INTRODUCTORY LECTURES ON AUTOMORPHIC FORMS

PUBLICATIONS OF THE MATHEMATICAL SOCIETY OF JAPAN

1. The Construction and Study of Certain Important Algebras. By Claude Chevalley.
2. Lie Groups and Differential Geometry. By Katsumi Nomizu.
3. Lectures on Ergodic Theory. By Paul R. Halmos.
4. Introduction to the Problem of Minimal Models in the Theory of Algebraic Surfaces. By Oscar Zariski.
5. Zur Reduktionstheorie Quadratischer Formen. Von Carl Ludwig Siegel.
6. Complex Multiplication of Abelian Varieties and its Applications to Number Theory. By Goro Shimura and Yutaka Taniyama.
7. Équations Différentielles Ordinaires du Premier Ordre dans le Champ Complexe. Par Masuo Hukuhara, Tosihusa Kimura et Mme Tizuko Matuda.
8. Theory of Q-varieties. By Teruhisa Matsusaka.
9. Stability Theory by Liapunov's Second Method. By Taro Yoshizawa.
10. Fonctions Entières et Transformées de Fourier. Application. Par Szolem Mandelbrojt.
11. Introduction to the Arithmetic Theory of Automorphic Functions. By Goro Shimura. (Kanô Memorial Lectures 1)
12. Introductory Lectures on Automorphic Forms. By Walter L. Baily, Jr. (Kanô Memorial Lectures 2)

PUBLICATIONS OF THE MATHEMATICAL SOCIETY OF JAPAN
12

INTRODUCTORY LECTURES ON AUTOMORPHIC FORMS

BY
Walter L. Baily, Jr.

KANÔ MEMORIAL LECTURES 2

Iwanami Shoten, Publishers
and
Princeton University Press
1973

LCC: 72-4034
ISBN: 0-691 08123-9
AMS (1971) 32.65

Kanô Memorial Lectures

In 1969, the Mathematical Society of Japan received an anonymous donation to encourage the publication of lectures in mathematics of distinguished quality in commemoration of the late Kôkichi Kanô (1865-1942).

K. Kanô was a remarkable scholar who lived through an era when Western mathematics and philosophy were first introduced to Japan. He began his career as a scholar by studying mathematics and remained a rationalist for his entire life, but enormously enlarged the domain of his interest to include philosophy and history.

In appreciating the sincere intentions of the donor, our Society has decided to publish a series of "Kanô Memorial Lectures" as a part of our Publications. This is the second volume in the series.

Publications of the Mathematical Society of Japan, volumes I through 10, should be ordered directly from the Mathematical Society of Japan. Volume 11 and subsequent volumes should be ordered from Princeton University Press, except in Japan, where they should be ordered from Iwanami Shoten, Publishers.

Co-published for
the Mathematical Society of Japan
by
Iwanami Shoten, Publishers
and
Princeton University Press

Printed in U.S.A.

INTRODUCTION

This book is based on lectures that I gave in Tokyo University in 1970 and 1971. Those lectures were given to a group most of whose members were graduate students, and were based on what seemed to me to be a reasonable introduction to the subject of automorphic forms on (domains equivalent to) bounded domains in $\boldsymbol{C}^n$, the space of n complex variables. The content of the lectures was based on the assumption that the hearer would seek out many of the details of proofs for himself elsewhere, especially in related areas such as those of algebraic groups and functional analysis. This book has been somewhat extended from the content of the lectures themselves by the addition of more examples and more details of proofs; however, the basic assumption remains that the interested reader will do the necessary additional research on background material for himself. Apart from this, however, it would be difficult to formulate any principles of precisely how it was decided to include some material and to exclude other material. It is hoped only that the book as a whole will serve some useful purpose as a sort of introductory guide to certain topics.

As for the subject matter itself, it is primarily that of complex analytic automorphic forms and functions on a (domain equivalent to a) bounded domain in a finite-dimensional, complex, vector space, most often denoted by $\boldsymbol{C}^n$. In other words, although, for example, we extensively reproduce certain relevant results of Harish-Chandra in this area, we do not attempt to go into the general subject of automorphic forms on a semi-simple Lie group. To the extent that our efforts do extend in this direction, it is mainly to prove certain theorems and lemmas that may be regarded as prerequisites to reading the first chapter of [26e], where general results are proved on the finite-dimensionality of spaces of automorphic forms on a semi-simple Lie group, which includes as a special case the situation we are interested in. This, in fact, was one of our objectives in this series of lectures. But our main concern has been with complex analytic functions. The reason, if one should be given, is that this is the context that seems most naturally related to algebraic geometry and problems of moduli of algebraico-geometric objects, apart from being the most classically oriented subdivision of the general topic of automorphic forms. If one is interested in the further number-theoretic connections of automorphic forms, it would appear essential to deal with the general situation of automorphic forms on a Lie group. Incidentally, it may seem (in spite of our alleged emphasis on complex analytic functions) that a large part of our effort is devoted to a development of representation theory. This seems quite natural, however, because of the obviously important role of that subject in connection with automorphic forms in any context.

We now turn to the discussion of the contents by part, chapter, and section. Part I deals mainly with the elementary theory of automorphic forms on a bounded domain D with respect to some discrete subgroup Γ of Hol(D), the full group of complex analytic self-transformations of D, with particular attention to the case when the orbit space is compact. A large part of the general theory here is due to H. Cartan. The chief result in the case when the orbit space D/Γ is compact is that that space is isomorphic (as a complex analytic space) to a projective algebraic variety, a fact which is proved in Chapter 5, section 2. Other than that, the table of contents is largely self-explanatory. In this section, very little use is made of any relationship between automorphic forms and harmonic analysis on the Lie group Hol(D).

By contrast, Part II treats the case of automorphic forms on a bounded symmetric domain, contains substantial sections devoted to basic facts from representation theory, and is dedicated very largely to applications of functional analysis on a Lie group to properties of automorphic forms. We begin by introducing the necessary material on algebraic groups. Because so much of this material is so technical and virtually no proofs are given, it was thought highly desirable to add a full chapter devoted entirely to examples; this has been accomplished by the insertion of Chapter 6. Chapter 7 is a sketchy account of the essentials needed from the general theory of algebraic Lie groups. Here we have included an account of the description of Harish-Chandra's realization of a bounded symmetric domain; the Iwasawa decomposition; and some of the results of Bruhat and Tits in the $\boldsymbol{p}$-adic case, which provide a $\boldsymbol{p}$-adic analog of the Iwasawa decomposition that is useful in the theory of

Eisenstein series. In Chapter 8, we review, with some proofs, some of the main results on compact groups: The Peter-Weyl theorem, the Frobenius reciprocity theorem, (both taken from the account in Weil's book [60a]), and the derivation of the Weyl character and dimension formulas (from [54: Exposé 21]). The latter find their place in the section dealing with the convergence of Fourier series in Chapter 9. As the title indicates, Chapter 9 is a collection of results of Harish-Chandra which are needed later, together with the proofs of those results as given by the same author [26a, b, d]. The main results we need are those used to prove the convergence of Poincaré series, the boundedness of Poincaré series "on the group", and the convergence of Fourier series, i.e., the expansion of an element of a representation space of a compact group in a series of the components of it obtained by orthogonal projections on the isotypic subspaces. Chapter 10 is mainly a collection of results from functional analysis, largely due to Godement, part of which, in addition to results given in Chapters 7 and 9, are prerequisite to reading, for example, [26e]. We also introduce the language of [52] for the study of automorphic forms on the domain by the functional analysis of their counterparts on the group Hol(D); much of this is due to Godement. Chapter 11 is concerned, finally, with the construction of automorphic forms through infinite series. In addition to using the results of Chapter 9 to demonstrate the convergence and boundedness on Hol(D) of Poincaré series, we also develop the convergence criterion of Godement for Eisenstein series. Together these give the Poincaré-Eisenstein series which are used in [3] to prove that the Satake compactification of D/Γ is a normal, complex analytic space and, as such, is isomorphic to a projective algebraic variety. To actually carry out the program of [3] would necessitate the introduction of reduction theory, the Satake topology, etc., for which we lack space. For details on these subjects, we refer the reader to [3; 6d]. We have limited ourselves to sketching an account, using certain ideas of Pyateckii-Shapiro on Fourier-Jacobi series [46], of how one may prove the finite-dimensionality of the spaces of automorphic forms. The idea is that one first proves the finite-dimensionality of the spaces of cusp forms following the ideas of [52; 26e]. For this, one proves Satake's lemma on characterization of cusp forms in relation to L^p spaces. Having the result for cusp forms, the general result is not difficult to obtain but we supply no further details here, other than to say that the main idea of the proof follows the lines of the references cited. However, it departs from the proof of Theorem 1 in [26e] in replacing certain facts about universal enveloping algebras by those concerning Fourier-Jacobi series. This is partly because these are somewhat specific to the complex analytic case, and partly because they have an interest in their own right. We conclude Chapter 11 with a sketch of the ideas behind the proof that the Satake compactification of D/Γ is an algebraic variety.

Part III consists of some special topics. Chapter 12 concerns itself with the arithmetic properties of the Fourier coefficients of Eisenstein series which seem independently important, especially in view of certain developments over recent years including [56d, f; 42; 36; 2i; 35; 58]. Chapter 13 contains a brief and somewhat incomplete account of certain matters introduced in Chapter 1. The main topic in Chapter 13 is theta functions and their relation to Eisenstein series via Siegel's main formula on definite quadratic forms.

Notation. No special attempt has been made to make notation uniform throughout this book. Therefore, the same letter may have different meanings in different places. The reader is advised to consult the beginning portions of any section or of any chapter to discover the local situation. The use of a dot to indicate multiplication within a group or operation of some mapping or group element on a space is not uniform. The dot may be used for the sake of emphasis in specific locations and suppressed in other entirely parallel situations.

Throughout, we use $\boldsymbol{Q}$, $\boldsymbol{R}$, $\boldsymbol{C}$, and $\boldsymbol{Z}$ to denote respectively the fields of rational numbers, real numbers, complex numbers, and the ring of rational integers. The reader's attention is directed to the supplementary notational references in the front of the book. Bracketed numerals refer to the bibliography.

Chicago, Autumn, 1972

W. L. Baily, Jr.

Acknowledgements

The author wishes most gratefully to acknowledge the help of Tokyo University in making available the facilities for giving these lectures; the generous support from the Mathematical Society of Japan; the assistance of Mr. M. Koike in taking notes of the lectures that served as a useful reference; the kind advice and encouragement of Prof. S. Iyanaga to write these lectures in book form; the proof-reading of portions of the manuscript by Messrs. M. Karel, M. Koike, and Prof. R. Narasimhan; and the careful typing of portions of the manuscript by Mr. F. Flowers.

The author also acknowledges his debt to the many other authors from whom he has borrowed heavily, but who, of course, cannot be held responsible for the present author's own oversights. In particular, the present author has had available to him notes of lectures given by Prof. A. Borel on the subject of automorphic forms, which have apparently not yet appeared in published form, and which served as a useful source of suggestions.

W. L. Baily, Jr.

CONTENTS

Part II
Automorphic forms on a bounded symmetric domain and analysis on a semi-simple Lie group

Supplementary notational references

Unless another meaning or notation is specified in a limited context, the following notational conventions are in use :

$\emptyset$ denotes the empty set.

G^0 denotes the identity component of the topological group G.

$\sum'$ denotes restricted direct sum.

A vertical bar $|$ will denote restriction of the function or mapping to the left to the set indicated to the right.

The identity of a group may be denoted by e or, if no confusion will result, by 1.

The Killing form of a Lie algebra is the bilinear form B defined by $B(X, Y) = \mathrm{tr}(ad\, X \cdot ad\, Y)$.

$[x]$ denotes the largest non-negative integer not greater than x.

In general, $e(\ )$ will denote the exponential function $e^{2\pi i(\ )}$.

If X is a complex manifold, then $\mathrm{Hol}(X)$ will denote the group of all one-to-one biholomorphic mappings of X onto itself.

If X is a symmetric $m \times m$ matrix, and M is $m \times n$, then $X[M] = {}^tMXM$.

PART I

ELEMENTARY THEORY OF AUTOMORPHIC FORMS ON A BOUNDED DOMAIN

CHAPTER 1
GENERAL NOTIONS AND EXAMPLES

§ 1. General notions

We begin by introducing the general context in which we shall consider automorphic forms and functions.

Let D be an open connected domain in the space $\boldsymbol{C}^n$ of n complex variables. Let $G=\mathrm{Hol}(D)$ be the group of all holomorphic one-to-one transformations of D onto itself, acting on the right. Denote by Γ a subgroup of G operating in properly discontinuous fashion on D (i.e., given two compact subsets A and B of D, the set $\Gamma_{A,B}=\{\gamma\in\Gamma \mid A\gamma\cap B \neq \emptyset$, the empty set$\}$ is finite). If $g\in G$, $Z\in D$, let $j(Z, g)$ denote the determinant of the functional (Jacobian) matrix of g at Z. We have the "cocycle relation":

$$j(Z, g_1g_2)=j(Z, g_1)\, j(Zg_1, g_2), \quad Z\in D,\ g_1,\ g_2\in G.$$

If d is an integer and f a meromorphic function on D such that

$$f(Z\gamma)\, j(Z, \gamma)^d \equiv f(Z) \tag{1}$$

for all $\gamma\in\Gamma$ and all $Z\in D$ for which both sides are defined, then we say that f is a meromorphic automorphic form of weight d with respect to Γ. If $d=0$, we call f an automorphic function, while (for any d) if f is analytic in D, we call f an automorphic form of weight d. It is clear from (1) that the poles and zeros of f, as well as their orders, are invariant under Γ. It turns out in many cases that for $d<0$, every (analytic) automorphic form is zero.

In the following sections, we describe a classical example intended to illustrate some of the features and applications of the theory of automorphic forms.

§ 2. Elliptic modular functions

Let

$$D=\{z\in \boldsymbol{C} \mid z=x+iy,\ y>0\};$$

then $G=PSL(2, \boldsymbol{R})^0$, i.e., is the quotient of

$$SL(2, \boldsymbol{R})=\left\{\begin{pmatrix} a & b \\ c & d \end{pmatrix} \middle| a, b, c, d \in \boldsymbol{R},\ ad-bc=1\right\}$$

by its center $\{\pm id.\}$. In what follows, we shall in practice denote an element of G by a pre-image of it in $SL(2, \boldsymbol{R})$. We take Γ to be the image of $SL(2, \boldsymbol{Z})$. An element $\begin{pmatrix} a & b \\ c & d \end{pmatrix}$ of G acts on D by a linear fractional transformation[1)]

$$z \mapsto \frac{az+b}{cz+d}$$

and the group Γ has a closed fundamental domain F of the well-known form:

$$F=\left\{z \in D \middle| |z| \geqq 1,\ |\operatorname{Re} z| \leqq \frac{1}{2}\right\};$$

that is to say, every orbit of Γ in D meets F, and two distinct points z_1 and z_2 of F are in the same orbit if and only if they both lie on the boundary ∂F of F and are related by $z_1=z_2 \pm 1$ or by $z_1=-1/z_2$ (or by both: $z_1=e^{2\pi i/3}$ and $z_2=e^{\pi i/3}$).

In what follows, we shall often use the abbreviated notation $\boldsymbol{e}(\ )=e^{2\pi i(\ )}$.

The condition (1) for a holomorphic function f on D to be an automorphic form of weight g becomes

$$f\left(\frac{az+b}{cz+d}\right)=(cz+d)^{2g} f(z) \tag{2}$$

for integers a, b, c, d such that $ad-bc=1$, and $z \in D$. In particular, $f(z+1)=f(z)$. Letting $\zeta=\boldsymbol{e}(z)$, we may write $f(z)=F(\zeta)$, where F is a function holomorphic in $0<|\zeta|<1$, as is easily verified. Writing down the Laurent expansion for F, one sees that f has a Fourier expansion

$$f(z)=\sum_{-\infty}^{+\infty} a_n \boldsymbol{e}(nz), \tag{3}$$

which converges absolutely and uniformly on compact subsets of D. We say that f has a certain kind of behaviour at ∞ if F has the same

1) In spite of our general agreement of §1 to let G operate on D on the right, we find it convenient here and in §3, in order not to conflict with firmly established traditional notation, to let G operate on the left.

kind of behaviour at 0. Then if $P=\infty$ or $P\in D$, let $w_P(f)$ be the order of f at P; in particular $w_\infty(f)=\inf_{a_n\neq 0} n$. As remarked in section 1, $w_P(f)$ is constant on orbits of Γ. Let N be a positive number >1, let λ_N be the contour in D consisting of the line $\left\{x+iy\,|\,y=N,\ -\frac{1}{2}\leqq x\leqq\frac{1}{2}\right\}$ and of that part of ∂F on which $\operatorname{Im} z\leqq N$, all described just once in the counter-clockwise direction. Applying the calculus of residues to the contour integral

$$\oint_{\lambda_N}\frac{df}{f}$$

(making suitable indentations in λ_N if necessary to avoid zeros of f) and letting $N\to\infty$ we obtain (assuming $f\not\equiv 0$)

$$w_\infty(f)+\frac{1}{2}w_i(f)+\frac{1}{3}w_\omega(f)+\Sigma^* w_P(f)=\frac{g}{6}, \tag{4}$$

where $\omega=\boldsymbol{e}\left(\frac{1}{3}\right)$ and Σ^* denotes a sum over representatives of distinct orbits of Γ in D, other than those containing i and ω; in particular an automorphic form of weight g on D holomorphic at ∞ has only finitely many zeros in F.

One way of constructing automorphic forms of weight g which are not identically zero is by forming the Eisenstein series

$$\boldsymbol{E}_g(z)=\Sigma\ (cz+d)^{-2g}, \tag{5}$$

where the sum is over a maximal set of mutually non-associate pairs of relatively prime integers (c, d). It is easily seen that the series in (5) converges for all $z\in D$ if $g>1$, and in fact converges absolutely and uniformly on any set of the form $|\operatorname{Re} z|\leqq A$, $\operatorname{Im} z>B$ for any $A, B>0$; therefore $\boldsymbol{E}_g$ is an automorphic form of weight g and $\boldsymbol{E}_g$ is regular at ∞. Put $z=iy$ and let $y\to+\infty$; one then sees that in the Fourier expansion

$$\boldsymbol{E}_g(z)=\sum_{n=0}^{\infty} a_n{}^g\boldsymbol{e}(nz) \tag{6}$$

we have $a_0{}^g=1$.

A remarkable fact about the numbers $a_n{}^g$ is that they are all rational numbers and, for fixed g, have bounded denominators. In fact, one may prove, as in [23] for instance, using the partial fraction expansions for the derivatives of the cotangent and the Fourier expansions for these derivatives in the upper half of the complex plane, that

$$\boldsymbol{E}_g(z)=1+\frac{(-1)^g 4g}{B_g}\sum_{n=1}^{\infty}\sigma_{2g-1}(n)\boldsymbol{e}(nz), \tag{7}$$

where B_g is the g-th non-vanishing Bernoulli number and

$$\sigma_k(n)=\sum_{\substack{d\mid n\\ d>0}} d^k.$$

Hence, $a_n{}^g\neq 0$ for all $n\geqq 0$. The properties just recited of the Fourier coefficients $a_n{}^g$ play an important role in the classical analytical formulation of complex multiplication.

Let M_g be the space of automorphic forms of weight g with respect to Γ which are regular at ∞. Define

$$\delta(z)=\boldsymbol{E}_2(z)^3-\boldsymbol{E}_3(z)^2.$$

Clearly $\delta\in M_6$ and since $\boldsymbol{E}_g(\infty)=1$, $\delta(\infty)=0$. On the other hand, it follows from (7) that

$$\delta(z)=1728\boldsymbol{e}(z)+\sum_{n>1} b_n\boldsymbol{e}(nz),$$

so that $\delta\not\equiv 0$ and the zero of δ at ∞ is simple. It follows then from (4) that δ has no zeros at any (finite) point of D. It also follows from (4) that any element of M_2, M_3, M_4, or of M_5 is a multiple of $\boldsymbol{E}_2$, $\boldsymbol{E}_3$, $\boldsymbol{E}_4$, or of $\boldsymbol{E}_5$, while M_1 must reduce to $\{0\}$, and M_0 consists of the constants. Now let $g\geqq 6$ and let $f\in M_g$. Clearly there exist integers $a, b\geqq 0$ such that $2a+3b=g$ and we define $f_1(z)=f(z)-\boldsymbol{E}_2(z)^a\boldsymbol{E}_3(z)^b f(\infty)$. Then $f_1\in M_g$ and $f_1(\infty)=0$, hence $f_1/\delta\in M_{g-6}$; it follows that $\dim M_g=\dim M_{g-6}+1$, and now by an easy induction that

$$\dim M_g=\begin{cases}\left[\dfrac{g}{6}\right]+1 & \text{if } g\not\equiv 1 \pmod 6,\\[2ex] \left[\dfrac{g}{6}\right] & \text{if } g\equiv 1 \pmod 6,\end{cases} \tag{8}$$

where $[r]$ denotes the biggest integer in r. (This derivation of this formula is to be found in [7].)

Let A be the Cartan matrix of the root system of the simple Lie algebra E_8 (cf. [11b]). Then A is a positive-definite, symmetric, integral matrix with determinant 1, and its diagonal entries are even. It is known that the class number $h(A)$ of the genus of quadratic forms containing A is one; i.e., given any other 8×8 positive-definite, symmetric, integral matrix B with determinant 1 and even diagonal entries, there exists a non-singular, unimodular, integral 8×8 matrix M such that ${}^tMAM=B$, where ${}^tM$ denotes the transpose of M. If $x=(x_1, \cdots, x_8)$ is an 8-tuple of numbers, let $A[x]={}^txAx$, where x is to be viewed as a column vector. Let $\boldsymbol{Z}$ denote the rational integers. If $z\in D$, the following series

$$\theta_A(z)=\sum_{n=(n_1,\cdots,n_8)\in \boldsymbol{Z}^8} \boldsymbol{e}\left(\frac{1}{2}zA[n]\right)$$

converges uniformly on compact subsets of D. Since the diagonal entries of A are even, $\theta_A(z+1)=\theta_A(z)$. Since A is unimodular and $A=AA^{-1}A$, A^{-1} is unimodularly equivalent to A. Using the Poisson summation formula (to be proved later: see Chapter 13, section 1) one then sees that $\theta(-z^{-1})=z^4\theta(z)$. Since $T: z\to z+1$ and $S: z\to -z^{-1}$ generate Γ, it is now clear that $\theta\in M_2$. Since dim $M_2=1$, we have $\theta=c\boldsymbol{E}_2$. Comparing the constant terms in the Fourier expansions one obtains $c=1$, so that $\theta=\boldsymbol{E}_2$. Thus one obtains the interesting (classical) result that for any positive integer, the number ν_m of ways of representing $2m$ in the form $2m=A[n]$, $n\in\boldsymbol{Z}^8$, is equal to 240 times the number theoretic function $\sum_{d\mid m} d^3$. (If we use A as the metric form on the space of root vectors for E_8, the roots themselves acquire length two.) By expressing the Fourier coefficients of $\boldsymbol{E}_2$ in another way, one also obtains (as we shall later in a more general situation——Chapter 12) that ν_m is the product of the $\boldsymbol{p}$-adic densities of representations of $2m$ by A, taken over all primes $\boldsymbol{p}$ and ∞. Thus one obtains a special case of Siegel's main formula for definite quadratic forms [56b]. It is expressed in this case, and characteristically is given, by an identity between a θ-series whose Fourier coefficients give global average densities and an Eisenstein series whose Fourier coefficients are Euler products of local densities.

Another question of importance is to find a common denominator for the Fourier coefficients of $\boldsymbol{E}_g$. Since, by (7), $B_g a_n{}^g\in\boldsymbol{Z}$, the denominators are bounded for any fixed g and, in any case, are closely connected with irregular primes [10; 33; 42; 56f]. A similar phenomenon occurs for $G=PSp(n, \boldsymbol{R})$, $\Gamma=PSp(n, \boldsymbol{Z})$, $D=\{Z=X+iY\mid$ where Z is $n\times n$, ${}^tZ=Z$, $Y\gg 0\}$, and it may be conjectured that the phenomenon is rather general (cf. [35] and [58]).

§ 3. The modular group and elliptic curves

Let L be a discrete subgroup of the additive group of $\boldsymbol{C}$ such that $T=\boldsymbol{C}/L$ is compact. Then L is a free Abelian group of rank 2 [60e, p. 37] and is called a lattice in $\boldsymbol{C}$; we let ω_1 and ω_2 be a pair of generators of L and put $\tau=\frac{\omega_1}{\omega_2}$. If L' is another lattice with generators ω_1' and ω_2', put $\tau'=\frac{\omega_1'}{\omega_2'}$, $T'=\boldsymbol{C}/L'$. We wish to determine necessary and sufficient conditions for T to be complex-analytically isomorphic to T'.

Suppose α is a complex analytic isomorphism of T' onto T. We may raise α to a complex analytic mapping $\tilde{\alpha}$, not identically constant, of the universal covering $\boldsymbol{C}$ of T' onto that of T, namely $\boldsymbol{C}$ itself; so $\tilde{\alpha}$ is an entire function, maps cosets of L' into cosets of L, and has a locally non-vanishing derivative; therefore, for any $l'\in L'$ there must exist $l\in L$ such that $\tilde{\alpha}(z+l')\equiv\tilde{\alpha}(z)+l$ (first of all, this holds in a neighborhood of some z_0 for fixed l, by the discreteness of L, and then for all z by analytic continuation). It follows that $\frac{d\tilde{\alpha}}{dz}$ is an entire function translationally invariant under L', hence is constant $\equiv\lambda\neq 0$, and $\lambda\cdot L'\subset L$; and since α is one-to-one, we have $\lambda\cdot L'=L$. Conversely, if $\lambda\neq 0$ is such that $\lambda\cdot L'=L$, we obtain an isomorphism of T' onto T. Thus $T\cong T'$ if and only if there exists a (non-zero) complex number λ such that $\lambda\cdot L'=L$; in terms of $\omega_1, \omega_2, \cdots$ this means there should exist $\begin{pmatrix} a & b \\ c & d \end{pmatrix}\in M_2(\boldsymbol{Z})$ with $ad-bc=\pm 1$ such that $\lambda\omega_1'=a\omega_1+b\omega_2$ and $\lambda\omega_2'=c\omega_1+d\omega_2$, thus $\tau'=\frac{a\tau+b}{c\tau+d}$. Without loss of generality we may assume $\operatorname{Im}\tau>0$, $\operatorname{Im}\tau'>0$, so that $ad-bc=1$. Therefore, the isomorphism classes of such elliptic curves $T=\boldsymbol{C}/L$ are in a natural one-to-one correspondence with the points of the orbit space D/Γ.

We let $G\times\boldsymbol{R}^2$ (semi-direct product) operate on $D\times\boldsymbol{C}$ by:

$$(g, (r_1, r_2))(\tau, z)=\left(g\tau, \frac{z+r_1\tau+r_2}{c\tau+d}\right),$$

$$g=\begin{pmatrix} a & b \\ c & d \end{pmatrix}\in G, \quad r_1, r_2\in\boldsymbol{R}.$$

It is readily verified that the discrete subgroup $\Delta=\Gamma\times\boldsymbol{Z}^2$ of $G\times\boldsymbol{R}^2$

acts in properly discontinuous fashion on the domain $D_1 = D \times \boldsymbol{C} \subset \boldsymbol{C}^2$. Let

$$\Delta(\boldsymbol{p}) = \left\{ (\gamma, (a_1, a_2)) \in \Delta \mid \gamma = \begin{pmatrix} a & b \\ c & d \end{pmatrix} \equiv \begin{pmatrix} 1 & 0 \\ 0 & 1 \end{pmatrix} \bmod \boldsymbol{p} \right\}$$

for an odd prime $\boldsymbol{p}$. Then, as is well-known and also easily verified directly, $\Delta(\boldsymbol{p})$ acts freely on D_1 and we have a natural mapping

$$\pi : D_1/\Delta(\boldsymbol{p}) \to D/\Gamma(\boldsymbol{p}),$$

where $\Gamma(\boldsymbol{p}) = \left\{ \gamma \in \Gamma \mid \gamma = \begin{pmatrix} a & b \\ c & d \end{pmatrix} \equiv \begin{pmatrix} 1 & 0 \\ 0 & 1 \end{pmatrix} \bmod \boldsymbol{p} \right\}$. If $x \in D/\Gamma(\boldsymbol{p})$, and if x is the canonical image of $\tau \in D$, then it is easy to see that $\pi^{-1}(x)$ is complex-analytically isomorphic to $\boldsymbol{C}/L_\tau$, $L_\tau = \boldsymbol{Z}\tau + \boldsymbol{Z}$. On the other hand, it will be verified later (Chapter 5) that $D_1/\Delta(\boldsymbol{p})$ and $D/\Gamma(\boldsymbol{p})$ both have structures of complex analytic spaces with respect to which π is a complex analytic mapping. Thus we have a "nice" fiber system of elliptic curves in which each isomorphism class of elliptic curves occurs at least once and at most a finite number ($\leqq [\Gamma : \Gamma(\boldsymbol{p})]$) of times. It can be proved [2d] that this fiber system may be compactified to an algebraic fiber system. This example is typical of one method of studying the moduli of Abelian varieties in relation to automorphic forms with respect to certain arithmetic groups.

Proceeding further one may treat in a similar way the connection between elliptic modular functions and complex multiplication; however, this would lead us too far from our main subject of automorphic forms. For this, the reader is referred to [55] and to other papers of the same author.

It is hoped the preceding examples will serve to illustrate some of the themes, parts of which we wish to develop in these notes.

CHAPTER 2
ANALYTIC FUNCTIONS AND ANALYTIC SPACES

§1. Power series and analytic functions

By a formal power series over a field $\mathfrak{k}$ (for our purposes usually the complex numbers $\boldsymbol{C}$), we mean a formal expression of the form

$$(1) \qquad \sum_{k_1 \geqq 0, \cdots, k_n \geqq 0} a_{k_1 \cdots k_n} z_1^{k_1} \cdot \cdots \cdot z_n^{k_n} = \sum_{(k) \geqq 0} a_{(k)} z^{(k)}$$

where the right side is a convenient abbreviation of the longer expression on the left and all $a_{(k)} \in \mathfrak{k}$. We may also write it as

$$(1') \qquad \sum_{(k) \geqq 0} a_{(k)} z^{(k)} = \sum_{l \geqq 0} \Big(\sum_{k_1 + \cdots + k_n = l} a_{(k)} z^{(k)} \Big) = \sum_{l \geqq 0} H_l(z),$$

where H_l denotes the homogeneous terms of degree l. We also use the notation: If $\alpha = (\alpha_1, \cdots, \alpha_n)$, then $(z-\alpha)^{(k)} = (z_1 - \alpha_1)^{k_1} \cdots (z_n - \alpha_n)^{k_n}$. When $\mathfrak{k}$ is a complete, valued field, then we may speak of convergence. A convergent power series means a power series that converges for some positive range of the absolute values of all the variables. The formal power series or the convergent power series form a ring with obvious rules for addition and multiplication. The same is true for power series of the form

$$(1'') \qquad \sum_{(k) \geqq 0} a_{(k)} (z-\alpha)^{(k)}$$

which we call "power series with center at α".

Now let the field $\mathfrak{k} = \boldsymbol{C}$ and suppose the series (1) converges at some point $z^0 = (z_1^0, \cdots, z_n^0)$, $|z_i^0| = R_i > 0$, $i = 1, \cdots, n$. Then there exists $M > 0$ such that $|a_{(k)}| R^{(k)} \leqq M$ for all $(k) \geqq 0$. So if $z = (z_1, \cdots, z_n)$ satisfies $|z_i| = r_i < R_i$, we have

$$|a_{(k)} z^{(k)}| \leqq M \Big(\frac{r}{R} \Big)^{(k)}$$

and so (1) converges absolutely like a multiple geometric series, and uniformly so in the region $\{|z_i| \leqq r_i \,|\, i = 1, \cdots, n\}$. Also, the partial derivatives of (1) of all orders are power series converging uniformly on the same region.

Let f be a complex-valued function in an open, connected subset, i.e., *domain*, D in $\boldsymbol{C}^n$. We say that f is analytic or holomorphic in D if for each $\alpha=(\alpha_1, \cdots, \alpha_n)\in D$, f is given in some neighborhood $\mathcal{U}$ of α by a power series with center at α converging throughout $\mathcal{U}$.

An alternative and equivalent characterization of analyticity is: f is analytic in D if continuously differentiable in the real and imaginary parts of $z_1, \cdots, z_n$, $z_j=x_j+iy_j$, there and if f satisfies the Cauchy-Riemann equations there

$$\frac{\partial f}{\partial x_j}=-i\frac{\partial f}{\partial y_j}, \quad j=1, \cdots, n, \tag{2}$$

or, using a customary notation,

$$\frac{\partial f}{\partial \bar{z}_j}=0, \quad j=1, \cdots, n. \tag{2'}$$

For it is clear that a function analytic in the power series sense satisfies (2); and the converse is settled by an easy generalization of Cauchy's integral formula:

$$(2\pi i)^{-n}\int_{C_n}\cdots\int_{C_1}\frac{f(\zeta)\, d\zeta_1 \cdots d\zeta_n}{(\zeta_1-z_1)\cdots(\zeta_n-z_n)}=(2\pi i)^{-n}\int_C \frac{f(\zeta)\, d\zeta}{(\zeta-z)^{(1)}}=f(z), \tag{3}$$

for z belonging to a product $\sigma=\sigma_1\times\cdots\times\sigma_n$ of discs with C_j the suitably oriented boundary of σ_j, and for f analytic according to the second definition on a neighborhood of $\bar{\sigma}=\bar{\sigma}_1\times\cdots\times\bar{\sigma}_n$; the rest of the notation in the middle term of (3) is adopted as a conventional abbreviation for that in the first term. From (3), one obtains in the usual way formulae for the coefficients of the power series for f with center at $\alpha=(\alpha_1, \cdots, \alpha_n)$, α_j being the center of σ_j. In particular, $a_{(0)}=f(\alpha)$ turns out to satisfy

$$|f(\alpha)|\leqq\frac{1}{vol(\sigma)}\int_\sigma |f(\zeta)|\, dv_\zeta, \tag{4}$$

dv_ζ being the Euclidean measure on σ. Applying Hölder's inequality, we have

$$|f(\alpha)|\leqq\left(\frac{1}{vol(\sigma)}\right)^{1/p}\left(\int_\sigma |f(\zeta)|^p\, dv_\zeta\right)^{1/p}, \quad p\geqq 1. \tag{5}$$

We need still another criterion for analyticity. The result we need, which will be applied at just one point later on, is

THEOREM 1 (*Hartogs* (*cf.* [5])). *Let f be a complex-valued function on the product $D\times P$, where D is a domain and P is an open polydisc. Suppose there exists an open polydisc p concentric with P such that $\bar{p}\subset P$. Assume the following:*

1) *f is analytic in $D\times p$.*

2) *for each $z\in D$, the restriction of f to $\{z\}\times P$ is analytic. Then f is analytic in $D\times P$.*

For the proof we refer the reader to [5 : pp. 137–139]. This theorem is used in proving Hartogs' theorem that analyticity in each variable implies analyticity in the sense we have defined previously.

We have the following classical result of Riemann :

THEOREM 2. *Let σ be a disc in $\boldsymbol{C}$ with center at α, and let f be analytic and bounded in $\sigma-\{\alpha\}$. Then f has a unique extension to an analytic function on σ.*

For the proof, see [4 : vol. I, p. 146].

This can be extended to several variables as we shall see later.

§ 2. Analytic sets

Let $D\subset \boldsymbol{C}^n$ be a domain, and let X be a closed subset of D. We say that X is an analytic subset of D if for each $a\in D$ there exists a neighborhood $\mathcal{U}\subset D$ of a and a finite number of analytic functions $f_1, \cdots, f_m$ on $\mathcal{U}$ such that

$$X\cap\mathcal{U}=\{z\in\mathcal{U}\,|\,f_1(z)=\cdots=f_m(z)=0\}. \tag{6}$$

If X is an analytic subset of D and $a\in X$, then a is called a "regular" or "simple" point of X if there exist a neighborhood $\mathcal{U}$ of a and analytic functions $f_1, \cdots, f_m$ on $\mathcal{U}$ satisfying (6) such that the matrix

$$\left(\frac{\partial f_\mu}{\partial z_\nu}\right)_{\mu=1,\cdots,m;\ \nu=1,\cdots,n} \tag{7}$$

has rank m at a—— of course then $m\leq n$. The simplest example of an analytic set all of whose points are regular is a so-called linear variety in $\boldsymbol{C}^n$, by which we mean a set L defined by a system of complex linear equations

$$\sum_{j=1}^{n} a_{ij} z_j + b_i = 0, \quad i=1, \cdots, s, \tag{8}$$

where it is assumed, of course, that the rank r of the matrix (a_{ij}) is the same as the number of linearly independent equations among those in (8). Then $\dim L = n - r$, and the fact that every point of L is regular follows from the fact that the system (8) is equivalent to any subsystem of r linearly independent equations.

If a is a simple point of X and $\mathcal{U}$ is a neighborhood of a and $f_1, \cdots, f_m$ are analytic functions satisfying (6) such that the matrix (7) has rank m at a, then the equations in (6) can be solved, by the implicit function theorem, to give m of the coordinates as analytic functions of $n-m$ others, and there is a biholomorphic mapping of a neighborhood $\mathcal{V}$ of a onto a domain $D' \subset \boldsymbol{C}^n$ such that the image of $X \cap \mathcal{V}$ is the intersection of D' with a linear variety of dimension $n-m$. Of course the dimension of X at the regular point a is defined to be $n-m$. It is obvious from the definition of regular point that the set X_{reg} of regular points of X is an open subset of X.

Let D and X be as above, let $x \in X$, let $\mathcal{U}$ be a neighborhood of x on X, and let f be a complex-valued function on $\mathcal{U}$; we say that f is analytic on $\mathcal{U}$ if for each $y \in \mathcal{U}$, there is a neighborhood $\mathcal{V}$ of y in D and an analytic function g on $\mathcal{V}$ such that $g|(\mathcal{V} \cap \mathcal{U}) = f|(\mathcal{V} \cap \mathcal{U})$. The families of analytic functions on the open subsets of X so-defined constitute, by definition, the canonical ringed structure of analytic functions on X.

In general, let X be a Hausdorff, locally compact space with a countable basis of neighborhoods. By a ringed structure on X, we mean, roughly speaking, that for each open subset $\mathcal{U}$ of X we are given a subring $\mathcal{R}_{\mathcal{U}}$ of the ring of continuous complex-valued functions on $\mathcal{U}$ containing the constant functions, such that where two open sets overlap, certain "reasonable" compatibility conditions between the two subrings are satisfied. It is left as an exercise for the reader to determine suitable "reasonable" compatibility conditions for the context in which we operate. The ringed structure defined by X and the assignments $\mathcal{U} \mapsto \mathcal{R}_{\mathcal{U}}$ is denoted by $(X, \mathcal{R})$. If $(X', \mathcal{R}')$ is another ringed space, and $\varphi: X \to X'$ is a continuous mapping, then we say φ is a morphism of ringed spaces if for every open

$\mathcal{U}'$ in X', the family $\mathcal{R}_{\mathcal{U}'}\circ\varphi$ of continuous functions on $\varphi^{-1}(\mathcal{U}')$ of the form $f\circ\varphi$, $f\in\mathcal{R}_{\mathcal{U}'}$, is a subring of $\mathcal{R}_{\varphi^{-1}(\mathcal{U}')}$. Under these conditions if φ is a homeomorphism of X onto a closed subspace of X' such that for each $a\in X$, neighborhood $\mathcal{U}$ of a, and $f\in\mathcal{R}_{\mathcal{U}}$, there exists a neighborhood $\mathcal{U}'$ of $f(a)$ and $f'\in\mathcal{R}'_{\mathcal{U}'}$ for which $f'\circ\varphi$ coincides with f on a neighborhood of a, then φ is called an injection. A ringed subspace of $(X, \mathcal{R})$ is a closed subspace X' of X with a ringed structure $\mathcal{R}'$ such that the identity mapping of X' into X is an injection of ringed spaces.

If D is a domain in $\boldsymbol{C}^n$, then D carries the ringed structure $\mathcal{O}$ of analytic functions, and our definition of the canonical ringed structure $\mathcal{A}$ of analytic functions on an analytic subset X of D amounts to saying that the inclusion of X into D is an injection $(X, \mathcal{A})\to(D, \mathcal{O})$ of ringed spaces.

By definition, a ringed space $(X, \mathcal{R})$ is called a (complex) analytic space if for each $a\in X$ there is a neighborhood $\mathcal{U}$ of a such that the ringed space $(\mathcal{U}, \mathcal{R}\,|\,\mathcal{U})$ (where $\mathcal{R}\,|\,\mathcal{U}$ is the restriction of $\mathcal{R}$ to open subsets of $\mathcal{U}$) is isomorphic to an open subset of some analytic set in a domain of $\boldsymbol{C}^n$ with its canonical ringed structure of analytic functions.

Clearly the notion of analytic space is a generalization of the notion of "complex manifold": a complex manifold is a complex analytic space all of whose points are regular. In fact it can be verified (by theorems on analytic sets which we shall state later) that if a neighborhood $\mathcal{U}$ of a point a of the analytic space X is isomorphic (as a ringed space) with an open set $\mathcal{V}$ on some analytic set Y and if b is the image of a in $\mathcal{V}$, then whether b is regular or not is independent of the particular choice of Y, etc. Hence the concept of regular point of an analytic space X is well-defined; the set of regular points of X is denoted by X_{reg} and any point of $X-X_{\text{reg}}$ is, by definition, a singular point of X. Put $X_{\text{sing}}=X-X_{\text{reg}}$. Then it may be proved that X_{reg} is a dense open subset of X, that X_{reg} is a union of connected, mutually disjoint complex manifolds, and that X_{sing} is an analytic subspace of X; for the proof of this it is sufficient to consider the case when X is an analytic subset of a domain D. For these facts we refer the reader to [44a : esp. pp. 56–68].

§ 3. Structure of local analytic sets

For the proofs of the facts cited here, we refer to [44a].

Let a be a point of the locally compact Hausdorff space X. Two subsets S and S' are called equivalent at a if there is a neighborhood $\mathcal{U}$ of a such that $S \cap \mathcal{U} = S' \cap \mathcal{U}$. Two functions f and f' defined in an open set containing a are called equivalent at a if there is a neighborhood $\mathcal{U}$ of a such that $f | \mathcal{U} = f' | \mathcal{U}$. In either case the equivalence classes are called "germs". The germ of a set S (resp. of a function f) at a will often be denoted S_a (resp. f_a). A germ (of functions or of sets) is called analytic at a if it has a representative which is analytic in some neighborhood of a. In many cases it will be convenient to denote a set or a function and its germ at some point by the same symbol——a convenience we shall avail ourselves of without warning.

If $a \in \boldsymbol{C}^n$, let $\mathcal{O}_a$ denote the ring of germs of holomorphic functions at a. If E is a set, the set

$$\mathfrak{J}(E)_a = \{f \in \mathcal{O}_a \,|\, f | E = 0\}$$

is an ideal of $\mathcal{O}_a$ and is equal to $\mathcal{O}_a$ if and only if a is not in the closure of E. If $\mathfrak{A}$ is an ideal in $\mathcal{O}_a$, then $\mathfrak{A}$ determines a unique germ $V(\mathfrak{A})_a$ of analytic sets at a on which all elements of $\mathfrak{A}$ vanish. We have

$$(9) \qquad \begin{aligned} &\mathfrak{J}(V(\mathfrak{A})_a) = \mathrm{rad}(\mathfrak{A}), \\ &V(\mathfrak{J}(E)_a)_a = \text{smallest germ of analytic sets at } a \text{ containing } E. \end{aligned}$$

If f is a formal power series in n indeterminates $X_1, \cdots, X_n$, then f is called regular of order p in X_n if

$$(10) \qquad f(0, \cdots, 0, X_n) = c \cdot X_n^p + \cdots, \quad c \neq 0.$$

Let $\boldsymbol{C}[[X_1, \cdots, X_n]]$ denote the formal power series in $X_1, \cdots, X_n$ and let $\boldsymbol{C}\{X_1, \cdots, X_n\}$ denote the convergent power series. One may always render a given power series regular in X_n by a linear change of coordinates. If $P \in \boldsymbol{C}[[X_1, \cdots, X_n]]$ and if

$$P = X_n^p + a_{p-1}(X_1, \cdots, X_{n-1}) X_n^{p-1} + \cdots + a_0(X_1, \cdots, X_{n-1}),$$

then P is called a distinguished polynomial in X_n if $a_{p-1}(0) = \cdots = a_0(0) = 0$; clearly P is then regular of order p.

THEOREM 3 (*Weierstrass*). *Let f be regular of order p in X_n and let $g \in \boldsymbol{C}[[X_1, \cdots, X_n]]$. Then there exist unique $Q \in \boldsymbol{C}[[X_1, \cdots, X_n]]$ and $R \in \boldsymbol{C}[[X_1, \cdots, X_{n-1}]][X_n]$,*

$$R(X_1, \cdots, X_n) = \sum_{k=0}^{p-1} R_k(X_1, \cdots, X_{n-1}) X_n^k$$

such that if we write

$$g(X) = \sum_{k=0}^{\infty} a_k(X_1, \cdots, X_{n-1}) X_n^k,$$

then

$$a_k(0) = R_k(0), \quad k = 0, \cdots, p-1, \quad \textit{and}$$
$$g = f \cdot Q + R.$$

In particular, if $g = X_n^p$, then

$$f \cdot Q = g - R = P$$

is a distinguished polynomial, and

$$f(0, \cdots, 0, X_n) = c \cdot X_n^p + \cdots, \quad P(0, \cdots, 0, X_n) = X_n^p,$$

so that $Q(0)$ is a constant $\neq 0$, i.e., Q is a unit.

If f and g are convergent, then Q and the coefficients of R are analytic in some neighborhood of the origin.

For the proofs, see [44a : Chap. II].

If X is a complex analytic space, a subset S of X will be called "thin" if 1) S is nowhere dense in X, and 2) if a is any point of X, then there exists a neighborhood $\mathcal{U}$ of a and an analytic function f on $\mathcal{U}$ such that $V(f) = \{x \in \mathcal{U} \mid f(x) = 0\}$ is nowhere dense in $\mathcal{U}$ and such that $S \cap \mathcal{U} \subset V(f)$. In particular, if D is a connected domain in $\boldsymbol{C}^n$, then any proper analytic subset of D is thin (identity theorem for analytic functions and analytic continuation). One result that may be obtained by applying the Weierstrass theorem in combination with Cauchy's integral formula is:

PROPOSITION 1. *Let $\boldsymbol{D}$ be a (connected) domain in $\boldsymbol{C}^n$, let S be a thin subset of D, and let f be a holomorphic function on $D - S$. Suppose f is locally bounded on D, i.e., suppose for each $a \in D$ there exists a neighborhood $\mathcal{U}$ of a in D such that $|f|$ is bounded on $\mathcal{U} \cap (D - S)$. Then f has a unique extension to a holomorphic function on D.*

For the proof, which is carried out in much the same way as the result of Riemann cited earlier, see [24 : pp. 19–20], where also the following easy consequence is derived:

COROLLARY. *Let D and S be as in the proposition. Then $D-S$ is connected.*

An analytic set X is called irreducible at a if its germ at a is not the proper union of two other (germs of) analytic sets at a: $X_a = X_{1a} \cup X_{2a}$, X_1, X_2 analytic $\Rightarrow X_a = X_{1a}$ or $X_a = X_{2a}$.

One other consequence of the Weierstrass theorem is that $\mathfrak{F} = \boldsymbol{C}[[X_1, \cdots, X_n]]$ and $\mathcal{O}_a$ are Noetherian rings. And every ideal in $\mathfrak{F}$ or in $\mathcal{O}_a$ is the intersection of a finite number of primary ideals. One verifies easily that if X is an analytic set at a, then X is irreducible at a if and only if $\mathfrak{J}(X)_a$ is a prime ideal in $\mathcal{O}_a$. And any (germ X of) analytic set(s) at a has a unique decomposition into non-redundant irreducible (germs of) analytic sets at a, called components of X.

We recall that if $\mathcal{R}$ and $\mathcal{R}'$ are commutative rings with unit, $\mathcal{R}' \subset \mathcal{R}$, then $\mathcal{R}$ is said to be integral over $\mathcal{R}'$ if every $x \in \mathcal{R}$ satisfies an integral equation

$$x^n + a_{n-1} x^{n-1} + \cdots + a_0 = 0, \quad \text{all } a_i \in \mathcal{R}'.$$

If $\mathcal{R}'$ is Noetherian, then for $\mathcal{R}$ to be integral over $\mathcal{R}'$, it is sufficient that $\mathcal{R}$ be an $\mathcal{R}'$-module of finite type [59 : vol. II].

Let X be an analytic set, $a \in X$, and assume (the germ of) X is irreducible at a. Then [44a : Chap. III] in some neighborhood of a, X_{reg} is a connected [44a : Prop. 11, p. 55] complex manifold of dimension d equal to the dimension of the prime ideal $\mathfrak{J}(X)_a$ (the length of a chain of prime ideals from $\mathfrak{J}(X)_a$ to the maximal ideal $\mathfrak{M}_a$ of $\mathcal{O}_a$ minus one). For simplicity only take $a=0$, let $\mathcal{O} = \mathcal{O}_0$, $\mathfrak{J} = \mathfrak{J}(X)_0$, $\mathcal{R} = \mathcal{O}/\mathfrak{J}$, and denote by π the canonical homomorphism of $\mathcal{O}$ onto $\mathcal{R}$. Then [44a : loc. cit.] one has

THEOREM 4. *One may choose the coordinates $z_1, \cdots, z_n$ on $\boldsymbol{C}^n$ such that, in some neighborhood of* 0,

a) *If $\mathcal{R}' = \boldsymbol{C}\{z_1, \cdots, z_d\}$, then $\mathcal{R}$ is integral over $\pi(\mathcal{R}')$ and*

$$\ker \pi \cap \mathcal{R}' = \{0\}.$$

b) *Let $\zeta_i = z_i | X$, $i = 1, \cdots, d$. Then $\zeta_1, \cdots, \zeta_d$ are local coordinates at every point on the complex manifold X_{reg}, outside a proper analytic* *subset.*

c) *0 is a regular point of X if and only if $\mathcal{R}=\pi(\mathcal{R}')$, i.e., if and only if $\mathcal{R}$ is the convergent power series ring in d variables.*

The integer d is called the dimension of X. We see that it is the dimension of every component of the germ of X at b for all $b\in X$ in a neighborhood of a and is characterized as being the dimension of X (as a complex manifold) at a dense open set of regular points. Part c) gives an algebraic characterization of regular points that makes clear the truth of our earlier assertion (section 2) that the regular points of an analytic space are characterized intrinsically. One then obtains

PROPOSITION 2. *Let X be a complex analytic space and let $X^{(d)}$ be the set of $x\in X$ such that X has a component of dimension d at x. Then $X=\bigcup_{d\geq 0} X^{(d)}$, and each $X^{(d)}$ is a complex analytic subspace of X* [44a : p. 67].

If $X=X^{(d)}$ for some $d\geqq 0$, we say X is of pure dimension d.

Let X and Y be complex analytic spaces and let $f\colon X\to Y$ be an analytic mapping (i.e., a morphism of the analytic ringed spaces). If Y' is an analytic subspace of Y, then $f^{-1}(Y')$ is an analytic subspace of X. If $x\in X$, we say that f is "light" at x if x is an isolated point of the fiber $f^{-1}(f(x))$. Suppose $y=f(x)$, let $\mathcal{R}_y$ resp. $\mathcal{R}_x$ be the ring of germs of analytic functions at y resp. at x, and assume $\dim_x X=\dim_y Y$. Let $f_x^*\colon \mathcal{R}_y\to\mathcal{R}_x$ be the homomorphism induced by f. Then we have

PROPOSITION 3. *$\mathcal{R}_x$ is integral over $f_x^*(\mathcal{R}_y)$ if and only if f is light at x.*

The proof follows at once from [44a : Thm. 1, p. 10 and Thm. 3, p. 44].

COROLLARY. *Let notation be as in Theorem 4, let $f_1, \cdots, f_m\in\mathcal{O}$, $f_1(0)=\cdots=f_m(0)=0$, let $\mathcal{A}=\mathcal{A}(f_1,\cdots,f_m)$ be the ring analytically generated*[1] *by $f_1,\cdots,f_m$, and let $\boldsymbol{I}$ be the ideal of analytic relations*[1]

1) The ring analytically generated by $f_1,\cdots,f_m$ is the ring of analytic functions of the form $g(f_1,\cdots,f_m)$ for some $g\in\boldsymbol{C}\{X_1,\cdots,X_m\}$; the ideal of analytic relations among $f_1,\cdots,f_m$ is the kernel in $\boldsymbol{C}\{X_1,\cdots,X_m\}$ of the assignment $g\mapsto g(f_1,\cdots,f_m)$, and $\mathcal{O}_0^{(m)}$ is the ring of germs of analytic functions at $0\in\boldsymbol{C}^m$.

among $f_1, \cdots, f_m$. *Denote by* f *the mapping of a neighborhood of* 0 *on* X *into a neighborhood of* $0 \in \boldsymbol{C}^m$ *whose coordinates are* $f_1, \cdots, f_m$. *Let* $Y = V(\boldsymbol{I})_0 \subset \boldsymbol{C}^m$. *Then* f *is an analytic mapping of* X *into* Y. *Moreover* $V(f_1, \cdots, f_m) \cap X = \{0\}$ *(in some neighborhood of* 0) *if and only if* f *is light at* 0, *which in turn is true if and only if* $\mathcal{R}$ *is integral over* $\pi(\mathcal{A}) = f_0^*(\mathcal{O}_0^{(m)})$; *and if these mutually equivalent conditions hold, then* f *is an open mapping of a neighborhood* $\mathcal{U}$ *of* 0 *on* X *onto a neighborhood of* 0 *on* Y *and (if* $\mathcal{U}$ *is small enough)* f *is light at every point of* $\mathcal{U}$.

The corollary follows from the proposition by employing the proposition, the references in [44a] cited in its proof, and [44a : Prop. 5, p. 71].

§ 4. The normalization theorem

Let X be an analytic space, $a \in X$, and let $\mathcal{R}_a$ be the ring of germs of analytic functions at a. If $\mathcal{R}$ is any commutative ring, the quotient ring $\tilde{\mathcal{R}}$ of $\mathcal{R}$ will mean the ring of all "quotients" of the form r/s where $s \in S$, the set of non-zero divisors of $\mathcal{R}$. Then X will be called normal at a if $\mathcal{R}_a$ is integrally closed in its quotient ring. If X is normal at each of its points, X is called, simply, "normal". If $\mathcal{R}_a$ is integrally closed in $\tilde{\mathcal{R}}_a$, then $\mathcal{R}_a$ is a domain of integrity [49 : vol. II, pp. 143-144]; in fact, if $X_a = X_1 \cup X_2$, then there exist analytic functions f_1 and f_2 at a such that f_i vanishes on X_i but not on any component of X_j, $j \neq i$, $i = 1, 2$, so that $f_1 + f_2 \in S$, and $x = f_1/(f_1 + f_2)$ satisfies $(x^2 - x) \mid X_a = 0$, thus the image $\bar{x}$ of x in $\tilde{\mathcal{R}}_a$ is integral over $\mathcal{R}_a$, but $\bar{x}$ is not in $\mathcal{R}_a$ since it cannot be continuous at a. Therefore, if X is normal at a, X is irreducible at a, consequently $\mathfrak{J}(X)_a$ is prime and $\mathcal{R}_a$ is a domain of integrity.

If X is an analytic space, and if (Y, f) is a pair consisting of a normal complex analytic space Y and an analytic mapping $f\colon Y \to X$, then (Y, f) is called a normalization of X if

(a) $f\colon Y \to X$ is proper and has finite fibers, and

(b) if S is the set of singular points of X and $A = f^{-1}(S)$, then $Y - A$ is dense in Y and $\pi \mid (Y - A)$ gives an analytic isomorphism of $Y - A$ onto $X - S$ [44a : p. 114].

THEOREM 5. *X has a normalization* (Y, f). *If* (Y', f') *is another* *normalization of X, then there exists a complex analytic isomorphism φ of Y onto Y' such that $f'\circ\varphi=f$.*

It is worth noting that if X is a normal analytic space, then the following further generalization of Riemann's theorem holds:

PROPOSITION 4. *Let X be a normal analytic space, S a thin subset of X, f a holomorphic function on $X-S$ which is locally bounded on X. Then f has a unique extension to a holomorphic function on X.*

[44a: Remark, p. 114].

It follows that if Y is another analytic space and if f is a one-to-one analytic mapping of Y onto X, then Y is normal and f is an analytic isomorphism of Y onto X.

§ 5. The Remmert-Stein theorem

THEOREM 6 (*Remmert-Stein* [47]). *Let D be a domain in $\boldsymbol{C}^n$, let X be a non-empty analytic subset of D and let Y be an analytic subset of $D-X$. Assume there exists a $d\geqq 0$ such that*

$$Y=\bigcup_{l\geqq d} Y^{(l)},$$

and

$$X=\bigcup_{l<d} X^{(l)}.$$

Then the closure $\bar{Y}$ of Y in D is an analytic subset of D and $\bar{Y}=\bigcup_{l\geqq d}\bar{Y}^{(l)}$.

§ 6. The quotient of C^n by a finite linear group

Let G be the group of all non-singular linear transformations of $\boldsymbol{C}^n$ and let Γ be a finite subgroup of G. The purpose of this section is to show that $X=\boldsymbol{C}^n/\Gamma$ has naturally the structure of a normal complex analytic space.

To begin with, there are two natural ringed structures that suggest themselves for the topological space X. For one, let $\pi:\boldsymbol{C}^n\to X$ be the natural mapping, and if f is a complex-valued function on an open subset $\mathcal{U}$ of X, define f to be analytic on $\mathcal{U}$ if $f\circ\pi$ is analytic on $\pi^{-1}(\mathcal{U})$ (of course, this implies f is continuous on $\mathcal{U}$); denote this ringed structure by $\mathcal{R}: \mathcal{U} \to \mathcal{R}_{\mathcal{U}}$. For the other, we claim first that the ring $\mathcal{P}^{\Gamma}$ of Γ-invariant polynomials has a finite set of generators as a C-algebra; this claim will be shortly justified below. Letting $Q_1, \cdots, Q_m$ be a finite set of such generators, such that each Q_j is a homogeneous polynomial, we define an analytic mapping

$$Q: C^n \to C^m$$

by $Q(z) = (Q_1(z), \cdots, Q_m(z))$. We also assert that the Γ-invariant polynomials separate the orbits of Γ in C^n. Granted this, which will also be proved later, one notes that Q induces an injection $\bar{Q}$ of the space X into C^m; it follows in particular that Q is light at every point of C^m. It will also be seen below that the ring $\mathcal{P}$ of polynomials in n variables is integral over $\mathcal{P}^{\Gamma}$, hence Q is a proper mapping. Therefore, by [44a: Corollary, p. 87], $\bar{Q}(X)$ is an analytic subset of C^m, and as such is a complex analytic space, of which the underlying space is homeomorphic to X. (In this case, since only polynomials are involved, a direct, purely algebraic proof of the same fact can be given, not using the analytical results of [44a], cf. [15b].) Transporting the analytic ringed structure on $\bar{Q}(X)$ back to X, we obtain a second ringed structure $\mathcal{R}'$ on X. The main step will be to show that $\mathcal{R}$ and $\mathcal{R}'$ are the same, and from this it will be easy to show that $(X, \mathcal{R})$ is a normal analytic space.

The essential steps (and most of the details) of the proof we give are to be found in [15b].

First we introduce the notation that if $a \in C^n$, then $\Gamma_a = \{\gamma \in \Gamma \mid a \cdot \gamma = a\}$. If f is analytic at a, let $\omega_a(f)$ denote the order (of the zero) of f at a. We use $\mathcal{P}$, as indicated above, to denote $C[X_1, \cdots, X_n]$, which will be identified with the polynomial functions on C^n, and for any subgroup Γ' of Γ, $\mathcal{P}^{\Gamma'}$ denotes the Γ'-invariant elements in $\mathcal{P}$. If $a = (a_1, \cdots, a_n) \in C^n$, then $\mathcal{P}^{\Gamma_a}$ is spanned by its homogeneous elements in $C[X_1 - a_1, \cdots, X_n - a_n]$.

LEMMA 1. *Let* $a, b \in C^n$, $a \notin b \cdot \Gamma$. *Let* $R_1 \in \mathcal{P}^{\Gamma_a}$, $R_2 \in \mathcal{P}^{\Gamma_b}$, *and let* r *be a positive integer. Then there exists* $P \in \mathcal{P}^{\Gamma}$ *such that* $\omega_a(P - R_1) > r$, $\omega_b(P - R_2) > r$.

PROOF. Let U_1 be a polynomial such that $\omega_a(U_1 - 1) > r$, $\omega_x(U_1) > r$ for $x \in a \cdot \Gamma \cup b \cdot \Gamma - \{a\}$, and let U_2 be a polynomial such that $\omega_b(U_2 - 1) > r$

and such that $\omega_x(U_2)>r$ for the other points x of $a\cdot\Gamma\cup b\cdot\Gamma$. Put $V_1=\prod_{\gamma\in\Gamma_a} U_1^\gamma$ and $V_2=\prod_{\gamma\in\Gamma_b} U_2^\gamma$ (where for any polynomial, P^γ denotes its image under the action of γ). Then V_1 (resp. V_2) has the same prescribed properties as U_1 (resp. U_2) and is Γ_a-(resp. Γ_b-)invariant. Moreover, $\omega_a(R_1-R_1V_1)>r$, $\omega_x(R_1V_1)>r$, $\omega_b(R_2-R_2V_2)>r$, $\omega_y(R_2V_2)>r$ for $x\in a\cdot\Gamma\cup b\cdot\Gamma-\{a\}$, $y\in a\cdot\Gamma\cup b\cdot\Gamma-\{b\}$, and R_1V_1 (resp. R_2V_2) is Γ_a-(resp. Γ_b-)invariant. Let Q_1 (resp. Q_2) be the sum $\sum_{\Gamma_a\backslash\Gamma\ni\gamma}(R_1V_1)^\gamma$ (resp. $\sum_{\Gamma_b\backslash\Gamma\ni\gamma}(R_2V_2)^\gamma$) and put $P=Q_1+Q_2$. It is routine to see that P satisfies the requirements of the lemma.

COROLLARY. *Let a and b be as in the lemma. Then there exists $P\in\mathcal{P}^\Gamma$ such that $P(a)=1$, $P(b)=0$.*

PROOF. $\mathcal{P}^\Gamma$ contains the constants. Then apply the lemma with $R_1\equiv 1$ and $R_2\equiv 0$.

LEMMA 2. *$\mathcal{P}$ is integral over $\mathcal{P}^\Gamma$ and $\mathcal{P}^\Gamma$ is finitely generated as a $\boldsymbol{C}$-algebra.*

PROOF. By the above corollary, it is seen that if $a\in\boldsymbol{C}^n-\{0\}$, there exists a homogeneous polynomial $P\in\mathcal{P}^\Gamma$ such that $P(a)\neq 0$, $P(0)=0$. Hence, by the Hilbert basis theorem, there exist finitely many elements $Q_1,\cdots,Q_l\in\mathcal{P}^\Gamma$ which are homogeneous and of positive degree having 0 as their only common zero. By a well-known theorem (Zariski), $\mathcal{P}$ is integral over the $\boldsymbol{C}$-algebra $\mathcal{Q}$ generated by $Q_1,\cdots,Q_l$, hence is all the more integral over $\mathcal{P}^\Gamma$. Now $\mathcal{Q}$ is a Noetherian ring, and since $\mathcal{P}$ is a finitely generated $\boldsymbol{C}$-algebra integral over $\mathcal{Q}$, $\mathcal{P}$ is a $\mathcal{Q}$-module of finite type, hence $\mathcal{P}^\Gamma=\mathcal{Q}\cdot Q_{l+1}+\cdots+\mathcal{Q}\cdot Q_m$. Thus $Q_1,\cdots,Q_m$ generate $\mathcal{P}^\Gamma$ as a $\boldsymbol{C}$-algebra.

The corollary and Lemma 2 contain the assertions made earlier without proof.

Now we want to prove the two ringed structures $\mathcal{R}$ and $\mathcal{R}'$ are the same. Since it is obvious that $\mathcal{R}\supset\mathcal{R}'$ (i.e., $\mathcal{R}_{\mathcal{U}}\supset\mathcal{R}'_{\mathcal{U}}$ for every open $\mathcal{U}$ in X), it remains to prove that $\mathcal{R}'\supset\mathcal{R}$. In other words we must prove that given $a=(a_1,\cdots,a_n)\in\boldsymbol{C}^n$, every convergent power series in $X_1-a_1,\cdots,X_n-a_n$ which is invariant under Γ_a can be written as $g(Q_1(X)-Q_1(a), \cdots, Q_m(X)-Q_m(a))$ for a convergent power series g in m variables. The proof can be divided into two parts, one formal, and one dealing with convergence.

Let $R_1, \cdots, R_t$ be homogeneous polynomials and let $d_i = \deg R_i$, $i=1, \cdots, t$. A polynomial in t variables $Y_1, \cdots, Y_t$ is called isobaric of weight p if every monomial appearing in it is of the form $Y_1^{n_1} \cdots Y_t^{n_t}$ with $\sum n_i d_i = p$.

So now suppose $R_1, \cdots, R_t$ are homogeneous polynomials in $X_1 - a_1, \cdots, X_n - a_n$ which generate the C-algebra $\mathcal{P}^{\Gamma_a}$. Then by Lemma 1, there exists for each $i=1, \cdots, t$, a polynomial $p_i(Y_1, \cdots, Y_m)$ such that

$$\omega_a(p_i(Q_1(X)-Q_1(a), \cdots, Q_m(X)-Q_m(a)) - R_i(X_1-a_1, \cdots, X_n-a_n)) > d_i.$$

We first show that each formal power series f invariant under Γ_a with center at a can be expressed as a formal power series in the t expressions

$$p_i(Q_1(X)-Q_1(a), \cdots, Q_m(X)-Q_m(a)), \quad i=1, \cdots, t.$$

Then we show that the formal power series in t variables can be taken as a convergent power series (the selection being quite non-constructive). But in fact, it is easily seen to be enough to consider the case $a=0$, if we formulate the two propositions to deal with the situation as is done below. It will then be apparent that the main reason for taking $a=0$ is one of convenience.

Let $a=0$. The set $Q_1, \cdots, Q_m$ of homogeneous polynomials is called a reduced set of generators of $\mathcal{P}^{\Gamma}$ if every *isobaric* polynomial $R(Y_1, \cdots, Y_m)$ such that $R(Q_1, \cdots, Q_m) \equiv 0$ has no linear terms. If $\{Q_1, \cdots, Q_m\}$ is not reduced, then there exists an isobaric polynomial R such that $R(Q_1, \cdots, Q_m) \equiv 0$ and such that the linear part L of R is non-zero. If Y_j appears with non-zero coefficient in L, then from considerations of degree, this is the only place in R where Y_j can appear, hence Q_j can be expressed as a polynomial in Q_i, $i \neq j$, and thus eliminated from the set of generators. Hence, we may assume $\{Q_1, \cdots, Q_m\}$ is a reduced set of generators of $\mathcal{P}^{\Gamma}$. Now denote the Γ-invariant formal power series (resp. convergent power series) by $\mathfrak{F}^{\Gamma}$ (resp. by $\mathcal{O}^{\Gamma}$).

PROPOSITION 5. *Let $Q_1, \cdots, Q_m$ be a reduced set of generators of*

$\mathcal{P}^r$, deg $Q_i = d_i$, *and let* $f_i \in \mathfrak{F}^r$ *be such that* $\omega(f_i - Q_i) > d_i$. *Then there exist m formal power series* $F_1, \cdots, F_m$ *such that*

(1) $F_i(Q_1, \cdots, Q_m) = f_i$, $i = 1, \cdots, m$ *and*

(2) *if* L_i *is the linear homogeneous part of* F_i, *then* $L_1, \cdots, L_m$ *are linearly independent.*

PROOF. We may assume $d_1 \leqq \cdots \leqq d_m$. From the definition of $Q_1, \cdots, Q_m$, it is obvious that there are m formal power series $F_1, \cdots, F_m$ such that $F_i(Q_1, \cdots, Q_m) = f_i$, $i = 1, \cdots, m$, and by hypothesis $\omega(f_i - Q_i) = \omega(F_i(Q_1, \cdots, Q_m) - Q_i) > d_i$, so for $d \leqq d_i$, the terms of weight d of $L_i(Y_1, \cdots, Y_m) - Y_i$ are 0. Therefore the matrix of the linear forms L_i is unipotent upper triangular. This proves the proposition.

COROLLARY. *Notation being as in Proposition 5, each* Q_i *can be expressed as a formal power series in* $f_1, \cdots, f_m$.

PROOF. This follows from the "formal implicit function theorem", since the linear forms $L_1, \cdots, L_m$ are linearly independent.

PROPOSITION 6. *Let the notation be as in Proposition 5. If* $f_1, \cdots, f_m$ *are convergent power series, then they analytically generate* $\mathcal{O}^r$ (i.e., *given* $f \in \mathcal{O}^r$, *there exists a convergent power series* $g(X_1, \cdots, X_m)$ *such that* $f = g(f_1, \cdots, f_m)$).

PROOF. Let $\mathcal{A}$ be the subring of $\mathcal{O}^r$ analytically generated by $f_1, \cdots, f_m$. The Krull topology on $\mathfrak{F} = \boldsymbol{C}[[X_1, \cdots, X_n]]$ is by definition the topology of $\mathfrak{F}$ as a topological ring in which the powers of the maximal ideal $\mathfrak{m}$ are a neighborhood basis of 0. Clearly $\mathcal{A}$ is Krull-dense in $\mathcal{P}^r$ by the preceding corollary. We want to show $\mathcal{A}$ is Krull-closed in $\mathcal{O}^r$. First of all, by the relation $\omega(f_i - Q_i) > d_i$, it is clear that $f_1, \cdots, f_m$ have 0 as their unique common zero, and so by the corollary of Proposition 3, $\boldsymbol{C}\{X_1, \cdots, X_n\} = \mathcal{O}$ is integral over $\mathcal{A}$, hence $\mathcal{O}^r$ is integral over $\mathcal{A}$. So if $x \in \mathcal{O}^r$, the $\boldsymbol{C}$-algebra $\mathcal{D}$ generated by x and $\mathcal{A}$ is an $\mathcal{A}$-module of finite type. Now $\mathfrak{m} \cap \mathcal{A}$ is the maximal ideal of $\mathcal{A}$ and is the set of convergent power series in $f_1, \cdots, f_m$ with zero constant term. Since $1 + y$ is invertible in $\mathcal{A}$ for all $y \in \mathfrak{m} \cap \mathcal{A}$, we may apply [65 : vol. II, Thm. 9, p. 262] and see that $\mathcal{A}$ is closed in $\mathcal{D}$, hence $\mathcal{D} = \mathcal{A}$, so $x \in \mathcal{A}$. Hence every element x of $\mathcal{O}^r$ is in $\mathcal{A}$, i.e., $\mathcal{O}^r = \mathcal{A}$, proving the proposition.

Thus $\mathcal{R}=\mathcal{R}'$.

Now for any finite group Γ of linear transformations operating on C^n, it is easily seen that $\mathcal{O}^\Gamma$ is integrally closed in its quotient field. Hence $\mathcal{R}_a \cong \mathcal{O}_a^{\Gamma_a}$ is integrally closed for any $a \in X$. Therefore $(X, \mathcal{R})$ is a normal analytic space. (The purely algebraic argument referred to earlier [15b] showing $Q(X)$ to be an affine algebraic variety shows in fact that $Q(X)$ is a normal affine variety.)

EXAMPLE. Let $n=2$ and let $\Gamma=\{\pm id.\}$. Then 0 is the only fixed point of $-id.$ and an analytic function at 0 is invariant under Γ if and only if in its power series expansion only terms of even degree occur. Any homogeneous polynomial of even degree is a polynomial in the three expressions z_1^2, z_1z_2, and z_2^2, these separate the orbits of Γ, and at any point different from (0, 0), a suitably chosen pair of them serve as local coordinates on the quotient space $C^2/\Gamma=X$. The quotient space X may be identified with the image of C^2 under the mapping

$$Q:(z_1, z_2) \mapsto (z_1^2, z_1z_2, z_2^2)=(\xi, \eta, \zeta) \in C^3,$$

and that image is the quadric surface $\eta^2-\xi\zeta=0$ whose unique singular point is at the origin.

CHAPTER 3
HOLOMORPHIC FUNCTIONS AND MAPPINGS ON A BOUNDED DOMAIN

§ 1. Semi-norms and norms

Let K be a field which is either $\boldsymbol{R}$ or $\boldsymbol{C}$ supplied with the usual absolute value. Let X be a linear space over K. A real-valued function p on X is called a semi-norm if

(1) $$p(x+y) \leqq p(x)+p(y), \text{ and}$$

(2) $$p(\alpha x) = |\alpha|\, p(x),$$

for all $x, y \in X$, $\alpha \in K$. These conditions imply, in addition, that

(3) $$p(x) \geqq 0 \text{ for all } x \in X.$$

A subset S of X is called [64 : p. 24]:

a) convex if for any two points x, $y \in S$, the real straight line segment between them is contained in S.

b) balanced if for every $x \in S$ and any $\alpha \in K$ such that $|\alpha| \leqq 1$ we have $\alpha x \in S$.

c) absorbing if for any $x \in X$, there exists $\alpha \in K$ such that $\alpha x \in S$.

Then if p is a semi-norm on X and $c > 0$, the set $U_{p,c} = \{x \in X \,|\, p(x) \leqq c\}$ is convex, balanced, and absorbing.

Let $\{p_\alpha\}_{\alpha \in A}$ be a family of semi-norms on X. This family is called separating if for each $x \in X$, $x \neq 0$, there exists $\alpha \in A$ such that $p_\alpha(x) \neq 0$. In this case, the topological linear space X on which a subbasis of neighborhoods of 0 is given by the family of sets $\{U_{p_\alpha, c}\}_{\alpha \in A,\, c > 0}$ is a Hausdorff, locally convex, topological, linear space. We shall call such a space a semi-normed linear space. We call the topology just described for X the weak topology on X. By weak convergence, we mean convergence in this topology, and by a bounded set, we mean a set on which each semi-norm p_α is bounded. Often "weak topology" refers to that given by the semi-norms defined as the absolute values of the members of a family of linear functionals.

If (X, p) is a semi-normed linear space for which the family of semi-norms consists of a single element p such that $p(x)=0$ only when $x=0$, then p is called a norm and (X, p), or simply X, is called a normed linear space. In this case, it is common to write $p(x)=||x||$. A normed linear space X is at the same time a metric space with metric function d given by $d(x, y)=||x-y||$. If X is complete in this metric, it is called a Banach space.

Let S be a Hausdorff, locally compact, topological space supplied with a measure μ. If f is a measurable function on S, we write $f\sim 0$ if $f=0$ except on a set of measure zero, and if f' is another measurable function, we write $f\sim f'$ if $f-f'\sim 0$. Henceforth, we do not distinguish between functions equivalent in this way and denote them by the same letter. If p is a real number $\geqq 1$, define $L^p(S, \mu)=\{f|f$ measurable, complex-valued on S, $\int_S |f|^p d\mu<+\infty\}$, and supply $L^p(S, \mu)$ with the norm $||f||_p=\left(\int_S |f|^p d\mu\right)^{1/p}$; if f is a measurable function on S, define $||f||_\infty$ as the infimum of the real positive numbers r such that $|f|\leqq r$ except on a set of measure zero, if such r exist, and define $||f||_\infty=\infty$ otherwise. The quantity $||f||_\infty$ is called the "essential supremum" of f. Let

$$L^\infty(S, \mu)=\{f|f \text{ measurable}, ||f||_\infty<\infty\}. \tag{4}$$

The space $L^p(S, \mu)$ with norm $|| \ ||_p$ is a Banach space, $1\leqq p\leqq\infty$. The space $L=L(S)$ of bounded, complex-valued continuous functions on S is topologized as a subspace of $L^\infty(S, \mu)$; it is evidently a closed subspace of $L^\infty(S, \mu)$.

We shall need later the following consequence of the Hahn-Banach Theorem [64 : p. 109]:

PROPOSITION 1. *Let X be a locally convex, topological, linear space and let M be a closed subspace of it. Let $x_0\in X-M$. Then there exists a continuous linear functional f on X such that $f(x_0)\neq 0$, $f|M=0$.*

Our general reference for the matters in this section is [64].

§ 2. Bounded families of holomorphic functions

Let D be a domain in $\boldsymbol{C}^n$ and let C be the family of all complex-valued continuous functions on D. For each compact subset A of D we define a semi-norm ν_A on C by

$$\nu_A(f) = \sup_{x \in A} |f(x)|. \tag{5}$$

These obviously make C into a semi-normed, topological, linear space, and since a uniform limit of holomorphic functions is holomorphic, it is clear that the holomorphic functions $\mathcal{O}(D)$ on D form a closed subspace of C.

Using Cauchy's integral formula in the same manner as in the proof of Montel's theorem, one proves that a weakly bounded subfamily of $\mathcal{O}(D)$ is uniformly equi-continuous on any compact subset of D. Let $\{f_n\}$ be a weakly bounded sequence of holomorphic functions on D. By the usual diagonalization process one obtains a subsequence that converges on a countable dense subset of D, hence, by the preceding remarks, that converges in the weak topology to a holomorphic function on D. In particular, any bounded sequence of holomorphic functions on D has a subsequence that converges to a holomorphic function on D.

Let V be a finite-dimensional complex vector space and suppose V is supplied with a positive-definite Hermitian form $(\ ,\ )$ that defines a metric and a norm $\|\ \|$ on V. By a holomorphic function from D to V we of course mean one that for some (and hence any) choice of basis on V has holomorphic coordinate functions. We may define again a family of semi-norms on the family of continuous functions from D to V: If A is a compact set in D, then $\nu_A(f) = \sup_{x \in A} \|f(x)\|$. Again, the family $\mathcal{O}(D, V)$ of holomorphic mappings from D to V is a closed subspace, and by applying the preceding results to each coordinate function, we see that a weakly bounded (and in particular a bounded) sequence in $\mathcal{O}(D, V)$ has a convergent subsequence.

Now in particular we may apply these results to holomorphic mappings of a bounded domain D into itself and obtain the result:

PROPOSITION 2. *Let D be a bounded domain in $\boldsymbol{C}^n$ and let $\{T_m\}$ be* *a sequence of holomorphic mappings of D into itself. Then $\{T_m\}$ has a convergent subsequence which has as limit a holomorphic mapping T of D into the closure $\bar{D}$ of D.*

Of course, the difficult point is to prove that, under some additional hypotheses, $T(D)\subset D$.

§ 3. The holomorphic automorphism group of D

Let D be a bounded domain in $\boldsymbol{C}^n$. Clearly the set of one-to-one biholomorphic mappings of D onto itself forms a group, which we denote by Hol(D), or more briefly, for present purposes, by G.

We supply G with the compact-open topology; this is the same as the topology that G receives as a subspace of the space of holomorphic mappings of D into $\boldsymbol{C}^n$, where the topology is that given by the semi-norms ν_A of the last section. Since D has a countable neighborhood base and is locally compact, and since, by definition, G operates effectively on D, it follows by standard arguments that G is Hausdorff and also has a countable neighborhood base.

LEMMA 1. *Let A_1 and A_2 be two compact subsets of D and define*

$$G_{A_1,A_2}=\{g\in G\mid A_1\cdot g\cap A_2\neq\emptyset\}.$$

Then G_{A_1,A_2} is compact.

PROOF. By what we already know of the topology on G, it is sufficient to show that G_{A_1,A_2} is sequentially compact. Let $\{T_m\}$ be a sequence in G_{A_1,A_2}; then, for each m there exists $(a_m, b_m)\in A_1\times A_2$ such that $T_m(a_m)=b_m$. Replacing $\{T_m\}$ by a subsequence we may assume by the preceding proposition that (a_m, b_m) converges to a limit $(a_0, b_0)\in A_1\times A_2$, that $\{T_m\}$ converges to a limit T_∞, and that $\{T_m^{-1}\}$ converges to a limit T_0, where T_0 and T_∞ are holomorphic mappings of D into $\bar{D}$. Clearly $T_0(b_0)=a_0$, $T_\infty(a_0)=b_0$. Let $\mathcal{N}$ be a neighborhood of b_0 with compact closure contained in D and let $\mathcal{M}$ be a neighborhood of a_0 with compact closure $\bar{\mathcal{M}}\subset D$ such that $T_\infty(\bar{\mathcal{M}})\subset\mathcal{N}$. Then T_0T_∞ is defined on $\bar{\mathcal{M}}$ and the composition $T_m{}^{-1}T_{m'}$ converges uniformly on $\bar{\mathcal{M}}$ to T_0T_∞. Since infinitely many elements of the sequence $T_m{}^{-1}T_{m'}$ are equal to the identity, it follows that T_0T_∞ is the identity on $\bar{\mathcal{M}}$. Similarly, there is a neighborhood $\mathcal{M}'$ of b_0 such that $T_\infty T_0$

is defined and equal to the identity there. It now remains to prove that the ranges of T_0 and of T_∞ lie in D. We treat the case of T_∞. Let $b \in D$ and let B be a compact neighborhood of b, $B = \bar{B} \subset D$. The compositum $T_l^{-1}T_m$ is defined for all l and m. We claim that $T_l^{-1}T_m$ converges weakly to the identity on D. We consider $\{T_l^{-1}T_m - id. = f_{l,m}\}$ as a bounded family of holomorphic mappings of D into $\boldsymbol{C}^n$; the sequence $\{f_{l,m}\}$ converges uniformly to zero on $\bar{\mathcal{M}}$ which has non-empty interior. Considering each of the coordinate functions of $f_{l,m}$ we see that to establish our claim it is sufficient to prove

LEMMA 2. *Let $\{f_k\}$ be a bounded sequence of holomorphic functions on a connected domain D and suppose this sequence tends uniformly to 0 on a subset $\bar{\mathcal{M}}$ of D with non-empty interior. Then $\{f_k\}$ converges weakly to 0 on D.*

PROOF OF LEMMA 2. Suppose B is a compact subset of D and suppose there exists an $\varepsilon > 0$ such that for every k_0 there exists $k \geqq k_0$ and $b_k \in B$ such that $|f_k(b_k)| \geqq \varepsilon$. Then replacing $\{f_k\}$ by a subsequence we may assume (1) f_k converges uniformly on every compact subset of D to a holomorphic function f, (2) $\{b_k\}$ converges to a point b of B, and (3) $\{f_k(b_k)\}$ converges to a limit c. By the weak convergence of $\{f_k\}$ we see that $f(b) = c$ and $|c| \geqq \varepsilon$, while $f = 0$ on all of $\mathcal{M}$, hence $f \equiv 0$. Clearly we have a contradiction. This proves Lemma 2.

Thus $T_l^{-1}T_m$ converges weakly to the identity. Let $b \in D$ and let B be a compact neighborhood of b, $B = \bar{B} \subset D$. There exists $N > 0$ such that $l, m > N$ imply $T_l^{-1}T_m$ differs from the identity by less than $\frac{1}{2} d(b, \partial B)$ on B, hence $T_l^{-1}T_m(b)$ remains in a compact neighborhood B_1 of b, $B_1 \Subset B$, for $l, m > N$. Choose and fix $l > N$. Then $T_l T_l^{-1} T_m(b) = T_m(b)$ remains in $T_l(B_1) \Subset T_l(B) \Subset D$ as $m \to \infty$, hence, $T_\infty(b) \in D$. Similarly the range of T_0 is contained in D. Combining this with the information that $T_0 T_\infty$ and $T_\infty T_0$ are equal to the identity on non-empty open sets we see that T_0 and T_∞ are in G. Hence G_{A_1, A_2} is compact.

COROLLARY 1. *G is locally compact.*

This follows from the definition of the compact-open topology.

COROLLARY 2. *If* $a \in D$, *then* $G_a = \{g \in G \mid a \cdot g = a\}$ *is compact.*

COROLLARY 3. *A subgroup* Γ *of* G *acts in properly discontinuous fashion on* D *(definition as in Chapter 1) if and only if* Γ *is a discrete subgroup of* G.

We may now apply Theorem 2, p. 208 of [43] to obtain:

THEOREM 7. G *is a Lie group.*

See also [15a] and [44c].

The purpose of our preceding discussion was to show how the theorem cited from [43] could be made applicable to Hol(D). However, Theorem 7 for this particular case is originally due to H. Cartan [15a] from whose work we have also borrowed the proof of Lemma 1.

§4. A uniqueness theorem of H. Cartan

Now we introduce another topology on the ring $\mathfrak{F} = \boldsymbol{C}[[X_1, \cdots, X_n]]$ of formal power series which is quite distinct from the Krull topology (though if $\boldsymbol{C}$ were replaced by a $\boldsymbol{p}$-adic field, for example, the two topologies would in some sense be complementary). Namely, we introduce the family of semi-norms $\{\nu_{(k)}\}_{(k)=(k_1, \cdots, k_m)}$ determined by $\nu_{(l)}(\sum_{(k)\geqq 0} a_{(k)} X^{(k)}) = |a_{(l)}|$.

By an endomorphism of the $\boldsymbol{C}$-algebra $\mathfrak{F}$, we mean a $\boldsymbol{C}$-linear ring endomorphism T of $\mathfrak{F}$ in the usual sense which is continuous in the Krull topology. Then $T(1) = 0$ or 1; in the first case $T \equiv 0$ and in the second case T is determined by n power series

$$T(X_i) = \sum_{(k)>0} a_{i,(k)} X^{(k)}, \tag{6}$$

having no constant terms. We introduce a system of semi-norms $\{\mu_{(k)}\}$ on End($\mathfrak{F}$) by defining $\mu_{(k)}(T) = \sup_i |a_{i,(k)}|$. By the linear part of T, we mean the transformation $L(T) \in \mathrm{End}(\mathfrak{F})$ defined by

$$L(T)(X_i) = \sum_{k_1+\cdots+k_n=1} a_{i,(k)} X^{(k)}. \tag{6_L}$$

THEOREM 8 (*H. Cartan*). *Let* T *be an automorphism of* $\mathfrak{F}$ *such that* $L(T)$ *is the identity and such that the family* $\{T, T^2, T^3, \cdots\}$ *of powers of* T *is weakly bounded. Then* T *is the identity.*

PROOF (cf. [5 : pp. 13-14]). We have $T(X_i) \equiv X_i \bmod \mathfrak{m}^2$, where $\mathfrak{m}$ is the maximal ideal of $\mathcal{O}$. If $T(X_i) \neq X_i$, let $\mathfrak{m}^r$ be the largest power of $\mathfrak{m}$ such that $T(X_i) \equiv X_i \bmod \mathfrak{m}^r$; we have $r \geqq 2$ and $T(X_i) \equiv X_i + A_{i,r}(X) \bmod \mathfrak{m}^{r+1}$, where $A_{i,r}$ is a homogeneous polynomial of degree r and not identically zero. We claim that $T^s(X_i) \equiv X_i + sA_{i,r}(X) \bmod \mathfrak{m}^{r+1}$ for all $s \geqq 1$. Since this is true for $s=1$, suppose it is true for $s-1$: $T^{s-1}(X_i) \equiv X_i + (s-1)A_{i,r}(X) \bmod \mathfrak{m}^{r+1}$; then $T^s(X_i) \equiv TX_i + (s-1)A_{i,r}(TX) \equiv X_i + sA_{i,r}(X) \bmod \mathfrak{m}^{r+1}$. But the sequence T^s is weakly bounded which is clearly impossible if $A_{i,r} \not\equiv 0$——a contradiction. Hence $T(X_i) = X_i$, $i=1, \cdots, n$, as claimed.

Now let D be a bounded domain, let $G = \mathrm{Hol}(D)$, and $a \in D$. For simplicity assume $a=0$. By Corollary 2 of Lemma 2, $G_0 = \{g \in G \mid 0 \cdot g = 0\}$ is compact, hence is weakly bounded (in either topology). The mapping $T \rightarrow L(T)$ is a homomorphism of G_0 into the general linear group of $\boldsymbol{C}^n$. By the preceding result, we see that $\ker(L) = \{e\}$, thus G_0 is isomorphic to a compact subgroup of the linear group of $\boldsymbol{C}^n$, and the isomorphism is given by viewing $\boldsymbol{C}^n$ as the tangent space to D at 0 and transferring the action of $T \in G_0$ to it in the natural way.

Now we show that there is a one-to-one biholomorphic mapping f of a neighborhood $\mathcal{U}$ of $0 \in D$ onto a neighborhood $\mathcal{V}$ of 0 such that the action of any $g \in G_0$ transferred to $\mathcal{V}$ via f is a linear transformation. Note that this is not achieved in the preceding result because that gives no change of coordinates at 0 to effect the transformation $T \rightarrow L(T)$. Let dg be the Haar measure on G_0 such that $\int_{G_0} 1 \cdot dg = 1$. Define f by

$$\zeta = f(z) = \int_{G_0} z \cdot g \cdot L(g)^{-1} \cdot dg \qquad \text{(action by } G \text{ on the right and viewing } z \in D \text{ as a row vector).}$$

The integral converges since the coordinate functions of g are continuous on the product of G_0 with any compact subset of D. A straight-forward calculation gives

$$\begin{aligned} \zeta(z) \cdot L(g') &= \int_{G_0} z \cdot g \cdot L(g)^{-1} L(g')\, dg \\ &= \int_{G_0} z \cdot g \cdot L(g'^{-1}g)^{-1}\, dg = \int_{G_0} z \cdot g' g L(g)^{-1}\, dg = \zeta(z \cdot g'), \end{aligned}$$

while the functional determinant of f at 0 is readily calculated to be unity, which gives us what we want:

PROPOSITION 3. *In a suitable system of coordinates in a neighborhood of* 0, G_0 *acts by linear transformations.*

CHAPTER 4
ANALYSIS ON DOMAINS IN C^n

§1. Measure theory

In this section, we set down without proof some further definitions and facts from measure theory for our present and future needs. Our direct reference is [11a] where chapter, section, and subsection will be cited as [B, Chap., §, no.]; most of the main ideas may also be found in [41].

Let X be a locally compact topological space, and F, a normed vector space over $\boldsymbol{R}$ with norm $|\ |$. $C_F(X)$ will denote the space of continuous F-valued functions on X. If A is a compact subset of X, define the semi-norm p_A on $C_F(X)$ by: $p_A(f)=\sup_{x\in A}|f(x)|$; we give $C_F(X)$ the topology supplied by the semi-norms p_A; when X is compact, this is equivalent to the topology defined by the single norm $|\ |_\infty$ defined by $|f|_\infty=\sup_{x\in X}|f(x)|$. Convergence of a sequence of functions in $C_F(X)$ with respect to the topology defined by the semi-norms p_A, A compact, will be called normal convergence; equivalently, the sequence will be said to converge normally. If $f\in C_F(X)$, let $\mathcal{S}(f)$ be the closure of $\{x\in X\,|\,f(x)\neq 0\}$. Let $\mathcal{K}_F(X)=\{f\in C_F(X)\,|\,\mathcal{S}(f)$ is compact$\}$, $\mathcal{K}(X)=\mathcal{K}_C(X)$, $\mathcal{K}_+(X)=\{f\in\mathcal{K}_R(X)\,|\,f(x)\geqq 0$ for all $x\in X\}$, $L_F(X)=\left\{f\in C_F(X)\,\middle|\,|f|_\infty=\sup_{x\in X}|f(x)|<+\infty\right\}$, $L=L_C$. Then the topology $L_F(X)$ receives as a subspace of $C_F(X)$ is equivalent to the topology supplied by the single norm $|\ |_\infty$.

To construct a measure on X, one begins with a linear functional on $\mathcal{K}_R(X)$ which is supposed to be continuous on each of the subspaces

(1) $$\mathcal{K}_A(X)=\{f\in\mathcal{K}_R(X)\,|\,\mathcal{S}(f)\subset A\},$$

where A is any compact subset of X. We supply the space $M(X)$ of such functionals with the topology it receives as a subspace of the dual space to $\mathcal{K}_R(X)$. Clearly, $M(X)$ is a $\mathcal{K}_R$-module in a natural way. An element μ of $M(X)$ is called positive, $\mu\geqq 0$, if $\mu(f)\geqq 0$ for all $f\in\mathcal{K}_+(X)$; μ is called bounded if it is continuous on all of $\mathcal{K}_R(X)$

topologized as a subspace of $C_R(X)$. The set of bounded μ is denoted by $M'(X)$; it consists of all continuous linear functionals on $\mathcal{K}_R(X)$ and is a Banach space when topologized as the dual of $\mathcal{K}_R(X)$. If $\mathcal{O}$ is an open set of X and $\mu \in M(X)$, let $\mu_{\mathcal{O}}$ be the restriction of μ to $\mathcal{K}_R(\mathcal{O})$. Define the support $\mathcal{S}(\mu)$ of μ by

$$\mathcal{S}(\mu) = X - \{\bigcup_{\mu_{\mathcal{O}}=0} \mathcal{O}\}.$$

The elements of $M(X)$ are called measures. Let $M_c(X)$ be the set of measures with compact support.

PROPOSITION 1. [B, Chap. III, §§ 3, 4]. *Every measure with compact support is bounded.*

Define the measure ε_x by $\varepsilon_x(f) = f(x)$. Then $\mathcal{S}(\varepsilon_x) = \{x\}$. If $\{x_\alpha\}_{\alpha \in A}$ is a discrete set in X without limit point, then $\sum_{\alpha \in A} \varepsilon_{x_\alpha}$ is also a measure, with support $\{x_\alpha \mid \alpha \in A\}$.

Let F be a Banach space and let $\mu \in M(X)$. One may show [B, Chap. III] that there is a unique linear mapping λ_μ of $\mathcal{K}_F(X)$ into F satisfying:

a) If $a \in F$, $g \in \mathcal{K}(X)$, then $\lambda_\mu(ag) = a\mu(g)$, and

b) If A is any compact subset of X, then the restriction of λ_μ to elements of $\mathcal{K}_F(X)$ with support in A is continuous.

We define $\int_X f d\mu = \lambda_\mu(f)$ for $f \in \mathcal{K}_F(X)$.

Let $I_+(X)$ be the space of functions f on X with values in $[0, +\infty]$ which are everywhere lower semi-continuous on X (i.e., for all $x \in X$, $\underline{\lim}_{y \to x} f(y) = f(x)$); if $f \in I_+(X)$, then let $H(f) = \{g \in \mathcal{K}_+(X) \mid g(y) \leqq f(y)$, all $y \in X\}$. One may prove that $f(x) = \sup_{g \in H(f)} g(x)$; for a measure μ and for $f \in I_+(X)$, define

$$\mu^*(f) = \sup_{g \in H(f)} \mu(g),$$

and if we have merely a function f satisfying $f \geqq 0$, define

$$\mu^*(f) = \inf_{h \geqq f,\, h \in I_+(X)} \mu^*(h),$$

while if A is a set, define (with φ_A the characteristic function of A)

$$\mu^*(A) = \mu^*(\varphi_A) = \text{outer measure of } A.$$

If $\alpha > 0$, $\mu^*(\alpha f) = \alpha\mu^*(f)$. A set N in X is called μ-negligible if $\mu^*(N) = 0$.

Again let F be a Banach space and fix a measure μ. If

$1 \leqq p < +\infty$, and if $f \in \mathcal{F}_F(X)$ = the set of all mappings of X into F, then define

$$N_p(f) = N_p(f, \mu) = \mu^*(|f|^p)^{1/p}$$

and let

$$\mathcal{F}_F{}^p(X) = \{f \in \mathcal{F}_F(X) \mid N_p(f) < +\infty\}.$$

The function N_p defines a (non-Hausdorff) topology on $\mathcal{F}_F{}^p(X)$ in the same way as a norm, except that $N_p(f)=0$ does not necessarily imply $f=0$. Let $L_F{}^p(X)'$ be the closure of $\mathcal{K}_F(X)$ in $\mathcal{F}_F{}^p(X)$, i.e., the set

$$\{f \in \mathcal{F}_F{}^p(X) \mid \text{for any } \varepsilon > 0 \text{ there exists } g \in \mathcal{K}_F(X) \text{ such that } N_p(f-g) < \varepsilon\}.$$

Then we denote by $L_F{}^p(X)$ the space obtained from $L_F{}^p(X)'$ by identifying functions that differ only on a μ-negligible set. In the future, we make no distinction between an element of $L_F{}^p(X)'$ and the equivalence class of $L_F{}^p(X)$ to which it belongs. If $F = \mathbf{C}$, put $L_C{}^p(X) = L^p(X)$. The elements of $L_F{}^1(X) = L_F{}^1(X, \mu)$ are called μ-integrable. A set A is called μ-integrable if φ_A (its characteristic function) is. We shall find it convenient to use the alternative notation $N_p(f) = \|f\|_p$.

If X is locally compact, $\mu \geqq 0$ a measure on X, then a mapping f of X into another topological space Y is called μ-measurable if for any compact set $K \subset X$, there exists a μ-negligible set $N \subset K$ such that $K - N$ is a countable disjoint union of compact sets K_n such that $f \mid K_n$ is continuous. Obviously every continuous mapping is measurable (let $N = \emptyset$). A set A is called measurable if $\varphi_A : X \to \{0, 1\}$ is a measurable mapping. To see the relation between this concept and the ordinary notion of Lebesgue measurability, we note that for Lebesgue measure λ, a set E is measurable if its inner and outer measures coincide, and that then E differs from the union of a countable family of closed sets which it contains by a set of measure zero.

A mapping φ of locally compact spaces, $\varphi : X \to Y$, is called μ-proper if it is μ-measurable and if for each compact subset K of Y, $\varphi^{-1}(K)$ is μ-integrable. If that is so, then there exists a unique measure ν on Y such that $\mu(f \circ \varphi) = \nu(f)$ for all $f \in K_F(Y)$. More precisely, $f \in \mathcal{F}_F(Y)$ is ν-integrable if and only if $f \circ \varphi$ is μ-integrable, and then $\mu(f \circ \varphi) = \nu(f)$.

Thus we note that a proper continuous mapping is μ-proper, and in particular, if X is a closed subspace of Y, then every measure on X naturally induces one on Y.

§ 2. L^p-spaces on a domain

Let D be a domain in $\boldsymbol{C}^n$ and let dv be the ordinary Euclidean measure there. Let $\mathcal{O}(D)$ be the $\boldsymbol{C}$-algebra of holomorphic functions on D, let $L^p(D)$, $p \geqq 1$, be the family of measurable, complex-valued functions f on D satisfying

$$|f|_p = \left(\int_D |f|^p \, dv\right)^{1/p} < +\infty,$$

$L^\infty(D)$ being the family of essentially bounded measurable functions on D, and let $\mathcal{O}^p(D) = \mathcal{O}(D) \cap L^p(D)$, $1 \leqq p \leqq +\infty$. Then $L^p(D)$ is a Banach space, and by the inequality (5) of Chapter 2, convergence of a sequence of holomorphic functions on D with respect to the norm $|\ |_p$ implies uniform convergence on any compact subset of D; hence, $\mathcal{O}^p(D)$ is a closed subspace of $L^p(D)$, $1 \leqq p \leqq +\infty$. If D is a bounded domain, then $v(D) = \int_D dv < +\infty$ and we have $L^p(D) \subset L^q(D)$ if $1 \leqq q \leqq p \leqq +\infty$.

§ 3. The Bergmann kernel function

With notation as in the preceding section, let $H = \mathcal{O}^2(D)$. As $L^2(D)$ is a complete separable Hilbert space with inner product (f, g) given by

$$(f, g) = \int_D f\bar{g} \, dv,$$

and norm $|f|_2 = (f, f)^{1/2}$, and as H is a closed subspace of $L^2(D)$, it follows that H is also a separable, complete, Hilbert space. Let $\{\varphi_\nu\}$ be an orthonormal basis of H and for $z, \zeta \in D$, define

$$\boldsymbol{K}_D(z, \zeta) = \boldsymbol{K}(z, \zeta) = \sum_\nu \varphi_\nu(z) \overline{\varphi_\nu(\zeta)}. \tag{2}$$

It may happen, for example if D is the whole of $\boldsymbol{C}^n$, that $H = \{0\}$. But in any case the series for $\boldsymbol{K}(z, \zeta)$ converges uniformly on compact subsets of $D \times D$, and if $f \in H$, then we have, following the usual Hilbert space arguments, that

$$f = \sum_\nu (f, \varphi_\nu) \cdot \varphi_\nu, \quad \text{or} \tag{3}$$

(4) $$f(z)=\int_D K(z,\zeta)f(\zeta)\,dv_\zeta,$$

where the usual Hilbert space a.e. equality in (3) and (4) becomes true equality everywhere because of our knowledge that convergence in L^2 for holomorphic functions implies uniform convergence on every compact set. As for the convergence of the series in (2), let

$$D^r=\{z\in D\,|\,\mathrm{dist}(z,C^n-D)>r\};$$

then, as in [5 : pp. 121–122], one readily verifies that

(5) $$\sum_\nu |\phi_\nu(z)|^2\leqq\omega_r^2,\quad z\in D^r,$$

where ω_r is a positive constant depending only on r; from this the asserted convergence follows easily.

Now it follows from general properties of Hilbert space that K is actually independent of the choice of orthonormal basis $\{\phi_\nu\}$. So let D' be another domain and assume φ is a biholomorphic homeomorphism of D' onto D. Then we have

(6) $$\int_D|f(z)|^2\,dv_z=\int_{D'}|f(\varphi(\zeta))|^2\,dv_{\varphi(\zeta)}=\int_{D'}|f(\varphi(\zeta))|^2\,|j(\zeta,\varphi)|^2\,dv_\zeta,$$

where $j(\zeta,\varphi)$ is the Jacobian determinant of φ at ζ. Hence, the mapping

(7) $$T_\varphi: f\mapsto f_\varphi : f_\varphi(\zeta)=f(\varphi(\zeta))\,j(\zeta,\varphi)$$

is an isometry of H onto $H'=\mathcal{O}^2(D')$. Therefore $\{\phi_{\nu\varphi}\}$ is an orthonormal basis of H' and so

(8) $$K_{D'}(\zeta_1,\zeta_2)=\sum_\nu \phi_{\nu\varphi}(\zeta_1)\overline{\phi_{\nu\varphi}(\zeta_2)}=j(\zeta_1,\varphi)\overline{j(\zeta_2,\varphi)}\,K_D(\varphi(\zeta_1),\varphi(\zeta_2)).$$

Now $K(z,z)$ is real-valued, and we have in particular from (8) that

(9) $$|j(\zeta,\varphi)|^2 K_D(\varphi(\zeta),\varphi(\zeta))=K_{D'}(\zeta,\zeta).$$

Henceforth write simply $K(z,z)=K(z)$. Then $K(z)$ is a real analytic function on the whole of D.

Suppose $K(z)\neq 0$ for every $z\in D$. Then $K(z)>0$ and is real analytic on all of D; hence, $\log K$ is real analytic on all of D and we define

(10) $$g_{\alpha\bar\beta}(z)=\frac{\partial^2}{\partial z_\alpha\partial\bar z_\beta}\log K(z).$$

We have, by easy calculation,

$$K^2 \frac{\partial^2}{\partial z_\alpha \partial \bar{z}_\beta} \log K = \sum_{\mu, \nu} \{\phi_\mu \bar{\phi}_\mu \phi_{\nu\alpha} \bar{\phi}_{\nu\beta} - \phi_\mu \bar{\phi}_\nu \bar{\phi}_{\mu\beta} \phi_{\nu\alpha}\}, \tag{11}$$

where $\phi_{\nu\alpha} = \partial\phi_\nu / \partial z_\alpha$. (Convergence may be verified from Cauchy's integral formula for the derivative.) It follows from (11) and from Schwarz's inequality that the Hermitian matrix $(g_{\alpha\bar{\beta}})$ is non-negative, semi-definite, and will be positive-definite at z if the vectors $\left(\frac{\partial \phi_\nu}{\partial z_1}, \cdots, \frac{\partial \phi_\nu}{\partial z_n}\right)$, $\nu = 1, 2, \cdots$, span C^n, i.e., the $\infty \times n$ matrix

$$(\partial\phi_\nu(z)/\partial z_j)_{j=1, \cdots, n;\ \nu=1, 2, \cdots} \tag{12}$$

has rank n. We henceforth assume D to be such that this is so for every $z \in D$. This will certainly be the case if D is a bounded domain, for then H contains all polynomials. Moreover, if $\varphi : D' \to D$ is biholomorphic, as above, then it follows from (9) and from the Cauchy-Riemann equations (equations (2′) of Chapter 2) that

$$\begin{aligned} g_{\alpha\bar{\beta}}(\varphi(\zeta)) &= \left(\frac{\partial^2}{\partial z_\alpha \partial \bar{z}_\beta} \log K_D\right)(\varphi(\zeta)) \\ &= \sum_{\gamma, \delta} \frac{\partial \zeta_\gamma}{\partial z_\alpha} \frac{\partial \bar{\zeta}_\delta}{\partial \bar{z}_\beta} \frac{\partial^2}{\partial \zeta_\gamma \partial \bar{\zeta}_\delta} (\log K_{D'}(\zeta) - 2 \log |j(\zeta, \varphi)|) \\ &= \sum_{\gamma, \delta} \left(\frac{\partial \zeta_\gamma}{\partial z_\alpha}\right) \left(\overline{\frac{\partial \zeta_\delta}{\partial z_\beta}}\right) g'_{\gamma\bar{\delta}}(\zeta). \end{aligned} \tag{13}$$

We pefine a Hermitian metric on D by $ds^2 = \sum g_{\alpha\bar{\beta}}\, dz_\alpha\, d\bar{z}_\beta$, and denote by ds'^2 the corresponding metric on D'. It follows from (13) that φ is (locally) an isometry with respect to these metrics. In the particular case when $D = D'$, one obtains then a Hermitian metric invariant under (every $g \in$) Hol(D). Likewise, if one defines a volume $d\beta$ in the customary way from ds^2, then $d\beta$ is invariant under Hol(D).

The function K, the metric ds^2, and the Hol(D)-invariant measure $d\beta$ on D are called the Bergmann kernel function, the Bergmann metric, and the Bergmann measure (or volume) on D.

§ 4. Holomorphic completions [5 : Chap. IV]

Let D be a domain in C^n. Suppose D_1 is also a (connected) domain in C^n with the properties (1) $D \subset D_1$, and (2) if f is a holomorphic function on D, then there exists a unique holomorphic function f_1 on

D_1 such that $f_1|D=f$. Let $D^-=\{\bar{z}|z\in D\}$ ($^-$ denoting complex conjugation) and define D_1^- similarly. If D' is another domain and if D_1' is related to it in the same way as D_1 is to D, let F be a holomorphic function on $D\times D'$. By hypothesis, if $z\in D$, then F_z defined on D' by $F_z(z')=F(z, z')$ has a unique holomorphic extension F_{z1} on D_1'; defining F_1 on $D\times D_1'$ by $F_1(z, z')=F_{z1}(z')$, it follows easily from Hartogs' theorem (Chapter 2, section 1) that the set of $w\in D\times D_1'$ such that F_1 is analytic in a neighborhood of w is (open and) closed, hence is all of $D\times D_1'$. Proceeding similarly on the first factor, one obtains finally a (unique) analytic function F_2 on $D_1\times D_1'$ such that $F_2|(D\times D')$ is analytic and coincides with F. Letting $D'=D^-$, we may take $D_1'=D_1^-$. Now $\boldsymbol{K}(z, \zeta)$ is a holomorphic function of $(z, \bar{\zeta})\in D\times D^-$, hence extends to a holomorphic function of $(z, \bar{\zeta})\in D_1\times D_1^-$, and by restricting it to the set of $(z, \bar{z})$, we see that $\boldsymbol{K}(z)$ may be prolonged to a real analytic function on D_1 (cf. [20]).

PROPOSITION 2. *Let D be a bounded domain. Assume there exists a compact subset A of D such that* $A\cdot \mathrm{Hol}(D)=D$. *Then* $\boldsymbol{K}$ *is unbounded on D in every neighborhood of any boundary point of D, i.e., any point of* $\bar{D}-D$.

PROOF. The fact that D is a bounded domain implies that $\boldsymbol{K}_D\neq 0$ everywhere on D (q.v. supra). Let D be a union of subdomains, $D=\bigcup_{n=1}^{\infty} D_n$, $\bar{D}_n=$(closure of $D_n)\subset D_{n+1}$, $v(D-D_n)=vol(D-D_n)<\frac{1}{n}$ (*vol* denotes Euclidean volume). Since A is compact and contained in D, there exists $\omega>0$ such that if $a\in A$, then there is a polydisc σ_a of volume ω and center a such that $\bar{\sigma}_a$ is contained in a fixed compact subset B of D. Let $G_n=G_{B,\bar{D}_n}$, where $G=\mathrm{Hol}(D)$. By Lemma 1, Chapter 3, section 3, G_n is compact, but G is not compact since $D=A\cdot G$ is not compact. Hence, if Δ is an infinite, discrete, closed subset of G, then $\Delta\cap(G-G_n)$ is non-empty for all n. Let $b\in\bar{D}-D$, $\{d_n\}\subset D$ be a sequence with $d_n\to b$, $\delta_n\in G$, $a_n\in A$ such that $a_n\cdot\delta_n=d_n$. By taking a subsequence if necessary, we may assume $\delta_n\in G-G_n$ (if $\delta_n\to\delta\in G$, then $a_n\cdot\delta_n$ must have a subsequence tending to a point of $A\cdot\delta$——a contradiction). Now we have

$$\frac{1}{n}>vol(B\cdot\delta_n)\geqq vol(\sigma_{a_n}\cdot\delta_n)$$

$$=\int_{\sigma_{a_n}} |j(\zeta, \delta_n)|^2 dv_\zeta \geqq vol(\sigma_{a_n}) |j(a_n, \delta_n)|^2 = \omega |j(a_n, \delta_n)|^2,$$

so $j(a_n, \delta_n) \to 0$ as $n \to \infty$, hence, by (9),

$$\boldsymbol{K}(a_n \cdot \delta_n) = |j(a_n, \delta_n)|^{-2} \boldsymbol{K}(a_n) \to \infty$$

as $n \to \infty$, while $a_n \cdot \delta_n \to b$.

Note that we have shown $|j(a, g)| \leqq \frac{\omega}{n}$ for $a \in A$, $g \in G - G_n$.

COROLLARY 1. *With the assumptions as in the proposition, D is a domain of holomorphy.*

PROOF. If every holomorphic function in D had an extension to a bigger domain, then, by known results, there is a point b of $\bar{D} - D$ such that every holomorphic function on D has an extension into a fixed neighborhood, say U, of b, hence by our earlier results $\boldsymbol{K}_D$ would be bounded in a neighborhood of b, which is impossible under the hypotheses of the proposition.

COROLLARY 2. *If D is a bounded homogeneous domain or if Γ is discrete in* Hol(D), *D is bounded, and D/Γ compact, then D is a domain of holomorphy.*

PROOF. In the first case, take A to be a single point of D. In the second, take A to be any compact subset of D such that $A \cdot \Gamma = D$.

(The author wishes to thank R. Narasimhan for a remark that somewhat simplified the statements and proofs of Proposition 2 and its corollaries. Cf. [44c : p. 127].)

§ 5. Finding an orthonormal basis of H

Let D be a bounded domain, let $a \in D$, and let $K = \{g \in \mathrm{Hol}(D) \mid a \cdot g = a\}$. If T_k, $k \in K$, is defined as in (7), section 3, then T_k is a unitary operator on H. As K is compact, this representation of K is fully reducible. If ρ_1 and ρ_2 are inequivalent, finite-dimensional, irreducible representations of K, and if H_1 and H_2 are the isotypic (definition as in Chapter 9, section 2) subspaces of H of types ρ_1 and ρ_2 respectively, then H_1 is orthogonal to H_2. If it happens that the isotypic subspace of H corresponding to a finite-dimensional irreducible representation ρ of K is finite-dimensional for every ρ, then one may find an orthonormal basis of H consisting of elements of such subspaces. In some cases (particularly in the case of a symmetric bounded domain) this makes possible the effective calculation of such a basis [28b] (cf. Chapter 9, section 7).

In fact, the representation $g \mapsto T_g$ defined by (7) is a unitary representation of $G = \mathrm{Hol}(D)$, as remarked there. This is also a useful observation in relating the theory of automorphic forms on D to the general ideas connected with unitary representations in Hilbert space.

CHAPTER 5
AUTOMORPHIC FORMS ON BOUNDED DOMAINS

§1. The quotient of a bounded domain by a discrete group

Let D be a bounded domain in $\boldsymbol{C}^n$ and let Γ be a discrete subgroup of $\mathrm{Hol}(D)$. Then Γ acts in properly discontinuous fashion on D, and the orbit space $X=D/\Gamma$ is a locally compact Hausdorff space. Let $\pi : D\to X$ be the canonical mapping. We define a ringed structure $\mathcal{R}$ on X as follows: If $\mathcal{U}$ is an open subset of X, and if f is a continuous complex-valued function on $\mathcal{U}$, then $f\in\mathcal{R}_{\mathcal{U}}$, by definition, if $f\circ\pi$ is analytic on $\pi^{-1}(\mathcal{U})\subset D$.

THEOREM 9. *$(X, \mathcal{R})$ is a normal complex analytic space.*

PROOF. The theorem is purely local in nature, so it will suffice to show that if $a\in X$, then there is a neighborhood $\mathcal{U}$ of a such that $(\mathcal{U}, \mathcal{R}|\mathcal{U})$ is a normal analytic space. We know that π is an open mapping because Γ acts in properly discontinuous fashion on D. Let $b\in\pi^{-1}(a)$. Since the action of Γ is discontinuous, Γ_b is finite and there exists a neighborhood $\mathcal{V}$ of b such that if $g\in\Gamma$, then $\mathcal{V}g\cap\mathcal{V}\neq\emptyset$ if and only if $g\in\Gamma_b$, and such that $\mathcal{V}g=\mathcal{V}$ for all $g\in\Gamma_b$. Then $\pi(\mathcal{V})=\mathcal{U}$ is a neighborhood of a. By Proposition 3, Chapter 3, section 4, there exists a biholomorphic mapping φ of a small neighborhood of b, which we may assume to be $\mathcal{V}$ itself, onto a neighborhood $\mathcal{N}$ of $0\in\boldsymbol{C}^n$ such that the transferred action of Γ_b on $\mathcal{N}$ is that of a finite group Γ_1 of linear transformations. Now it is clear that $(\mathcal{U}, \mathcal{R}|\mathcal{U})$ is isomorphic to $\mathcal{N}/\Gamma_1$, supplied with the ringed structure described in Chapter 2, section 6, and by the results of that section is therefore a normal analytic space, which completes the proof of the theorem.

§2. Automorphic forms and Poincaré series

Again let D be a bounded domain and let Γ be a discrete subgroup of $G=\mathrm{Hol}(D)$. We define automorphic forms of weight d

with respect to Γ as in Chapter 1, section 1. It is now our purpose to exhibit a way of constructing non-trivial automorphic forms.

Let f be a bounded holomorphic function on D (for example, any polynomial), let d be a positive integer, and form the series

$$\sum_{\gamma \in \Gamma} f(z\cdot\gamma)\, j(z, \gamma)^d, \quad z \in D, \tag{1}$$

whose value will be denoted by $\mathscr{P}_f(z)$ if it converges.

PROPOSITION 1. *The series (1) converges uniformly and absolutely on each compact subset of D if $d \geqq 2$.*

PROOF. Let A be a compact subset of D and let B be a compact subset of D containing A in its interior, and $\omega > 0$ be as in the proof of Proposition 2, Chapter 4, section 4, and for each $a \in A$, let σ_a be a polydisc with center a such that $\sigma_a \subset B$ and $vol(\sigma_a) = \omega$. Let m be the number of elements in $G_{B,B} \cap \Gamma$ and let $M > 0$ be a bound on $|f|$ on D. We first treat the case $d = 2$. Then for $a \in A$ we have

$$|j(a, \gamma)|^2 \leqq \omega^{-1} \int_{\sigma_a} |j(z, \gamma)|^2\, dv_z = \omega^{-1}\, vol(\sigma_a \cdot \gamma). \tag{2}$$

Therefore, if $z \in A$,

$$\begin{aligned} \sum_{\gamma \in \Gamma} |f(z\cdot\gamma)|\, |j(z, \gamma)|^2 &\leqq \sum_{\gamma \in \Gamma} M |j(z, \gamma)|^2 \\ &\leqq M\omega^{-1} \sum_{\gamma \in \Gamma} vol(\sigma_z \cdot \gamma). \end{aligned} \tag{3}$$

If $\sigma_z \cdot \gamma \cap \sigma_z \cdot \gamma' \neq \emptyset$, then $\gamma'\gamma^{-1} \in G_{B,B}$, so that for given z and γ, the number of sets of the form $\sigma_z \cdot \gamma'$ which overlap $\sigma_z \cdot \gamma$ is at most m, hence the collection of sets $\{\sigma_z \cdot \gamma\}_{\gamma \in \Gamma}$ covers any point of D at most m times; therefore,

$$\sum_{\gamma \in \Gamma} |f(z\cdot\gamma)|\, |j(z, \gamma)|^2 \leqq \omega^{-1} Mm\, vol(D), \tag{4}$$

which settles the case $d = 2$. But then $|j(z, \gamma)| < 1$ for $z \in A$ and for all but a finite number of $\gamma \in \Gamma$; and for $|j(z, \gamma)| < 1$, $|j(z, \gamma)|^d$ is a decreasing function of d, and from this we obtain the desired result for all $d \geqq 2$.

Now from the "cocycle relation" for $j(z, g)$, it follows that $\mathscr{P}_f(z\cdot\gamma)\, j(z, \gamma)^d = \mathscr{P}_f(z)$ for all $z \in D$, i.e., $\mathscr{P}_f$ is an automorphic form of weight d with respect to Γ.

Let $a \in D$. Let $\mathcal{N}$ be a neighborhood of 0 in another copy of $\boldsymbol{C}^n$ and let λ be a biholomorphic isomorphism of $\mathcal{N}$ onto a neighborhood $\mathcal{U}$ of a stable under Γ_a such that λ maps 0 onto a and such that $|j(0, \lambda)|=1$. If $\zeta \in \mathcal{N}$ and z is its image under λ, we write $\zeta \cdot \lambda = z$. We assume $\mathcal{N}$, $\mathcal{U}$, and λ chosen such that 1) $\gamma \in \Gamma$, $\mathcal{U} \cdot \gamma \cap \mathcal{U} \neq \emptyset$ imply $\gamma \in \Gamma_a$ (so $\mathcal{U}\gamma = \mathcal{U}$ by the earlier assumption), and 2) $\Gamma_1 = \lambda \Gamma_a \lambda^{-1}$ acts on $\mathcal{N}$ by linear transformations. If f is a holomorphic function on a neighborhood of a satisfying $j(z, \gamma)^d f(z \cdot \gamma) \equiv f(z)$ for all $\gamma \in \Gamma_a$ when $z \cdot \Gamma_a$ is contained in the domain $def(f)$ of f, then we say f is a local automorphic form of weight d (with respect to Γ_a, of course). For such a function, define ${}^\lambda f$ by

$$(5) \qquad {}^\lambda f(\zeta) = j(\zeta, \lambda)^d f(\zeta \cdot \lambda), \quad \zeta \in \mathcal{N} \cap (def(f))\lambda^{-1}.$$

Then application of the cocycle relation shows that

$$ {}^\lambda f(\zeta \cdot \bar{\gamma})\, j(\zeta, \bar{\gamma})^d = f(\zeta), \quad \bar{\gamma} \in \Gamma_1.$$

Since $\bar{\gamma}$ is linear for $\bar{\gamma} \in \Gamma_1$, $j(\zeta, \bar{\gamma})$ is a constant N-th root of unity if $N = \mathrm{ord}(\Gamma_a)$. Let $\mathcal{A}(d)_a$ denote the linear space of germs of local automorphic forms of weight d with respect to Γ_a at a. It is clear from (5) that the assignment $\lambda^* : f \mapsto {}^\lambda f$ is such that for *each* d divisible by N it is an isomorphism of $\mathcal{A}(d)_a$ onto the ring $\mathcal{O}^{\Gamma_1}$ of germs of Γ_1-invariant holomorphic functions at 0, and it is clear that it maps the intersection $\mathfrak{m}'$ of $\mathcal{A}(d)_a$ with the maximal ideal $\mathfrak{m}_a$ of $\mathcal{O}_a$ onto the maximal ideal $\mathfrak{m}_1$ of $\mathcal{O}^{\Gamma_1}$, i.e., it is bicontinuous in the Krull topologies. It is also important to notice that $\mathfrak{m}_a$ is invariant under Γ_a, and that $\mathfrak{m}$, the maximal ideal in $\mathcal{O} = \mathcal{O}_0$, is invariant under Γ_1. Let $\lambda^*(l)$ be the canonical isomorphism of $\mathcal{A}(d, l)_a = \mathcal{A}(d)_a/(\mathfrak{m}')^l$ onto $\mathcal{O}^{\Gamma_1}/(\mathfrak{m}_1)^l$ for all $l > 0$, both modules being finite-dimensional vector spaces over $\boldsymbol{C}$. Let $f \in \mathcal{O}^{\Gamma_1}$ be chosen, and let $f' = \lambda^{*-1}(f)$. Let $l > 0$ be given and let P_l be a polynomial function on D such that $f' - P_l \in \mathfrak{m}_a^{l+1}$; if we form

$$\bar{P}_{l,d} = N^{-1} \sum_{\gamma \in \Gamma_a} (P_l \circ \gamma)\, j(\ ,\gamma)^d,$$

then $\bar{P}_{l,d} \equiv f'(\mathfrak{m}_a^{l+1})$. Let $\bar{\sigma}$ be a compact neighborhood of a, assume $\bar{\sigma} \subset (\mathcal{N})\lambda$, let $\eta = \min_{\zeta \in \lambda^{-1}(\bar{\sigma})} |j(\zeta, \lambda)|^{-1}$, and let $\Gamma(\bar{\sigma}) = \{\gamma \in \Gamma \,\big|\, |j(z, \gamma)| > \frac{1}{2}\eta$, some $z \in \bar{\sigma}\}$. From the proof of Proposition 1, it is clear that $\Gamma(\bar{\sigma})$ is finite. Let $A = a \cdot \Gamma(\bar{\sigma})$. Clearly $a \in A$ and $\Gamma_a \subset \Gamma(\bar{\sigma})$. We now suppose that P_l vanishes to order $l+1$ at every point of $A - \{a\}$ and form

$$P'_{l,d}=N^{-1}\cdot \sum_{\gamma\in\Gamma(\bar\sigma)} (P_l\circ\gamma)\,j(\ ,\gamma)^d;$$

then $P'_{l,d}\equiv f' \bmod \mathfrak{m}_a^{l+1}$, that is, the terms in the power series developments of $P'_{l,d}$ and of f' up to and including terms of order l are the same. On the other hand, let

$$P''_{l,d}=N^{-1}\cdot \sum_{\gamma\in\Gamma-\Gamma(\bar\sigma)} (P_l\circ\gamma)\,j(\ ,\gamma)^d.$$

Then $|P''_{l,d}|=O\left(\left(\frac{\eta}{2}\right)^d\right)$ on $\bar\sigma$ as $d\to\infty$, and $\mathcal{P}_{l,d}$ defined by

$$\mathcal{P}_{l,d}(z)=N^{-1}\cdot \sum_{\gamma\in\Gamma} P_l(z\cdot\gamma)\,j(z,\gamma)^d$$

is an automorphic form given by a series (1). We have $\lambda^*(\mathcal{P}_{l,d}) =\lambda^*(P'_{l,d})+\lambda^*(P''_{l,d})$, and

α) $\lambda^*(P'_{l,d})\equiv f \bmod \mathfrak{m}_0^{l+1}$;

β) $|\lambda^*(P''_{l,d})|=O(2^{-d})$ on $(\bar\sigma)\lambda^{-1}$.

Now f is fixed. but as $d\to\infty$, each power series coefficient of $\lambda^*(P''_{l,d})$ goes to zero by Cauchy's integral formula applied to a fixed, small polycylindrical neighborhood of 0. The same thing can be done independently at any finite number of points $\{a_1, \cdots, a_k\}$ of D, as long as they lie in different orbits of Γ——one only need find the polynomial P_l subject to a finite number of conditions for each of the orbits $a_i\cdot\Gamma$, $i=1, \cdots, k$, and replace the integer $N=\text{ord}(\Gamma_a)$ by the least common multiple of $\{\text{ord}(\Gamma_{a_i})\}_{i=1,\cdots,k}$. Thus we have proved

THEOREM 10 (*Giraud, H. Cartan* [15c]). *Let $a_1, \cdots, a_k\in D$ belong to distinct orbits of Γ and let a positive integer l be given. Adopt notation as above for each point a_i, attaching the subscript i to those objects associated to a_i. Let $f_i\in\mathcal{O}^{\Gamma_{1i}}$, $i=1, \cdots, k$. Let $\varepsilon>0$ be given. Then there exists $d>0$ and a series (1) such that the power series coefficients of $\lambda_i^*(\mathcal{P}_f)$ differ from those of f_i, up to terms of order l, by not more than ε in absolute value, $i=1, \cdots, k$.*

But now, by forming linear combinations of $\mathcal{P}_f$ and observing that $\mathcal{P}_f+\mathcal{P}_{f'}=\mathcal{P}_{f+f'}$, we obtain:

COROLLARY. *Under the same hypotheses as in the theorem, the same conclusion holds with $\varepsilon=0$; i.e., there exists a series (1) such* *that* $\lambda_i^*(\mathcal{P}_f)\equiv f_i \bmod \mathfrak{m}_{0i}^{l+1}$, $i=1, \cdots, k$.

Again let us go back to our notation for the single point a. We know from section 6 of Chapter 2 that there exist a finite number of homogeneous Γ_1-invariant polynomials $Q_1, \cdots, Q_m$ of positive degrees such that the mapping

$$Q: \zeta \mapsto (Q_1(\zeta), \cdots, Q_m(\zeta))$$

gives a mapping of a neighborhood $\mathcal{N}$ of 0 onto a normal analytic set Y which is isomorphic to $\mathcal{N}/\Gamma_1$ with its natural ringed structure as an analytic space. Moreover, if $f_1, \cdots, f_m \in \mathcal{O}^{\Gamma_1}$ are such that $f_i \equiv Q_i$ (mod $\mathfrak{m}_0^l$) for sufficiently large l, $i=1, \cdots, m$, then the same thing is true of the mapping

$$f: \zeta \mapsto (f_1(\zeta), \cdots, f_m(\zeta)),$$

possibly on a smaller neighborhood than $\mathcal{N}$. Now if we add to the set $\{f_i\}_{i=1,\cdots,m}$ another element $f_0 \in \mathcal{O}^{\Gamma_1}$, $f_0 \equiv 1$ (mod $\mathfrak{m}_0^l$), then we may replace the set $(f_1, \cdots, f_m)$ by the set $(f_1/f_0, \cdots, f_m/f_0)$, or, changing our point of view slightly, we may take the mapping

$$f^{\sim}: \zeta \mapsto [f_0(\zeta) : \cdots : f_m(\zeta)]$$

of a neighborhood of 0 into a neighborhood of the point $[1 : 0 : \cdots : 0]$ in m-dimensional complex projective space $\boldsymbol{CP}^m$.

The series (1) are called Poincaré series of weight d. Now applying the above Corollary we see that given any $a \in D$, there exists a Poincaré series which does not vanish at a. If A is a compact set in D, it follows that there exist a finite number of Poincaré series $\mathcal{P}_1, \cdots, \mathcal{P}_t$ such that for any $a \in A$ there exists an index i, $1 \leqq i \leqq t$, such that $\mathcal{P}_i(a) \neq 0$. Let the weight of $\mathcal{P}_i$ be d_i, $d = \text{l.c.m.}(d_i)_{i=1,\cdots,t}$. Then $\mathcal{P}_1^{d/d_1}, \cdots, \mathcal{P}_t^{d/d_t}$ are automorphic forms of weight d without common zeros on A. Likewise, with the help of the Corollary, given two points a and b of D in distinct orbits of Γ, there exist Poincaré series $\mathcal{P}$ and $\mathcal{P}'$ such that $\mathcal{P}(a) = \mathcal{P}'(b) = 0$ and $\mathcal{P}'(a)\,\mathcal{P}(b) \neq 0$.

Now let A be a compact subset of D and let $\psi_0, \cdots, \psi_s$ be automorphic forms of the same weight d having no common zeros on A. Define a mapping into s-dimensional projective space $\boldsymbol{CP}^s$ by

$$\psi(z) = [\psi_0(z) : \cdots : \psi_s(z)].$$

Then ψ is a well-defined mapping on A and if $a, b \in A$ belong to the same orbit of Γ, then $\psi(a) = \psi(b)$. Suppose $a, b \in A$ belong to different orbits of Γ but $\psi(a)=\psi(b)$. Let $\mathcal{P}$ and $\mathcal{P}'$ be automorphic forms such that $\mathcal{P}(a)=\mathcal{P}'(b)=0$, $\mathcal{P}'(a)\,\mathcal{P}(b)\neq 0$, let d' be the l.c.m. of the weights of $\psi_0, \cdots, \psi_s, \mathcal{P}$, and $\mathcal{P}'$, and let $\psi_0', \cdots, \psi_{s'}'$, be all monomials in these of weight d'. Then ψ' with coordinates $\{\psi_i'\}$ is also a well-defined mapping of A into $\boldsymbol{CP}^{s'}$, $\psi'(c)\neq\psi'(d)$ if $\psi(c)\neq\psi(d)$, and $\psi'(a)\neq\psi'(b)$. We say of two such mappings ψ and ψ', that ψ' dominates ψ, $\psi'\succ\psi$, if ψ' is defined wherever ψ is defined and if $\psi(c)\neq\psi(d)$ implies $\psi'(c)\neq\psi'(d)$ for all c and d in $def(\psi)$. We may now form a sequence $\{\psi^{(k)}\}$ such that $\psi^{(k+1)}\succ\psi^{(k)}$, such that each $\psi^{(k)}$ is defined at every point of A, and such that if $\psi^{(k)}$ is not one-to-one on orbit equivalence classes within A, then there exist a and $b\in A$, $a\notin b\cdot\Gamma$, such that $\psi^{(k)}(a)=\psi^{(k)}(b)$ but $\psi^{(k+1)}(a)\neq\psi^{(k+1)}(b)$. The set Δ_k of pairs $(x, y)\in D^k\times D^k$, where D^k is the open set on which $\psi^{(k)}$ is well-defined, such that $\psi^{(k)}(x)=\psi^{(k)}(y)$ is an analytic subset of $D^k\times D^k$ and $\Delta_k\supset\Delta_{k+1}\cap(D^k\times D^k)$. Applying the descending chain condition on analytic subsets of $D^k\times D^k$ meeting a given compact set, one obtains a mapping $\psi^{(0)}$ which is one-to-one on orbits of Γ meeting A. Again applying the discussion just following the Corollary, one may enrich the mapping $\psi^{(0)}$, if necessary, to obtain a mapping $\psi^* : z\mapsto[\psi_0^*(z) : \cdots : \psi_M^*(z)]$ such that for each $a\in A$, there exists a neighborhood $\mathcal{U}$ of a stable under Γ_a such that ψ^* maps $\mathcal{U}$ onto an analytic set Z in a neighborhood of $\psi^*(a)$ in $\boldsymbol{CP}^M$ and induces an isomorphism of the analytic space Z onto $\mathcal{U}/\Gamma_a$. If, finally, we understand A/Γ to mean the space of orbits of Γ meeting A, then ψ^* is one-to-one on A/Γ and induces an injection with respect to its natural complex structure.

Suppose now that D/Γ is compact. Then there is a compact subset A of D such that $A\cdot\Gamma=D$. Applying the above, we obtain

THEOREM 11. *Let D/Γ be compact. Then there exist a finite number of automorphic forms $\psi_0, \cdots, \psi_M$ of the same weight with respect to Γ such that*

$$\psi : z\mapsto[\psi_0(z) : \cdots : \psi_M(z)]$$

induces an injection of the complex space D/Γ into $\boldsymbol{CP}^M$.

It is almost a trivial consequence of the Remmert-Stein theorem (section 5 of Chapter 2) that a closed analytic subset of complex projective space is an algebraic variety (Chow's theorem) [44a : pp. 125–127]. Hence, Theorem 11 can be rephrased by saying that under the given hypotheses, D/Γ is an algebraic projective variety.

This form of the proof, based on Poincaré series, is due to H. Cartan [51]. Another quite different way of proving the same result is essentially based on a criterion of Kodaira : One uses the fact that the 2-form associated to the Bergmann metric is in the characteristic class of the negative of the canonical bundle on the V-manifold D/Γ. More details may be found in [2a, b].

PART II

AUTOMORPHIC FORMS ON A BOUNDED SYMMETRIC DOMAIN AND ANALYSIS ON A SEMI-SIMPLE LIE GROUP

CHAPTER 6
EXAMPLES FOR ALGEBRAIC GROUPS

§1. Definitions for algebraic groups and arithmetic subgroups

If S is a compact Riemann surface of genus >1, the universal covering surface of S is complex-analytically isomorphic to the unit disc σ, the fundamental group Γ of S acts in properly discontinuous fashion on σ, and the orbit space σ/Γ is compact and isomorphic to S. Aside from groups that arise in this way and those derived from them in trivial fashion by taking products, etc., about the only known groups Γ which operate discontinuously on a bounded domain D such that D/Γ is compact, or else has a reasonable compactification, are obtained as arithmetic subgroups of algebraic groups.

In the next chapter, we shall present known results on algebraic groups largely without proofs. Since these results are rather technical, we have decided to insert this chapter to give some examples first, at the same time providing some of the more elementary definitions as well.

If A is an associative algebra, let A^* be the group of its invertible elements. If V is a finite-dimensional vector space over a field k, let $\mathrm{End}(V)$ be the k-algebra of its linear endomorphisms and denote by $GL(V)=\mathrm{End}(V)^*$ the group of non-singular linear transformations of V. Ordinarily the field k will be a subfield of the field $\boldsymbol{C}$ of complex numbers, but all the general results of this chapter and the next hold as long as k is of characteristic zero. By a linear algebraic group we mean a subgroup G of $GL(V)$, for some such vector space V and field k, with the property that G is the intersection of $GL(V)$ with an affine algebraic subvariety W of $\mathrm{End}(V)$. If K is a field, we say that G (or W) is defined over K if the ideal $\mathfrak{J}(W)$ associated to W in the ring of polynomial functions on $\mathrm{End}(V)$ is generated by a finite number of polynomials with coefficients in K (all this, of course, with respect to some choice of basis for V). If L is a field containing K, we denote by G_L the group of L-rational points in G, i.e., the group of matrices in G (with respect to the given basis) having all entries in L. Suppose that $K=\boldsymbol{Q}$ and that Λ is a lattice in V_R (i.e., a discrete subgroup of the real vector space V_R such that V_R/Λ is compact [60e]) contained in V_Q; let

$$G_\Lambda=\{g\in G\,|\,g\cdot\Lambda=\Lambda\}. \tag{1}$$

A subgroup Γ of G_R is called arithmetic if it is commensurable with G_Λ, i.e., if $G_\Lambda\cap\Gamma$ is of finite index in Γ and in G_Λ. Clearly $G_\Lambda\subset G_Q$; and if Λ' is another such lattice, then $G_{\Lambda'}$ is also arithmetic, so that the definition of arithmetic group does not depend on the choice of lattice. It is obvious that an arithmetic subgroup of G_R is also a discrete subgroup of G_R. Less obvious is the following [6b : § 5] :

PROPOSITION 1. *Let ρ be a rational homomorphism (for the definition see Chapter 7) defined over $\boldsymbol{Q}$ of G onto an algebraic subgroup G' of $GL(V')$. If Γ is an arithmetic subgroup of G_R, then $\rho(\Gamma)$ is an arithmetic subgroup of G'_R.*

§ 2. **Some examples** (cf. [6d])

If we take $W=\mathrm{End}(V)$, then $G=GL(V)$ is defined over the prime field $\boldsymbol{Q}$. Now we suppose that V is the n-dimensional vector space $\boldsymbol{C}^n$ over $\boldsymbol{C}$, so that $GL(V)$ becomes the group $GL(n, \boldsymbol{C})$ of non-singular, complex, $n\times n$ matrices $g=(g_{ij})_{i,j=1,\cdots,n}$. Clearly G is connected. The subgroup $SL(n, \boldsymbol{C})=\{g\in G\,|\det g=1\}$ is also connected, as it is generated by all the unipotent elements g for which all $g_{ii}=1$, $1\leqq i\leqq n$, and just one g_{ij} with $i\neq j$ is non-zero. Let $B=\{g\in G\,|\,g_{ij}=0 \text{ for } i>j\}$, the group of upper triangular matrices in G, $U=\{g\in B\,|\,g_{ii}=1,\ i=1,\cdots,n\}$, and $T=\{g\in G\,|\,g_{ij}=0 \text{ for } i\neq j\}$. Then U and T are connected, closed subgroups of B, U is a normal subgroup of B, and B is the semi-direct product of T and U. Let $\boldsymbol{N}=\boldsymbol{N}(T)$ be the normalizer of T in G, and let $Z=Z(T)$ be the centralizer of T. It is clear that $Z=T$. The matrix representation of $g=(g_{ij})$ is with respect to a basis $\{v_i\}$ of V consisting of eigenvectors of T; and the normalizer of T must permute the corresponding one-dimensional eigenspaces $\boldsymbol{C}v_i$, $i=1,\cdots,n$; therefore, every $g\in\boldsymbol{N}$ may be written in the form $t\cdot\pi$ where $t\in T$ and π is a permutation matrix. Conversely, every such element belongs to $\boldsymbol{N}$; hence, $\boldsymbol{N}=T\cdot S_n$ (semi-direct product), where S_n is viewed as the group of $n \times n$ permutation matrices. Thus the group $W = \boldsymbol{N}/T$ is isomorphic to S_n. If $t \in T$ has diagonal entries $t_1, \cdots, t_n$, then $t(v_i) = t_i \cdot v_i$. Define $\chi_i(t) = t_i$; χ_i is a rational function of t. Every rational homomorphism (i.e., homomorphism given by a rational function of $t_1, \cdots, t_n$) of T into $\boldsymbol{C}^*$ is an element of the free Abelian group $\boldsymbol{X}^*(T)$ generated by $\chi_1, \cdots, \chi_n$. G is a Lie group and its Lie algebra $\mathfrak{g}$ is the Lie algebra of $n \times n$ complex matrices with bracket $[A, B] = AB - BA$. T acts on $\mathfrak{g}$ by the adjoint representation and the non-trivial characters of that representation are $\{\chi_i \chi_j^{-1} \mid i \neq j\}$. The subgroup U of B is generated by its one-dimensional unipotent subgroups $U_{ij} = \exp(\mathfrak{u}_{ij})$, $i < j$, where $\mathfrak{u}_{ij}$ is the one-dimensional eigenspace of T in $\mathfrak{g}$ corresponding to the character $\chi_i \chi_j^{-1}$, which we denote by χ_{ij}.

By a flag of subspaces of V, we mean a set $V_0, \cdots, V_t$ of subspaces of V such that $V_i \subsetneqq V_{i+1}$, $i = 0, \cdots, t-1$; then a maximal flag consists of $n+1$ subspaces:

$$\{0\} = V_0 \subset V_1 \subset \cdots \subset V_n = V, \quad \dim V_i = i.$$

Let F_0 be the particular maximal flag for which $V_i = \boldsymbol{C}v_1 + \cdots + \boldsymbol{C}v_i$, $i > 0$. If $g \in G$, then $F = gF_0$ is another maximal flag, $F = \{V_0 \subset V_1' \subset \cdots \subset V_n' = V\}$, $V_i' = \boldsymbol{C}v_1' + \cdots + \boldsymbol{C}v_i'$, $v_i' = gv_i$. Let j_1 be the last index such that v_{j_1} occurs with a non-vanishing coefficient in v_1'; then there is an element $b_1 \in B$ such that $b_1 v_1' = v_{j_1}$. Now let j_2 be the last index different from j_1 such that v_{j_2} occurs with a non-vanishing coefficient in v_2'; then there is an element $b_2 \in B$ such that $b_2 v_{j_1} = v_{j_1}$ and $b_2 v_2' = v_{j_2} + c v_{j_1}$. Proceeding inductively, we find an element $b \in B$ and indices $j_1, \cdots, j_n$ such that $bv_i' = v_{j_i} + \sum_{k<i} c_{ik} v_{j_k}$. Thus $bF = \{V_0 \subset V_1'' \subset V_2'' \subset \cdots \subset V_n'' = V\}$, where V_i'' is spanned by v_{j_k}, $k \leqq i$. By applying an appropriate permutation matrix π such that $\pi v_{j_i} = v_i$, we have $\pi b F = F_0$, hence $\pi b g = b' \in B$, or $g = b^{-1} \pi^{-1} b'$. It follows that G is the union of the double cosets $B\pi B$, $\pi \in S_n$. If $B\pi B = B\pi' B$ for $\pi \neq \pi'$, then we would have $\pi b {}^t\pi' = b'$ for some $b, b' \in B$, where ${}^t\pi' = (\pi')^{-1}$ is the transpose of π' (π' is an orthogonal matrix); thus the element in the i-th row and j-th column of b' would be $b_{\pi^{-1}(i)\pi'^{-1}(j)}$; but if $i = j$ and $\pi^{-1}(i) > \pi'^{-1}(i)$ (which must be the case for some i if $\pi \neq \pi'$), we have $0 \neq b'_{ii} = b_{\pi^{-1}(i)\pi'^{-1}(i)} = 0$, which is a contradiction. Hence $\pi \neq \pi'$ implies that $B\pi B \neq B\pi' B$; therefore, the indicated cosets are disjoint and we have

$$G = \bigcup_{w \in W} BwB \quad \text{(disjoint)}. \tag{2}$$

Let H be any subgroup of G containing B. Then $BHB=H$, so that H may be written as a union of double cosets of B,

$$H=\bigcup_{w\in W_H} BwB, \tag{3}$$

and it is obvious that W_H is a subgroup of W since the double coset representatives have been normalized so as to form a subgroup of G. However, not every subgroup of $W=S_n$ may occur as W_H in this way. The group W is generated by its transpositions r_{ij} (in fact, by the "fundamental" transpositions $r_{i,i+1}$), and if we define $r_{ij}(\chi_{ij})(t)=\chi_{ij}(r_{ij}tr_{ij}^{-1})$, then $r_{ij}(\chi_{ij})=\chi_{ji}=\chi_{ij}^{-1}$. Now one may show that if $H\supset B$, then H is automatically closed and connected and is its own normalizer [53 : Exp. 12]. Let $\mathfrak{h}$ be the Lie algebra of H. Then one may also show that $\mathfrak{h}$ is the vector space direct sum of the Lie algebra of T and of the subspaces $\mathfrak{u}_{ij}$ of $\mathfrak{g}$ which it contains, that W_H contains a "reflection" r_{ij} precisely when $\mathfrak{h}$ contains both $\mathfrak{u}_{ij}$ and $\mathfrak{u}_{ji}$ (which are interchanged by $Ad\, r_{ij}$), and that then W_H is generated by the "fundamental reflections" $r_{i,i+1}$ which it contains. A pair of reflections r_{ij} and r_{kl} are called perpendicular if they commute, i.e., if the 2-cycles (ij) and (kl) are disjoint; a set of reflections is called connected if it is not the union of disjoint, non-empty, mutually perpendicular subsets; then the set $\mathcal{R}_H$ of fundamental reflections in W_H may be decomposed into mutually perpendicular components. The latter amounts to saying that there exist integers $0<a_1\leqq b_1<a_2\leqq b_2<\cdots<a_t\leqq b_t\leqq n$ (where the possibility that $t=1:\ 0<a_1\leqq b_1\leqq n$ is not excluded), such that $\mathcal{R}_H$ consists of the fundamental reflections $r_{i,i+1}$ for which $a_j\leqq i\leqq b_j-1$ for some j, $1\leqq j\leqq t$. If $t=1$ and $a_1=b_1=n$, then $\mathcal{R}_H=\emptyset$, $W_H=\{id.\}$, and $H=B$. In matrix notation this means that for $g\in G$ we have $g=(g_{ij})\in H$ if and only if $i>j$ always implies $g_{ij}=0$ unless $a_k\leqq j<i\leqq b_k$ for some $k\in\{1,\cdots,t\}$. Let U_H be the connected subgroup of H whose Lie algebra is the direct sum of $\mathfrak{u}_{ji}$ where i,j are any pair with $i>j$ that do *not* satisfy $a_k\leqq j<i\leqq b_k$ for any $k\in\{1,\cdots,t\}$, and let T_H be the connected component of the identity in the set of $t\in T$ which satisfy $t_i=t_{i+1}$ whenever $r_{i,i+1}\in\mathcal{R}_H$. Then direct matric calculations show that $R_H=U_H\cdot T_H$ (semi-direct product) is a connected, solvable (precise definition later), normal subgroup of H, that U_H is a connected, unipotent (definition later), normal subgroup of H, and that H is the semi-direct product of U_H and of the

centralizer $Z(T_H)$ in G of T_H; and as for $Z(T_H)$ itself, it is the product of T_H and of a product L_H of groups $SL(b_k-a_k+1, \boldsymbol{C})$, $k=1, \cdots, t$, such that $T_H \cap L_H$ is finite. Virtually all of the above facts about $GL(n, \boldsymbol{C})$ can be, and here in part have been, verified by direct matric calculation, which gives this example the advantage that everything about it can be easily explained in elementary fashion.

Now, as above, let $G_1 = SL(n, \boldsymbol{C})$. Since the determinant is a homogeneous polynomial of degree n in $\{g_{ij}\}$, $1 \leqq i, j \leqq n$, $SL(n, \boldsymbol{C})$ is also a linear algebraic group. Then we define $T_1 = T \cap G_1$, $B_1 = B \cap G_1$, $\mathfrak{g}_1 =$ Lie algebra of $G_1 = \{g \in \mathfrak{g} \mid \mathrm{tr}(g) = 0\}$, $\boldsymbol{N}_1 = \boldsymbol{N}_{G_1}(T_1)$, $Z_1 = Z_{G_1}(T_1) = T_1$, $W_1 = \boldsymbol{N}_1/Z_1$. However, it is easy to see, in fact, that every element of W has a representative in $\boldsymbol{N}_1$, changing if necessary one of the 1's in some π to -1, and $Z_1 = T \cap G_1$, so that $W_1 \cong W$, although it is no longer possible to pick a set of coset representatives of W_1 forming a subgroup of G. Therefore, we again have

$$G_1 = \bigcup_{w \in W_1} B_1 w B_1 \quad \text{(disjoint)} \tag{4}$$

and each subgroup H_1 of G_1 containing B_1 is of the form $H \cap G_1$, where H is a subgroup of G containing B. The characters $\{\chi_i\}$ of T described earlier, restricted to T_1, generate the rational character group $\boldsymbol{X}^*(T_1)$ of T_1 and are now related by $\chi_1 \cdots \chi_n \equiv 1$; and the characters χ_{ij} (described earlier) restricted to T_1 generate a subgroup of $\boldsymbol{X}^*(T_1)$ of finite index.

The groups T and T_1 are themselves evidently algebraic groups (as are also, incidentally, U, B, B_1, H, etc.) and are examples of what are called tori.

DEFINITION. *Let D be a linear algebraic subgroup of $GL(V)$. Then D is called diagonalizable if it is represented by diagonal matrices with respect to some basis of V. D is called a torus if it is connected, commutative, and consists of semi-simple elements.*

THEOREM 12. [6a]. *A linear algebraic group is a torus if and only if it is diagonalizable and connected.*

The algebraic group $G_m = GL(1, \boldsymbol{C})$ is clearly a torus defined over the prime field $\boldsymbol{Q}$, and a direct product of tori is a torus, hence G_m^n is a torus for any positive integer n. Conversely, it can be proved that any torus contained in $GL(l, \boldsymbol{C})$ is isomorphic, over a suitable subfield of $\boldsymbol{C}$, to G_m^n for some n. This is a common alternative definition for a torus.

It is clear that T is a torus in G and T_1 is a torus in G_1; and since T is its own centralizer in G, and T_1 is so in G_1, it follows that T is a maximal torus in G and T_1 is a maximal torus in G_1.

§3. Further examples——the orthogonal group

With V an n-dimensional vector space over $\boldsymbol{C}$, let Q be a non-degenerate quadratic form on V. Identifying V with $\boldsymbol{C}^n$ by a choice of basis in V, Q is given by a symmetric matrix (q_{ij}) with non-zero determinant.

If Q and Q' are non-singular, $n\times n$, symmetric matrices over $\boldsymbol{C}$, then it is known that there exists $a\in GL(n, \boldsymbol{C})$ such that ${}^t aQa=Q'$ (${}^t a$ $=$transpose of a), i.e., Q and Q' are congruent.

Define

$$O(Q,\ V)=\{g\in GL(V)\,|\,{}^t gQg=Q\}.$$

If Q' is another quadratic form and if ${}^t aQa=Q'$, it is clear that

$$O(Q',\ V)=a^{-1}O(Q,\ V)a.$$

So from the complex point of view, it makes little difference which Q we consider. However, as we are interested in some of the different real forms of $O(Q, V)$, we shall discuss different real cases for Q.

If r and s are non-negative integers such that $2r+s=n$, let

$$Q_{r,s}=\begin{pmatrix} 0 & 0 & I_r \\ 0 & E_s & 0 \\ I_r & 0 & 0 \end{pmatrix}, \tag{5}$$

where I_r is the $r\times r$ matrix $(\delta_{i,r-j+1})=\begin{pmatrix} 0 & & \cdot\, 1 \\ & \cdot^{\cdot^{\cdot}} & \\ 1\,\cdot & & 0 \end{pmatrix}$, while if p and q are non-negative integers such that $p+q=n$, let

$$Q(p,\ q)=\begin{pmatrix} E_p & 0 \\ 0 & -E_q \end{pmatrix}. \tag{6}$$

Thus (p, q) is the signature of Q, while $Q(p, q)$ is congruent over the reals to $Q_{r,s}$ if and only if $p=r+s$, $q=r$. Let $O_{r,s}=O(V, Q_{r,s})$, $O(p, q)$ $=O(V, Q(p, q))$.

If $g \in O_{r,s}$, we write

$$g=\begin{pmatrix} A_{11} & A_{12} & A_{13} \\ A_{21} & A_{22} & A_{23} \\ A_{31} & A_{32} & A_{33} \end{pmatrix},$$

where A_{11} and A_{33} are $r \times r$, A_{22} is $s \times s$, and the sizes of the other rectangular blocks are determined accordingly. The equations for g to belong to $O_{r,s}$ become

$$(7) \qquad {}^tA_{11}IA_{31}+{}^tA_{21}A_{21}+{}^tA_{31}IA_{11}=0,$$

$$(8) \qquad {}^tA_{11}IA_{32}+{}^tA_{21}A_{22}+{}^tA_{31}IA_{12}=0,$$

$$(9) \qquad {}^tA_{11}IA_{33}+{}^tA_{21}A_{23}+{}^tA_{31}IA_{13}=I,$$

$$(10) \qquad {}^tA_{12}IA_{32}+{}^tA_{22}A_{22}+{}^tA_{32}IA_{12}=E,$$

$$(11) \qquad {}^tA_{12}IA_{33}+{}^tA_{22}A_{23}+{}^tA_{32}IA_{13}=0, \quad \text{and}$$

$$(12) \qquad {}^tA_{13}IA_{33}+{}^tA_{23}A_{23}+{}^tA_{33}IA_{13}=0,$$

where $I=I_r$, $E=E_s$. If we are interested only in the structure relative to $\boldsymbol{C}$, then we might as well take the form $Q_{r,0}$ or $Q_{r,1}$ (according as n is even or odd) since that has maximal index over the reals.

So, letting $s=0$, or 1, let $T_{r,s}$ be the intersection of $O_{r,s}$ with the group of all diagonal matrices in $GL(n, \boldsymbol{C})$. Then for $t=g\in T_{r,s}$, (7), (8), (11), and (12) are identically satisfied while (9) and (10) become ${}^tA_{11}IA_{33}=I$ and $a_{22}^2=1$ ($s=1$) or nothing ($s=0$), and if A_{11} is a diagonal matrix with entries $t_1, \cdots, t_r$, then A_{33} is the diagonal matrix with entries $t_r^{-1}, \cdots, t_1^{-1}$. It is easy to see that $T_{r,s}$ is its own centralizer in $O_{r,s}$ ($s=0$ or 1). Now let B be the subgroup of $g \in O_{r,s}$ for which A_{21}, A_{31}, and A_{32} are all zero, A_{11} is an upper triangular $r \times r$ matrix (so that $A_{33}=I^tA_{11}^{-1}I$ is also upper triangular) and $a_{22}=+1$ (if $s=1$). Let U be the subgroup of $g \in B$ such that the diagonal entries of A_{11} are all 1. The group $O_{r,s}$ is not connected since the continuous function det takes the values ± 1 on it. Let $SO_{r,s}=\{g\in O_{r,s} \mid \det(g)=+1\}$. For immediate purposes, let G be this group and let T now be its intersection with $T_{r,s}$. Again one may easily verify that U is normal in B, that T (which is a maximal torus in G because it is its own centralizer there) normalizes U, and that $B=T\cdot U$ (semi-direct product). Now, an element of the normalizer $\boldsymbol{N}$ of T in G which stabilizes all the eigenspaces in the representation of T on V must be diagonal, hence must be in T itself. Any $g \in \boldsymbol{N}$ must permute those eigenspaces and one soon sees that the permissible permutations are those which permute the diagonal entries of A_{11} (and correspondingly those of A_{33}) together with those that exchange diagonal entries in A_{11} with those in A_{33} in appropriate reversed order, subject to the condition that the element $g \in \boldsymbol{N}$ must have $\det(g) = +1$. If $s=1$, this is no restriction since we can always put -1 in the middle of the permutation matrix if necessary, and one obtains $\text{card}(W) = 2^r r!$; but if $s=0$, then the number of exchanges of diagonal elements of A_{11} with corresponding diagonal elements of A_{33} in reverse position must be *even*, cutting our possibilities in half to get $\text{card}(W) = 2^{r-1} r!$ (the remaining automorphisms of G normalizing T are "outer"; they interchange the short branches on the Dynkin diagram of D_r :

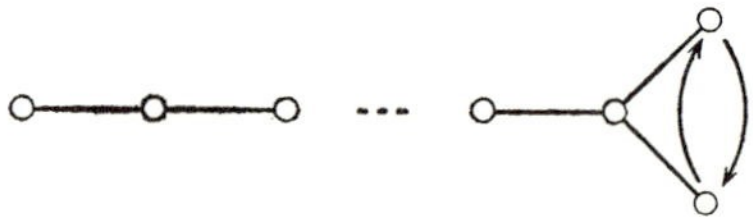

and they come from a coset modulo $\boldsymbol{N}$ of a permutation matrix in $GL(n, \boldsymbol{C})$ (not in $SL(n, \boldsymbol{C})$)). As before one has a decomposition

(13) $$G = \bigcup_{w \in W} BwB.$$

To prove it, one must use a modified concept of flag. A subspace V' of V is called Q-isotropic if $Q(v_1, v_2) = 0$ for all $v_1, v_2 \in V'$. Then by a maximal flag in this case we mean a sequence of isotropic subspaces of V: $V_0' = \{0\} \subset V_1' \subset \cdots \subset V_r'$, where $\dim V_i' = i$. Using this notion, the proof of (13) is much the same as that of (2). The details are left as an exercise. It is clear that B is the intersection of G with the group of upper triangular matrices in $GL(n, \boldsymbol{C})$ and one may verify as before that if H is a subgroup of G containing B, then H is closed, connected, is its own normalizer, and has a double coset decomposition

(14) $$H = \bigcup_{w \in W_H} BwB;$$

in fact, H is of the form $H_1 \cap G$, where H_1 is a subgroup of $GL(n, \boldsymbol{C})$ containing the upper triangular matrices, and each of the double cosets in (14) is the intersection with G of one of the double cosets (2).

As for the structure of the unipotent group U, we note that if A_{21}, A_{31}, and A_{32} are all zero and $a_{22}=1$ (when $s=1$) then the equations (7)-(12) reduce to

$$(15)\qquad \begin{aligned} &{}^tA_{11}IA_{33}=I, \quad {}^tA_{12}IA_{33}+A_{23}=0, \quad \text{and} \\ &{}^tA_{13}IA_{33}+{}^tA_{33}IA_{13}+{}^tA_{23}A_{23}=0; \end{aligned}$$

now A_{23} is $1\times r$ ($s=1$) or nothing ($s=0$). The equation $A_{23}=-{}^tA_{12}IA_{33}$ determines A_{23} from A_{12} and A_{33}, and the equation

$${}^tA_{13}IA_{33}+{}^tA_{33}IA_{13}+{}^tA_{23}A_{23}=0$$

becomes

$${}^tA_{13}IA_{33}+{}^tA_{33}IA_{13}+{}^tA_{33}IA_{12}{}^tA_{12}IA_{33}=0;$$

if $A_{33}=E_r$, the latter becomes

$${}^tA_{13}I+IA_{13}+IA_{12}{}^tA_{12}I=0,$$

which means that the symmetric part of IA_{13} is determined by A_{12}, while its skew-symmetric part is quite arbitrary. If $t=(A_{ij})\in T$, denote the diagonal entries of A_{11} by $t_1, \cdots, t_r$; we obtain then as characters in the adjoint representation of T on the Lie algebra of U (i.e., as the positive roots), the set

$$(16)\qquad \{t_it_j^{-1},\ i<j;\ t_it_j,\ i\neq j;\ \text{and if } s=1, \text{ i.e., if } n \text{ is odd, the characters } t_i,\ i=1, \cdots, r\}.$$

Let U^- be the group of *lower* triangular unipotent matrices in G and let $\mathfrak{g}$, $\mathfrak{u}$, $\mathfrak{u}^-$ and $\mathfrak{t}$ be the Lie algebras of G, U, U^-, and T, respectively. Then

$$(17)\qquad \mathfrak{g}=\mathfrak{u}\oplus\mathfrak{u}^-\oplus\mathfrak{t},$$

and if $\mathfrak{g}_R$ is the Lie algebra of all real matrices in G, then

$$(18)\qquad \mathfrak{g}_R=\mathfrak{u}_R\oplus\mathfrak{u}_R^-\oplus\mathfrak{t}_R.$$

At the opposite extreme is the case $s=n$, $r=0$. In this case, we let $O(n)=O_{0,n}=O(n, 0)$, $G=SO(n, \boldsymbol{C})$; then $\mathfrak{g}_R$ has no nilpotent elements $\neq 0$ and G_R, no unipotent elements different from the identity.

We now consider certain intermediate cases. As remarked previously, there would be no point in considering any other cases unless we chose some field different from $\boldsymbol{C}$ over which to consider the different "forms" of $SO(Q)$. So now we focus attention on V_R, the real points of V. In the examples we consider, the real Lie group G_R in question is to be the group of real points of $SO(Q_{r,s})$, $r=1$ or 2. However, to avoid an awkward disposition of algebraic signs we replace $Q_{r,s}$ by $Q_{r,-s}$ in which the $s\times s$ identity matrix E_s is to be replaced by $-E_s$. This does not change the abstract Lie group, but changes at least some outward aspects of the geometry. We denote the equations defining $SO(Q_{r,-s})$ corresponding to (7)-(12) by (7′)-(12′) respectively.

A. First let $r=1$, $s=n-1$ (so $\dim V=n+1$), and let $Q=Q_{1,-(n-1)}$. So if $x=(x_1,\cdots,x_{n+1})\in V$, then $Q(x)=2x_1x_{n+1}-x_2^2-\cdots-x_n^2$. We consider the real algebraic manifold

$$\mathcal{Q}=\{x\in V_R\,|\,Q(x)=1\}, \tag{19}$$

and denote by $\mathcal{Q}_1$ the open subset of it on which $x_1>0$. It is clear that $\mathcal{Q}$ is the union of this and of the subset on which $x_1<0$. Let $G=SO(Q)=SO(Q,V)$. We first calculate the identity component of the subgroup of upper triangular matrices in G_R. If $g=(A_{ij})$ is such a matrix, then $A_{11}=a$, $A_{33}=a^{-1}$, and $A_{13}=y$ are all real numbers and the equations (7′)-(12′) show that (since $A_{ij}=0$ for $i>j$) A_{22} is an orthogonal upper triangular matrix, hence is diagonal with ±1's on the diagonal and $\det A_{22}=+1$. Let $A_{12}=v$. The same equations show then that $y=\frac{1}{2}a^{-1}v\,{}^tv$ and $A_{23}=a^{-1}\,{}^tv$. Since we are interested only in the identity component, we see that then $A_{22}=E_s$ and that that identity component is the semi-direct product of

$$A=\{g=(A_{ij})\,|\,A_{ij}=0,\ i\neq j,\ A_{22}=E_s,\ a_{11}=a>0,\ a_{33}=a^{-1}\} \tag{20}$$

and of

$$N=\{(A_{ij})\,|\,a_{11}=a_{33}=1,\ A_{22}=E_s,\ A_{ij}=0,\ i>j,\ A_{12}=v,\ a_{13}=\frac{1}{2}v\,{}^tv,\ A_{23}={}^tv\}. \tag{21}$$

Let $\omega=2^{-1/2}$ (now and henceforth) and put $x_0=(\omega,0,\cdots,0,\omega)\in V$; clearly $x_0\in\mathcal{Q}$ and its image under au where $a\in A$ and $u\in N$ are of the forms just described, is

$$x=\left(a\omega\left(1+\frac{1}{2}v\,{}^tv\right),\ \omega v_2,\ \cdots,\ \omega v_n,\ a^{-1}\omega\right) \tag{22}$$

(where we elect to number the coordinates of v from 2 to n); conversely if $x'=(x'_1, \cdots, x'_{n+1})\in Q_1$, and if u is of the form described above, then $u\cdot {}^t x' = {}^t(x''_1, x'_2+v_2x'_{n+1}, \cdots, x'_n+v_nx'_{n+1}, x'_{n+1})$, and since $v_2, \cdots, v_n$ are arbitrary and $x'_{n+1}\neq 0$, we may choose u such that $u\cdot {}^t x'$ is of the form ${}^t(\alpha, 0, \cdots, 0, \beta)$, $\alpha\beta=\frac{1}{2}$, $\alpha>0$, which is clearly the image of x_0 under some $a\in A$. It follows that AN is transitive on Q_1. Let

$$\tau=\begin{pmatrix} \omega & 0\cdots 0 & \omega \\ 0 & & 0 \\ \vdots & E_{n-1} & \vdots \\ 0 & & 0 \\ \omega & 0\cdots 0 & -\omega \end{pmatrix}. \tag{23}$$

Then τ transforms $Q_{1,-(n-1)}$ into the quadratic form $x_1^2-x_2^2-\cdots-x_{n+1}^2$, with matrix that we denote by $Q^-=Q(1, n)$, the group G_R into the special orthogonal group G_R^- of Q^-, the quadric domain Q_1 into the domain $\{Q^-=1, x_1>0\}$, and x_0 into the point $(1, 0, \cdots, 0)$ whose stabilizer in G_R^- is

$$\begin{pmatrix} 1 & 0 \\ 0 & SO(n) \end{pmatrix}_R,$$

as may be seen by direct calculation. This is just the identity component of $G_R^-\cap SO(n+1)$. Now $\tau=\tau^{-1}={}^t\tau\in O(n+1)$, which normalizes $SO(n+1)$, hence the stabilizer of x_0 in G_R is the identity component K^0 of $G_R\cap SO(n+1)$. Since G_R stabilizes Q as a set, and since Q_1 is a connected component of Q on which G_R^0 is transitive, it follows that

$$G_R^0=ANK^0=NAK^0=K^0AN$$

(since, evidently, A normalizes N), where G_R^0 is the identity component of G_R. Note that $G_R^-\cap SO(n+1)$ is not connected (it contains $\begin{pmatrix} -1 & 0 & O \\ 0 & -1 & \\ O & & E \end{pmatrix}$), hence $G_R\cap SO(n+1)$ is not connected. Since the union of the "orbits" of two cosets of G_R^0 (in G_R) covers all of Q, which consists of the components on which $x_1>0$ and on which $x_1<0$, it follows that if we put $K=G_R\cap SO(n+1)$, we have

$$G_R=KAN \quad \text{(topological product)}. \tag{24}$$

The open set $x_1^2 - x_2^2 - \cdots - x_{n+1}^2 > 0$ consists of two components. We see that $(G_R^-)^0$ extended by the multiplicative group of positive, real, scalar multiplications is transitive on either one, say on the one, Q^+, where $x_1 > 0$.

B. Now let $r=2$, $s=n-2$ (so $\dim V = n+2$), and let $Q = Q_{2,-(n-2)}$. Then $Q(x) = 2(x_1 x_{n+2} + x_2 x_{n+1}) - \sum_{i=3}^{n} x_i^2$. Let $G = SO(Q)$. We are going to construct a certain domain in $\boldsymbol{C}^n$ and show how G_R^0 may be made to act on it by biholomorphic transformations. Our procedure will be to follow the ideas in Harish-Chandra's general theory. But the results we give here may all be easily verified independently of the general theory, and thus serve as an illustration of general theorems to be presented later without proof.

We note that the equations (7′)-(12′) for $g = (A_{ij})_{i,j=1,2,3}$ to belong to G (where A_{11} and A_{33} are two-by-two, A_{22} is $(n-2)\times(n-2)$, etc.) may be obtained from (7)-(12) by changing the sign of the middle term on the left side in each equation and by replacing E on the right by $-E$, when it occurs. Let $\mathfrak{g}$ be the Lie algebra of G. If $X \in \mathfrak{g}$, we again write $X = (A_{ij})$, the rectangular blocks A_{ij} being the same size as before. Since $Q = Q^{-1}$, one sees immediately that G is stable under the transformation $g \to {}^t g^{-1}$ and that $\mathfrak{g}$ is stable under $\iota : X \to -{}^t X$. Since $X \in \mathfrak{g}$ if and only if ${}^t XQ + QX = 0$, it follows that if we write $\mathfrak{g} = \mathfrak{k} \oplus \mathfrak{p}$, where $\mathfrak{k}$ and $\mathfrak{p}$ are the $(+1)$- and (-1)-eigenspaces of ι, the matrices X in $\mathfrak{k}$ are the matrices of the form

$$(25)\quad X = \begin{pmatrix} \mathcal{A} & \mathcal{B} & C \\ -{}^t\mathcal{B} & \mathcal{D} & {}^t\mathcal{B}I \\ ICI & -I\mathcal{B} & I\mathcal{A}I \end{pmatrix},\ {}^tC = -ICI = \begin{pmatrix} c & 0 \\ 0 & -c \end{pmatrix},\ \mathcal{A} = -{}^t\mathcal{A} = \begin{pmatrix} 0 & a \\ -a & 0 \end{pmatrix},$$
$$\mathcal{D} = -{}^t\mathcal{D}, \text{ and } \mathcal{B} \text{ arbitrary.}$$

Thus, the dimension of $\mathfrak{k}$ is $\frac{1}{2}(n^2 - n + 2)$; moreover $X \in \mathfrak{p}$ if and only if

$$(26)\qquad X = \begin{pmatrix} \mathcal{A} & \mathcal{B} & C \\ {}^t\mathcal{B} & 0 & {}^t\mathcal{B}I \\ C & I\mathcal{B} & -I\mathcal{A}I \end{pmatrix},\ C = \begin{pmatrix} c & 0 \\ 0 & -c \end{pmatrix},\ \mathcal{A} = {}^t\mathcal{A},$$
$$\mathcal{B} \text{ arbitrary;}$$

thus, $\dim \mathfrak{p} = 2n$; finally, the center of $\mathfrak{k}$ consists of those X of the form (25), where $\mathscr{B}=0$, $\mathscr{D}=0$, $\mathscr{A}=\begin{pmatrix} 0 & \beta \\ -\beta & 0 \end{pmatrix}$, $C=\begin{pmatrix} \beta & 0 \\ 0 & -\beta \end{pmatrix}$, and we denote such an X by λ_β. Direct calculation shows that the matrix λ_β (which clearly is semi-simple) has eigenvalues $2i\beta$ and $-2i\beta$ (each with multiplicity one) and 0 (with multiplicity n), that $ad\,\lambda_\beta$ normalizes $\mathfrak{p}$ and that $ad\,\lambda_\beta|\mathfrak{p}$ has eigenvalues $2i\beta$ and $-2i\beta$; we let $\mathfrak{p}^+$ and $\mathfrak{p}^-$ be the subspaces of $\mathfrak{p}$ on which $ad\,\lambda_\beta$ has respectively the eigenvalues $2i\beta$ and $-2i\beta$.

We now let

$$(27)\qquad c=\begin{pmatrix} \frac{1}{2} & \frac{i}{2} & 0 & \frac{i}{2} & \frac{1}{2} \\ \frac{i}{2} & \frac{1}{2} & 0 & -\frac{1}{2} & -\frac{i}{2} \\ 0 & 0 & E & 0 & 0 \\ \frac{i}{2} & -\frac{1}{2} & 0 & \frac{1}{2} & -\frac{i}{2} \\ \frac{1}{2} & -\frac{i}{2} & 0 & -\frac{i}{2} & \frac{1}{2} \end{pmatrix} = \exp\begin{pmatrix} 0 & i\alpha & 0 & i\alpha & 0 \\ i\alpha & 0 & 0 & 0 & -i\alpha \\ 0 & 0 & 0 & 0 & 0 \\ i\alpha & 0 & 0 & 0 & -i\alpha \\ 0 & -i\alpha & 0 & -i\alpha & 0 \end{pmatrix}_{\alpha=\frac{\pi}{4}},$$

where the 0's stand for rectangular zero matrices of various sizes, the entries $\pm i\alpha$ being one-by-one. Clearly c is symmetric, unitary, and belongs to $\exp\mathfrak{p}\subset G$. Direct calculation shows that $c\lambda_\beta c^{-1}$ has $-2i\beta$ in the upper left-hand corner, $2i\beta$ in the lower right-hand corner, and zeros elsewhere. Moreover, $Ad\,c(\mathfrak{p}^+)=c\mathfrak{p}^+c^{-1}$ consists of the matrices $X=(A_{ij})$ for which $A_{11}=\begin{pmatrix} 0 & 0 \\ 2a_1 & 0 \end{pmatrix}$, $A_{33}=\begin{pmatrix} 0 & 0 \\ -2a_1 & 0 \end{pmatrix}$, $A_{31}=\begin{pmatrix} 2a_2 & 0 \\ 0 & -2a_2 \end{pmatrix}$, $A_{21}=(b_j\ \ 0)$ (where b_j and 0 represent column vectors of length $n-2$), $A_{32}=\begin{pmatrix} 0 \\ {}^t b_j \end{pmatrix}$, and all other $A_{ij}=0$, so that $\exp X=p$, where

$$(28)\qquad p=p_x=\begin{pmatrix} 1 & 0 & 0 \\ x & E & 0 \\ -\frac{1}{2}Q'(x) & {}^t x' & 1 \end{pmatrix},\quad Q'=Q_{1,-(n-2)},$$

x being a column vector of length n from which x' is obtained by interchanging the first and last entries of x and changing the signs of those entries, while $Ad\,c(\mathfrak{p}^-)$ and its image under exp consist of the matrices which are transposes of the corresponding matrices for $\mathfrak{p}^+$. Let $Ad\, c(\mathfrak{p}^\pm)=\mathfrak{p}_1^\pm$, $\exp(\mathfrak{p}_1^\pm)=P_1^\pm$, and let $Ad\, c(K_C)=K_{1C}$. Then K_{1C} consists of the matrices $k\in G_C$ of the form

$$k=\begin{pmatrix}\alpha & 0 & 0\\ 0 & A & 0\\ 0 & 0 & \alpha^{-1}\end{pmatrix}, \tag{29}$$

where $A\in O(Q_{1,-(n-2)})$. Now let $g=(A_{ij})\in G_R$. We want to find $p\in P_1^+$ (as in (28)), $p^-\in P_1^-$, and $k\in K_{1C}$ (as in (29)) such that

$$c\cdot g=p^-kp^+. \tag{30}$$

Let $\boldsymbol{\alpha}$, $\boldsymbol{\beta}$, $\boldsymbol{\gamma}$, and $\boldsymbol{\delta}$ be the two-by-two matrices appearing in the upper left-hand, upper right-hand, lower left-hand, and lower right-hand corners of c respectively. If $p^+=p_x$, $p^-={}^t(p_y)$, and k is as in (29), then

$$p^-kp^+=(B_{rs})_{r,s=1,2,3}, \tag{31}$$

where $B_{11}=\alpha+{}^tyAx+\alpha^{-1}Q'(x)\,Q'(y)$ and $B_{33}=\alpha^{-1}$ are one-by-one (i.e., numbers), B_{22} is $n\times n$ and $B_{22}=A+\alpha^{-1}y'\,{}^tx'$, the size of the other B_{rs} being determined accordingly, with, in particular, $B_{32}=\alpha^{-1}\,{}^tx$, $B_{31}=-\dfrac{1}{2}\alpha^{-1}Q'(x)$. The last two rows of $c\cdot g$ may be written in blocks as

$$(\boldsymbol{\gamma}A_{11}+\boldsymbol{\delta}A_{31},\ \boldsymbol{\gamma}A_{12}+\boldsymbol{\delta}A_{32},\ \boldsymbol{\gamma}A_{13}+\boldsymbol{\delta}A_{33}).$$

The main matter to be resolved in showing that (30) may be solved (and uniquely so) for x, y, A, and α is to show that the lower right-hand corner of $c\cdot g$ is different from zero for every $g\in G_R$. Using the equation

$$0={}^tA_{13}IA_{33}+{}^tA_{33}IA_{13}-{}^tA_{23}A_{23}, \tag{12'}$$

one sees that if that entry were zero, setting its real and imaginary parts equal to zero would imply that the sum of the squares of the elements in the last column of g would be zero, which is impossible. Thus, α^{-1} is a well-determined, non-zero complex number. With this knowledge it is now easy to solve in turn for x' (that is to say, for x), for y, and A; and elementary calculations based on the equations that must be satisfied by all $(A_{ij})\in G$ show that when c, y, A, and α are determined in this way, then the matrix (B_{rs}) with $B_{33}=\alpha^{-1}$, $B_{32}=\alpha^{-1}\,{}^tx$, etc., is an element of G that can be no other than $c\cdot g$.

Now, for any $g \in G_R$, let $\tau = \tau(g) = (\tau_1, \tau_2, \cdots, \tau_{n+2})$ be the last row of $c \cdot g$. If g is the identity, then this is the last row of c: $2\tau(e) = (1, -i, 0, \cdots, 0, -i, 1)$. We have $\overline{\tau(e)}Q^t\tau(e) > 0$ and $\tau(e)Q^t\tau(e) = 0$. Since g is a real matrix and is in the orthogonal group of Q, it follows that $\overline{\tau(g)}Q^t\tau(g) > 0$ and $\tau(g)Q^t\tau(g) = 0$ for all $g \in G_R$. So let $(\tau_1, \cdots, \tau_{n+2}) = \tau$ satisfy ${}^t\bar{\tau}Q\tau > 0$, ${}^t\tau Q\tau = 0$, $\tau_{n+2} \neq 0$. Then we may put $\zeta_j = -\tau_j\tau_{n+2}^{-1}$, $j = 1, \cdots, n+1$, and write $\zeta_j = \xi_j + i\eta_j$, $\xi_j, \eta_j \in \boldsymbol{R}$. We have then

$$2\,\mathrm{Re}(-\bar{\zeta}_1 + \bar{\zeta}_2\zeta_{n+1}) - \sum_{j=3}^{n} |\zeta_j|^2 > 0 \tag{32}$$

and

$$2(-\zeta_1 + \zeta_2\zeta_{n+1}) - \sum_{j=3}^{n} \zeta_j^2 = 0; \tag{33}$$

taking real parts in the two equations, we obtain

$$2(-\xi_1 + \xi_2\xi_{n+1} + \eta_2\eta_{n+1}) - \sum_{j=3}^{n} (\xi_j^2 + \eta_j^2) > 0 \tag{34}$$

and

$$2(-\xi_1 + \xi_2\xi_{n+1} - \eta_2\eta_{n+1}) - \sum_{j=3}^{n} (\xi_j^2 - \eta_j^2) = 0, \tag{35}$$

which combined give

$$2\eta_2\eta_{n+1} - \sum_{j=3}^{n} \eta_j^2 > 0. \tag{36}$$

From the last equation, it is clear that η_2 and η_{n+1} are different from zero. Now, if $g \in G_R$, the last row of $c \cdot g$ is $\left(-\frac{1}{2}\alpha^{-1}Q'(x), \alpha^{-1}\,{}^tx', \alpha^{-1}\right) = (\tau_1(g), \cdots, \tau_{n+2}(g))$, so that in our notation above $x_2(g) = \zeta_{n+1}(g)$, $x_j(g) = -\zeta_j(g)$, $j = 3, \cdots, n$, $x_{n+1}(g) = \zeta_2(g)$, and (36) translated says that

$$2\,\mathrm{Im}\,\zeta_2 \cdot \mathrm{Im}\,\zeta_{n+1} - \sum_{j=3}^{n} (\mathrm{Im}\,\zeta_j)^2 > 0. \tag{37}$$

Since $\zeta_2(g)$ is a continuous function of g, $\zeta_2(e) = i$, and $\mathrm{Im}\,\zeta_2(g) = \eta_2(g) \neq 0$ for $g \in G_R^0$, it follows that $\eta_2(g) > 0$ for $g \in G_R^0$. Therefore we conclude that if D' is the set of all $\zeta = (\zeta_2, \cdots, \zeta_{n+1}) \in \boldsymbol{C}^n$ such that, with $\zeta = -x'$, p_x appears in a decomposition (30) for some $g \in G_R^0$, then D' is contained in the domain

$$D = \{\zeta \mid 2\eta_2\eta_{n+1} - \sum_{j=3}^{n} \eta_j^2 > 0,\ \eta_2 > 0\}. \tag{38}$$

We shall shortly show that D' is all of D. If $\zeta(g) \in D'$ and if $g_1 \in G_R^0$, define an action of G_R^0 on D' by $\zeta(g)\cdot g_1=\zeta(gg_1)$. It is trivial to verify that this action is well-defined.

Let A_1 be the group of diagonal matrices in G_R^0 with positive diagonal entries; then,

$$(39)\qquad A_1=\left\{a=\begin{pmatrix} a_1 & 0 & 0 & 0 & 0 \\ 0 & a_2 & 0 & 0 & 0 \\ 0 & 0 & E_{n-2} & 0 & 0 \\ 0 & 0 & 0 & a_2^{-1} & 0 \\ 0 & 0 & 0 & 0 & a_1^{-1} \end{pmatrix}\right\}.$$

Write $a=(a_1, a_2)$ for the element in large parentheses. Let N_1 denote the group of unipotent upper triangular matrices in G_R. Then N_1 contains as a subgroup the group N discussed in part A. (with n replaced by $n-1$), by way of the natural injection

$$(40)\qquad N\ni u\mapsto\begin{pmatrix} 1 & 0 & 0 \\ 0 & u & 0 \\ 0 & 0 & 1 \end{pmatrix}\in N_1,$$

and P_{1R}^+ is a normal subgroup of N_1 such that $N_1=N\cdot P_{1R}^+$ (semi-direct product). If A also denotes the subgroup of $SO(Q_{1,-(n-2)})$ discussed in A., then it may be identified with the set of $a\in A_1$ as in (39) such that $a_1=1$. Let $a'n'\in A_1N_1$. We may write $a'=(a_1, 1)a$, $a\in A$, and $n'=u\cdot p_x$, $x\in \boldsymbol{R}^n$, p_x as in (28). Then $\tau(e)\cdot a'n'=\tau(a'n')=(a_1, iy_2, \cdots, iy_{n+1}, a_1^{-1})$, where

$$(41)\qquad 2y_2y_{n+1}-y_3^2-\cdots-y_n^2>0, \quad y_2>0,$$

and

$$(42)\qquad \tau(a'up_x)=(*, -x_2+iy_2, x_3+iy_3, \cdots, -x_{n+1}+iy_{n+1}, a_1^{-1}),$$

or $\zeta(a'up_x)=-a_1(z_2, \cdots, z_{n+1})$ with $z_j=-x_j+iy_j$, $j=2, n+1$ and $z_k=x_k+iy_k$, $k=3, \cdots, n$. Since AN extended by positive real scalar multiplications is transitive on

$$Q_1^+=\{x\,|\,2x_2x_{n+1}-x_3^2-\cdots-x_n^2>0,\ x_2>0,\ x_j\in \boldsymbol{R}\},$$

it follows that we may obtain every point of D in the form $\zeta(a'up_x)$ and a', u and x are readily seen to be unique. Thus G_R^0 is transitive on D, as asserted earlier. To determine the subgroup of G_R^0 leaving

$\zeta(e)$ fixed, it is expedient to denote by Ω the $(n+2)\times(n+2)$ matrix

$$\Omega=\begin{pmatrix}\omega E_2 & 0 & \omega I \\ 0 & E & 0 \\ \omega I & 0 & -\omega E_2\end{pmatrix},$$

and to transform the quadratic form Q and group G by Ω. We have $\Omega^2=E_{n+2}$, ${}^t\Omega=\Omega=\Omega^{-1}$ and $\tau(e)\Omega=(1, -i, 0, \cdots, 0)=x_0'$, of which the stabilizer in $\Omega G_R^0\Omega$ is seen to be the group of matrices

$$\begin{pmatrix}E_2 & 0 \\ 0 & SO(n)\end{pmatrix}_R. \tag{43}$$

Since $(-i, 1)\begin{pmatrix}\cos\theta & \sin\theta \\ -\sin\theta & \cos\theta\end{pmatrix}=e^{-i\theta}(-i, 1)$, it follows that the *projective* stabilizer of x_0' in $\Omega G_R^0\Omega$ is the larger group of matrices

$$\begin{pmatrix}SO(2) & 0 \\ 0 & SO(n)\end{pmatrix}_R, \tag{44}$$

i.e., the projective stabilizer of x_0' (as a point in complex projective space) is the identity component of the intersection of $\Omega G_R^0\Omega$ with $SO(n+2)$, so that the projective stabilizer of $\tau(e)$ in G_R^0 is the identity component K_1^0 of $K_1=G_R^0\cap SO(n+2)$. Now since the last coordinate of $\tau(g)$ is always $\neq 0$ for $g\in G_R^0$, it follows that the first coordinate is determined by the last $n+1$ coordinates, hence the point of D corresponding to $\tau(g)$ is uniquely determined by the projective image of $\tau(g)$ and conversely (in fact, D may be viewed as an open orbit of G_R^0 on the complex projective quadric ${}^t\tau Q\tau=0$); hence, K_1^0 is the stabilizer of $\zeta(e)\in D$. Since A_1N_1 is simply transitive on D, we obtain

$$G_R^0=K_1^0A_1N_1 \quad \text{(topological product)}, \tag{45}$$

and since A_1N_1 has no proper compact subgroups, as one readily sees, it follows that $K_1=G_R^0\cap SO(n+2)$ is connected and therefore equal to K_1^0.

We have seen that $c\cdot G_R^0\subset P_1^-K_{1C}P_1^+$. We shall now show that $cG_R^0c^{-1}\subset P_1^-K_{1C}P_1^+$, and hence that $G_R^0\subset P^-K_CP^+$. Suppose $g\in G_R^0$ and that we have (30) with $p^-={}^t(p_y)$, $p^+=p_x$, and k as in (29). Then our problem is to solve for x', y', and $k'\in K_C$ such that

$$c\cdot g\cdot c^{-1}=p^-kp^+\cdot c^{-1}={}^t(p_{y'})k'p_{x'}, \tag{46}$$

and since K_{1c} normalizes P_1^-, it is clear that solving (46) is equivalent to solving the equation

$$ {}^t p_{y''} k_1 p_{x''} = p_x c^{-1} \tag{47} $$

for x'', y'', and k_1. It is seen that the last row of $p_x c^{-1}$ is

$$ \mu = (\rho_1, \rho_2, x_3, \cdots, x_n, \rho_{n+1}, \rho_{n+2}), $$

where $\rho_1 = a+ib$, $\rho_2 = c-id$, $\rho_{n+1} = -c-id$, $\rho_{n+2} = a-ib$, and a, b, c, and d are defined by $2a = \gamma+1$, $2b = x_2 + x_{n+1}$, $2c = x_2 - x_{n+1}$, $2d = \gamma - 1$ with $\gamma = -\frac{1}{2}Q'(x)$. Since $p_x c^{-1}$ is in the (complex) orthogonal group of Q, one has

$$ 2(\rho_1\rho_{n+2} + \rho_2\rho_{n+1}) - x_3^2 - \cdots - x_n^2 = 0. \tag{48} $$

On the other hand, if λ is the last row of p_x, it follows from the fact that x corresponds (in the manner described previously) to a point of D that

$$ \bar{\lambda} Q {}^t\lambda > 0, \tag{49} $$

and hence

$$ \bar{\mu} Q' {}^t\mu > 0, \tag{50} $$

where $Q' = cQ\bar{c}$ is the $(n+2)\times(n+2)$ diagonal matrix of which the first and last diagonal elements are 1 and the others are -1. Therefore

$$ |\rho_1|^2 + |\rho_{n+2}|^2 - |\rho_2|^2 - |\rho_{n+1}|^2 - |x_3|^2 - \cdots - |x_n|^2 > 0. \tag{51} $$

Now (48) is equivalent to

$$ 2(a^2+b^2) - 2(c^2+d^2) - x_3^2 - \cdots - x_n^2 = 0 \tag{52} $$

and (51), to

$$ 2(|a|^2+|b|^2) - 2(|c|^2+|d|^2) - |x_3|^2 - \cdots - |x_n|^2 > 0. \tag{53} $$

If $z_1, \cdots, z_m$, $m > 3$, are complex numbers such that $|z_1|^2 + |z_2|^2 - |z_3|^2 - \cdots - |z_m|^2 > 0$ and $z_1^2 + z_2^2 - z_3^2 - \cdots - z_m^2 = 0$, then it is impossible for z_1 to be a real multiple of z_2; it follows that both z_1 and z_2 are non-zero and that $\operatorname{Im}\frac{z_1}{z_2} \neq 0$. Now, if $\operatorname{Im}\frac{z_1}{z_2} > 0$, we have $z_1 + iz_2 \neq 0$, and in that case, dividing the quadratic equation by $(z_1+iz_2)^2$, we obtain

$$ \left(\frac{z_3}{z_1+iz_2}\right)^2 + \cdots + \left(\frac{z_m}{z_1+iz_2}\right)^2 = \frac{z_1 - iz_2}{z_1+iz_2}, \tag{54} $$

while division of the inequality by $|z_1+iz_2|^2$ yields

$$(55) \qquad \left|\frac{z_3}{z_1+iz_2}\right|^2+\cdots+\left|\frac{z_m}{z_1+iz_2}\right|^2<\frac{1}{2}\left(1+\left|\frac{z_1-iz_2}{z_1+iz_2}\right|^2\right).$$

Clearly $|z_1-iz_2|<|z_1+iz_2|$. Put

$$\zeta_1=\frac{z_3-iz_4}{z_1+iz_2}, \quad \zeta_2=\frac{z_3+iz_4}{z_1+iz_2}, \quad \zeta_k=\frac{z_{k+2}}{z_1+iz_2}, \qquad k>2.$$

Combining the above information, one obtains [46 : pp. 87–89] :

$$(56) \qquad |\zeta_1|^2+|\zeta_2|^2+2|\zeta_3|^2+\cdots+2|\zeta_{m-2}|^2 < 1+|\zeta_1\zeta_2+\zeta_3^2+\cdots+\zeta_{m-2}^2|<2.$$

Returning to equation (52) and inequality (53), we see that $\operatorname{Im}\left(\frac{a}{b}\right)$ is non-zero and is of constant sign as x varies on a connected domain or as g varies on G_R^0; since $x_2(g)=\zeta_{n+1}(g)$, $x_{n+1}(g)=\zeta_2(g)$, $\operatorname{Im}\zeta_{n+1}>0$, $\operatorname{Im}\zeta_2>0$, with $\zeta_2(e)=\zeta_{n+1}(e)=i$, and $-\frac{1}{2}Q'(x(e))=+1$, it follows that $\operatorname{Im}\left(\frac{a}{b}\right)<0$ and thus $0\leqq|\rho_1|<|\rho_{n+2}|$. It follows from the preceding lines, then, that as x runs over D, the solutions of (47) for x'' describe a bounded domain in (the Lie algebra $\mathfrak{p}_1^+$ of) P_1^+. Therefore, $G_R^0\subset P^-K_CP^+$ and the projection of G_R^0 on the third factor is a bounded domain.

Thus we have illustrated one of Harish-Chandra's theorems [26c]. We have chosen the example of the bounded domain associated to $SO(n, 2)$ because it has not been treated so often in the literature (cf. [28a] and [56e]) and because of its relation in a special case [63] to the group of invariants of the infinitesimal form of Maxwell's equations and of the metric of signature $(n-1, 1)$. More generally, the central quotient of $SO(n, 2)^0$ is the identity component of the group of conformal transformations of the n-dimensional real vector space supplied with the metric form $dx_1^2-dx_2^2-\cdots-dx_n^2$.

In the next Chapter we give the general formulation of Harish-Chandra's realization [26c; 27: Chap. VIII] of a Hermitian symmetric space of non-compact type as a bounded domain. In the next sections, we repeat this construction for the Siegel upper half-space of degree n, in order to explain partial Cayley transformations, and for the exceptional irreducible domain of dimension 27, in order to prepare the ground for the study of its Eisenstein series and related arithmetical problems.

We conclude this section by remarking that much of the calculation presented here in the description and properties of the symmetric domains of type IV (Siegel's classification) is based on [56e : § 48] and [46 : § 8].

§ 4. Again $GL(n)$ and $SL(n)$

If $g \in GL(n, \boldsymbol{R})$ and if Q is a real, positive-definite, symmetric $n \times n$ matrix, let $Q \cdot g = {}^t g Q g$. For fixed Q, we obtain every real, positive-definite, symmetric $n \times n$ matrix in the form $Q \cdot g$ for some $g \in GL(n, \boldsymbol{R})$. On the other hand, by the Babylonian reduction theorem [56a], there is a unique, positive, diagonal matrix a and a unique unipotent, upper triangular matrix u such that $Q \cdot ua = E$. The set of all $k \in GL(n, \boldsymbol{R})$ such that $E \cdot k = E$ is just $O(n)$. If A denotes the group of all positive, diagonal, real matrices, and N, the group of all real, upper triangular, unipotent matrices, then we have

$$GL(n, \boldsymbol{R}) = O(n, \boldsymbol{R}) \cdot A \cdot N \quad \text{(topological product)}. \tag{57}$$

Let $A_1 = A \cap SL(n, \boldsymbol{R})$. Then also

$$SL(n, \boldsymbol{R}) = SO(n, \boldsymbol{R}) \cdot A_1 \cdot N. \tag{58}$$

These results and those like them in the preceding section are special cases of the so-called Iwasawa decomposition which will be discussed in the next chapter.

§ 5. Examples continued——the symplectic group

Let G be the group $Sp(n, \boldsymbol{C})$ of $2n \times 2n$ complex matrices g such that ${}^t g J g = J$, where J is a fixed, real, skew-symmetric matrix.

We choose $J = \begin{pmatrix} 0 & E \\ -E & 0 \end{pmatrix}$, E being the $n \times n$ identity, and write $g \in G$ in the form $g = \begin{pmatrix} A & B \\ C & D \end{pmatrix}$, where A, B, C, and D are $n \times n$ square matrices. It is seen that the algebraic conditions for g to belong to G are

$$ {}^t AC - {}^t CA = {}^t BD - {}^t DB = 0, \quad {}^t AD - {}^t CB = E, \tag{59}$$

which are equivalent to those obtained by replacing g by ${}^t g$ since $g \in G$ if and only if ${}^t g \in G$. In particular, $g \mapsto {}^t g^{-1}$ is an automorphism of G. Let $\mathfrak{g}$ be the Lie algebra of G and write any $2n \times 2n$ matrix X in the form $X = \begin{pmatrix} A & B \\ C & D \end{pmatrix}$; the conditions for $X \in \mathfrak{g}$ are just

$$ {}^t XJ + JX = 0, \tag{60} $$

and $X \mapsto -{}^t X$ is an automorphism of $\mathfrak{g}$ which we denote by φ. It is a semi-simple transformation of $\mathfrak{g}$ having $+1$ and -1 as its eigenvalues; let $\mathfrak{k}$ be the $(+1)$-eigenspace of φ and $\mathfrak{p}$, the (-1)-eigen space. Then

$$ \mathfrak{k} = \left\{ \begin{pmatrix} A & B \\ -B & A \end{pmatrix} \mid A \text{ skew-symmetric, } B \text{ symmetric} \right\}, \tag{61} $$

$$ \mathfrak{p} = \left\{ \begin{pmatrix} A & B \\ B & -A \end{pmatrix} \mid A \text{ and } B \text{ both symmetric} \right\}, \tag{62} $$

while the center of $\mathfrak{k}$ consists of the elements

$$ \lambda_\theta = \begin{pmatrix} 0 & \theta E \\ -\theta E & 0 \end{pmatrix}, \quad \theta \in \boldsymbol{C}. \tag{63} $$

If $X = \begin{pmatrix} A & B \\ B & -A \end{pmatrix} \in \mathfrak{p}$, then $[\lambda_\theta, X] = 2\begin{pmatrix} \theta B & -\theta A \\ -\theta A & \theta B \end{pmatrix}$. Let $\tau = Ad(\exp \lambda_{\pi/4})$. Then τ has eigenvalues $\pm i$ on $\mathfrak{p}$ (which it stabilizes) and the respective eigenspaces are given by

$$ \left.\begin{aligned} \mathfrak{p}_{+i} &= \mathfrak{p}^+ = \left\{ \begin{pmatrix} A & iA \\ iA & -A \end{pmatrix} \right\} \\ \mathfrak{p}_{-i} &= \mathfrak{p}^- = \left\{ \begin{pmatrix} A & -iA \\ -iA & -A \end{pmatrix} \right\} \end{aligned} \right\}, \quad A \text{ symmetric}, \tag{64} $$

and both are normalized by $ad\ Y$, $Y \in \mathfrak{k}$, hence by $Ad\ K_C$, where K_C is the connected analytic subgroup of G with Lie algebra $\mathfrak{k}$. As in the preceding example of $SO(n, 2)$, we define an element $c \in \exp \mathfrak{p}$, in this case putting

$$ c = \exp \begin{pmatrix} 0 & i\alpha E \\ i\alpha E & 0 \end{pmatrix} \Big|_{\alpha = \pi/4} = \begin{pmatrix} \omega E & i\omega E \\ i\omega E & \omega E \end{pmatrix}, \quad \omega = 2^{-1/2}. \tag{65} $$

Then define $P_1^\pm = c \cdot \exp \mathfrak{p}^\pm \cdot c^{-1}$, $K_{1C} = c \cdot \exp \mathfrak{k} \cdot c^{-1}$; one sees that P_1^+ consists of the matrices of the form

$$ p_Z = \begin{pmatrix} E & 0 \\ Z & E \end{pmatrix}, \quad {}^t Z = Z, \tag{66} $$

P_1^- consists of those of the form

$$p_{Z'}^- = \begin{pmatrix} E & Z' \\ 0 & E \end{pmatrix}, \quad {}^tZ' = Z', \tag{67}$$

and K_{1C}, of those of the form

$$k_M = \begin{pmatrix} M & 0 \\ 0 & {}^tM^{-1} \end{pmatrix}, \quad \det M \neq 0. \tag{68}$$

Then it is easy to show that

$$c \cdot G_R \subset P_1^- K_{1C} P_1^+; \tag{69}$$

specifically, if $g = \begin{pmatrix} A & B \\ C & D \end{pmatrix} \in G_R$, then one proves, utilizing (59), that $Bi + D$ is non-singular, and that $c \cdot g = p_{Z'}^- k_M p_Z$ with $M = \omega^{-1}\, {}^t(Bi+D)^{-1}$ and $Z = (Bi+D)^{-1}(Ai+C)$. Let $\tau(g)$ be the $n \times 2n$ complex matrix $(\omega Ai + \omega C,\ \omega Bi + \omega D)$; we have $\tau(e) = (i\omega E,\ \omega E)$, so that $i\overline{\tau(e)} J\, {}^t\tau(e) = 2E > 0$, i.e., is a positive-definite Hermitian matrix. Hence $i\overline{\tau(g)} J\, {}^t\tau(g) > 0$ for all $g \in G_R^0$; but we may dispense with any distinction between G_R and its identity component since G_R is connected. A simple proof of this results from the fact that G is generated by all the elements p_Z and p_Z^-, and G_R, by those with real Z. If $g = \begin{pmatrix} A & B \\ C & D \end{pmatrix} \in G_R$, let $\zeta(g)$ be the Z such that $c \cdot g = p_{Z'}^- k_M p_Z$; then from what we have just seen, and utilizing (59), it is easy to prove that $\zeta(g) = Z = X + iY$ is symmetric, and Y is positive-definite. As in the preceding example, we let G_R operate on the set D' of $\zeta(g)$, $g \in G_R$, by $\zeta(g) \cdot g' = \zeta(gg')$, $g' \in G_R$, and we define

$$D = \{Z \mid Z = X + iY \text{ is } n \times n \text{ symmetric, } Y > 0\}. \tag{70}$$

If $Z = X$ is real symmetric, then

$$\zeta(g) \cdot p_X = \zeta(g) + X, \tag{71}$$

while if M is real, $n \times n$, non-singular, then

$$\zeta(g) \cdot k_M = {}^tM \zeta(g) M. \tag{72}$$

From the preceding section, we know that the group of k_M for which M is upper triangular is transitive on the cone of positive-definite, real, symmetric matrices Y, when we define

$$Y \cdot k_M = {}^tMYM. \tag{73}$$

It now follows in the same manner as in the preceding example that

G_R is actually transitive on D, i.e., $D=D'$. The proofs of some of the results analogous to those of the example for $SO(n, 2)$, e.g., the determination of the stability group in G_R of iE (it is $U(n)$) and the proof of the Iwasawa decomposition for G_R are left as exercises for the reader. One notes that since $Sp(1, \boldsymbol{R}) \cong SL(2, \boldsymbol{R})$, the series of examples just considered includes that of the upper half-plane in one variable.

If $D=\{z=x+iy \mid y>0\}$ is the upper half-plane in the space $\boldsymbol{C}$ of one complex variable, the mapping

$$z \mapsto \frac{z-i}{-iz+1} \tag{74}$$

sends D onto the unit disc $\{z \mid |z|<1\}$. This mapping, known as the inverse Cayley transformation, can be extended to the case of the domain D defined in (70) for any n. For a given n, there are n transformations of the type (74) called "partial Cayley transformations", and the n images of D under these may be described in a simple way. If r is a positive integer $\leqq n$, let $s=n-r$, let E' denote the $r \times r$ identity, and let E'' be the $s \times s$ identity. Define

$$c=c_r=\begin{pmatrix} \omega E' & 0 & i\omega E' & 0 \\ 0 & E'' & 0 & 0 \\ i\omega E' & 0 & \omega E' & 0 \\ 0 & 0 & 0 & E'' \end{pmatrix}.$$

If $\tau=(ZE)$ is the $n \times 2n$ matrix made of the last n rows of p_Z, where Z is such that $Z \in D$, then we have

$$-i\tau J\,{}^t\bar{\tau}>0,$$

i.e., $-i\tau J\,{}^t\bar{\tau}$ is a positive-definite Hermitian matrix, so that also

$$-i\tau c_r^{-1}(c_r J \bar{c}_r)\bar{c}_r^{-1}\,{}^t\bar{\tau}>0 \tag{75}$$

(noting that ${}^tc=c$). Now by analogy with equation (47) of section 3 we seek to solve the equation

$$p^- k p_\zeta^+ = p_Z^+ \cdot c_r^{-1}. \tag{76}$$

Let $(M_1 M_2)$ be the $n \times 2n$ matrix composed of the last n rows of $p_Z^+ \cdot c_r^{-1}$, where each M_j is $n \times n$. Then, as in earlier cases, one shows easily that M_2 is non-singular and that accordingly (76) has a unique solution with $\zeta = M_2^{-1}M_1$. If we write

$$\zeta = \begin{pmatrix} \zeta_{11} & \zeta_{12} \\ {}^t\zeta_{12} & \zeta_{22} \end{pmatrix},$$

where ζ_{11} is $r\times r$, ζ_{22} is $s\times s$, and ζ_{12} is $r\times s$, then ζ_{jj} is symmetric. $j=1, 2$, and if

$$Z = \begin{pmatrix} Z_{11} & Z_{12} \\ {}^tZ_{12} & Z_{22} \end{pmatrix}$$

is a like decomposition of Z, then we have

$$\begin{aligned} \zeta_{11} &= (E'-iZ_{11})^{-1}(Z_{11}-iE'), \\ \zeta_{12} &= \omega^{-1}(E'-iZ_{11})^{-1}Z_{12}, \quad \text{and} \\ \zeta_{22} &= i\,{}^tZ_{12}(E'-iZ_{11})^{-1}Z_{12}+Z_{22}. \end{aligned} \tag{77}$$

Moreover, it follows from (75) that

$$\begin{pmatrix} E'-\zeta_{11}\bar{\zeta}_{11} & -i\zeta_{12}-\zeta_{11}\bar{\zeta}_{12} \\ i{}^t\bar{\zeta}_{12}-{}^t\zeta_{12}\bar{\zeta}_{11} & 2\operatorname{Im}\zeta_{22}-{}^t\zeta_{12}\bar{\zeta}_{12} \end{pmatrix} > 0. \tag{78}$$

Now a Hermitian matrix

$$\begin{pmatrix} A & B \\ {}^t\bar{B} & C \end{pmatrix}$$

will be positive-definite if and only if $A>0$ and $C-{}^t\bar{B}A^{-1}B>0$. The first condition becomes

$$E'-\zeta_{11}\bar{\zeta}_{11}>0, \tag{79}$$

and the second,

$$2\operatorname{Im}\zeta_{22}-{}^t\zeta_{12}\bar{\zeta}_{12}+(i{}^t\bar{\zeta}_{12}-{}^t\zeta_{12}\bar{\zeta}_{11})(E'-\zeta_{11}\bar{\zeta}_{11})^{-1}(i\zeta_{12}+\zeta_{11}\bar{\zeta}_{12})>0. \tag{80}$$

Define

$$L_{\zeta_{11}}(\zeta_{12}, \zeta'_{12}) = {}^t\zeta_{12}\bar{\zeta}'_{12}-(i{}^t\bar{\zeta}_{12}-{}^t\zeta_{12}\bar{\zeta}_{11})(E'-\zeta_{11}\bar{\zeta}_{11})^{-1}(i\zeta'_{12}+\zeta_{11}\bar{\zeta}'_{12}). \tag{81}$$

Then (80) becomes

$$2\operatorname{Im}\zeta_{22}-L_{\zeta_{11}}(\zeta_{12}, \zeta_{12})>0, \tag{82}$$

and the image of D under the mapping $Z\mapsto\zeta$ becomes the set of ζ satisfying (79) and (82), since the condition $\operatorname{Im} Z>0$ is clearly equivalent to the conditions (79) and (82), as the chain of calculations reducing the one condition to the others is reversible. We denote the set of $r\times r$ symmetric matrices ζ_{11} satisfying $E'-\zeta_{11}\bar{\zeta}_{11}>0$ by F.

The domain described by the conditions (79) and (82) is an example of what is called a Siegel domain of the third kind. We shall return to this subject in Chapters 7 and 11.

We conclude this section by letting Γ be the subgroup $Sp(n, \boldsymbol{Z})$ of integral unimodular matrices in G. It is a discrete subgroup of G_R and is known [25; 1] to be maximal discrete there. Its image in $\mathrm{Hol}(D)$ operates in properly discontinuous fashion, so that D/Γ has a natural structure of a complex analytic space (and, in fact, is a V-manifold in the sense of Satake [50a]).

§ 6. An exceptional domain (cf. [2i])

We now describe a symmetric tube domain $\mathfrak{T}$, equivalent to a bounded domain D, such that $\mathrm{Hol}(D)$ is a real form of the exceptional Lie group E_7. First we give some algebraic preliminaries necessary for the description of this domain. Most of these may be found in [17] and [19b], the latter being the basic reference for the description of the real form of E_7.

By the (real) Cayley algebra, we mean an eight-dimensional non-associative algebra $\mathfrak{C}$ over $\boldsymbol{R}$ with basis $e_0, \cdots, e_7$ and law of multiplication given [17] by the rules:

(α) $xe_0 = e_0x = x$ for all $x \in \mathfrak{C}$,

(β) $e_i^2 = -e_0$, $i = 1, \cdots, 7$,

(γ) $e_1e_2e_4 = e_2e_3e_5 = e_3e_4e_6 = e_4e_5e_7 = e_5e_6e_1 = e_6e_7e_2 = e_7e_1e_3 = -e_0$

(the triads are associative [17]);

for simplicity, we write $e_0 = 1$, and we view $\boldsymbol{R}$ as imbedded in $\mathfrak{C}$ by identifying it with $\boldsymbol{R}\cdot e_0$. We define an involution $x \mapsto \bar{x}$ of $\mathfrak{C}$ by: $x = x_0 + x_1e_1 + \cdots + x_7e_7 \mapsto \bar{x} = x_0 - x_1e_1 - \cdots - x_7e_7$, $x_i \in \boldsymbol{R}$; furthermore, we define $T: \mathfrak{C} \to \boldsymbol{R}$ by $T(a) = a + \bar{a}$, $N: \mathfrak{C} \to \boldsymbol{R}$ by $N(a) = a\bar{a}$, and $B: \mathfrak{C} \times \mathfrak{C} \to \boldsymbol{R}$ by $B(a, b) = T(a\bar{b}) = T(\bar{a}b)$. One defines [17] a subring o of the ring

$$\mathfrak{C}_{\boldsymbol{Q}} = \{x = x_0 + \cdots + x_7e_7 \in \mathfrak{C} \mid x_i \in \boldsymbol{Q},\ i = 0, \cdots, 7\}$$

such that o is a lattice in the real vector space $\mathfrak{C}$, which is self-dual with respect to the inner product $B(\ ,\ )$. This lattice o, which in [17] is denoted by J, is isometric to the root lattice of the Lie algebra E_8 if we give each root the length 2.

Denote by $M_3(\mathfrak{C})$ the set of three-by-three matrices over $\mathfrak{C}$ and let $\mathfrak{J}$ be the real subspace of all $X=(x_{ij})\in M_3(\mathfrak{C})$ satisfying $x_{ij}=\bar{x}_{ji}$; in particular, x_{ii} is a real number that we denote by ξ_i. Clearly $\dim_{\boldsymbol{R}}\mathfrak{J}=27$. We supply $\mathfrak{J}$ with a product by $X\circ Y=\frac{1}{2}(XY+YX)$, where XY is the ordinary matrix product; with this product, $\mathfrak{J}$ becomes a real Jordan algebra. If $X=(x_{ij})\in\mathfrak{J}$, define $\operatorname{tr}(X)=\xi_1+\xi_2+\xi_3$, and define an inner product $(\ ,\ )$ on $\mathfrak{J}$ by $(X, Y)=\operatorname{tr}(X\circ Y)$. Moreover, we define a homogeneous cubic polynomial function det on $\mathfrak{J}$ by

$$\det X=\xi_1\xi_2\xi_3-\xi_1N(x_{23})-\xi_2N(x_{13})-\xi_3N(x_{12})+T((x_{12}x_{23})\bar{x}_{13}),$$

and define a symmetric trilinear form $(\ ,\ ,\ )$ on $\mathfrak{J}\times\mathfrak{J}\times\mathfrak{J}$ by requiring the identity $(X, X, X)=\det X$ to hold. Then we define a bilinear mapping $(X, Y)\mapsto X\times Y$ of $\mathfrak{J}\times\mathfrak{J}$ into $\mathfrak{J}$ by stipulating that $(X\times Y, Z)=3(X, Y, Z)$ hold identically. Finally, we denote by $\mathfrak{R}$ the set of squares $X\circ X$ of elements of $\mathfrak{J}$, and by $\mathfrak{R}^+$, the interior of $\mathfrak{R}$; then $\mathfrak{R}^+$ is a convex open cone in $\mathfrak{J}$. We let $\mathfrak{J}_o=\{X=(x_{ij})\in\mathfrak{J}\,|\,x_{ij}\in o,\ i, j=1, 2, 3$ (in particular, $\xi_i\in o\cap\boldsymbol{Q}=\boldsymbol{Z})\}$; this lattice in $\mathfrak{J}$ is self-dual with respect to the inner product $(\ ,\ )$.

Let S' be the three-dimensional torus of triples $s=(s_1, s_2, s_3)$, $s_i\in\boldsymbol{C}^*$, supplied with componentwise multiplication as the group law. If $s\in S'_{\boldsymbol{R}}$, let s operate on $\mathfrak{J}$ by $s\cdot X=s\cdot(x_{ij})=(s_is_jx_{ij})$; this representation θ of $S'_{\boldsymbol{R}}$ is an isogeny of it into $GL(\mathfrak{J})$. Extending this to a complex representation of S' in the obvious manner, let $S=\theta(S')$. It is clear that S and S' are isomorphic over $\boldsymbol{Q}$ to $(\boldsymbol{C}^*)^3$. We define two subgroups $\mathcal{S}$ and $\mathcal{J}$ of $GL(\mathfrak{J})$:

$$\mathcal{S}=\{g\,|\,g\in GL(\mathfrak{J}),\ \det(gX)\equiv\nu(g)\det X,$$
$$\text{where } \nu(g) \text{ is a non-zero constant, independent of } X\},$$

$$\mathcal{J}=\text{the kernel of } \nu \text{ in } \mathcal{S}.$$

Clearly $S\subset\mathcal{S}$, and if $s=(s_1, s_2, s_3)\in S'$, then $\nu(\theta(s))=(s_1s_2s_3)^2$. Define $S_0=\mathcal{S}\cap\mathcal{J}$ and use S'_0 to denote the identity component of $\theta^{-1}(S_0)$.

Let $\boldsymbol{X}$ and $\boldsymbol{X'}$ be two real vector spaces, each isomorphic to $\mathfrak{J}$, and let Ξ and Ξ' be two one-dimensional real vector spaces. Denote by $\boldsymbol{W}$ the direct sum $\boldsymbol{X}\oplus\Xi\oplus\boldsymbol{X'}\oplus\Xi'$, and if $w=(X, \xi, X', \xi')\in\boldsymbol{W}_C$, the complexification of $\boldsymbol{W}$, and $s=(s_1, s_2, s_3)\in S'$, define

$$\theta_1(s)\cdot w=(\theta(s)\cdot X,\ |s|^{-1}\xi,\ \theta(s)^{-1}X',\ |s|\xi'), \tag{83}$$

where $|s| = \nu(\theta(s)) = (s_1 s_2 s_3)^2$. Then θ_1 is a homomorphism of S' into $GL(\boldsymbol{W}_C)$ and we let $S_1 = \theta_1(S')$. Define [19b] a quartic form Q on $\boldsymbol{W}_C$ by

$$(84)\quad Q(w) = (X \times X, X' \times X') - \xi \det X - \xi' \det X' - \frac{1}{4}((X, X') - \xi\xi')^2,$$

and use $\{\,,\}$ to denote the skew-symmetric bilinear form on $\boldsymbol{W} \times \boldsymbol{W}$ given by

$$(85)\qquad \{w_1, w_2\} = (X_1, X_2') - (X_2, X_1') + \xi_1\xi_2' - \xi_2\xi_1'.$$

Now let $\mathcal{G}$ be the group of all $g \in GL(\boldsymbol{W})$ leaving Q and $\{\,,\}$ invariant. The group $\mathcal{G}$ can easily be shown to be connected. It contains the following subgroups:

1. The real torus S_{1R};
2. the group $\mathcal{J}$, where the action of $g \in \mathcal{J}$ on $\boldsymbol{W}$ is given by

$$g(X, \xi, X', \xi') = (g \cdot X, \xi, g^* \cdot X', \xi'),$$

g^* being the inverse adjoint of g on $\mathfrak{J}$ defined by the identity $(g \cdot X, g^* \cdot Y) \equiv (X, Y)$;

3. the group $\boldsymbol{P}^+$ consisting of all the transformations p_B', $B \in \mathfrak{J}$, where p_B' is given by $p_B'(X, \xi, X', \xi') = (X_1, \xi_1, X_1', \xi_1')$ and

$$X_1 = X + 2B \times X' + \xi B \times B, \quad X_1' = X' + \xi B,$$
$$\xi_1' = \xi' + (B, X) + (B, B \times X') + \xi \det B, \quad \xi_1 = \xi;$$

and, finally,

4. the group $\boldsymbol{P}^- = \iota^{-1}\boldsymbol{P}^+\iota$, where one defines $\iota \in \mathcal{G}$ by

$$\iota(X, \xi, X', \xi') = (-X', -\xi', X, \xi).$$

The latter group consists of the elements p_B, $B \in \mathfrak{J}$, where p_B is obtained from p_B' by interchanging the primed and unprimed variables. One way of seeing that $\mathcal{G}$ is connected is to show that it is generated by its subgroups $\boldsymbol{P}^+$ and $\boldsymbol{P}^-$, each of which is obviously connected.

The group $\mathcal{G}$, and each of its subgroups 1.–4., is algebraic and has a complexification $\mathcal{G}_C$, resp. S_1, $\mathcal{J}_C$, $\boldsymbol{P}_C^+ = P^+$, and $\boldsymbol{P}_C^- = P^-$, operating on $\boldsymbol{W}_C$. Denote by $\mathcal{S}_C$ the subgroup of $\mathcal{G}_C$ generated by $\mathcal{J}_C$ and S_1; it is seen to be isomorphic to the complexification of the group $\mathcal{S}$ defined earlier.

Identifying $\boldsymbol{C}^{27}$ with $\mathfrak{J}_C$, we define the tube domain $\mathfrak{T}$ in $\boldsymbol{C}^{27}$ by

$$(86)\qquad \mathfrak{T} = \{Z = X + iY \mid X \in \mathfrak{J}, Y \in \mathfrak{R}^+\}.$$

We now describe the action of $\mathcal{G}$ on $\mathfrak{T}$ in fashion parallel to that in examples of preceding sections. The subgroup of $\mathcal{G}$ generated by $\mathcal{J}$ and S_{1R} is to be identified with $\mathcal{S}$ by restriction of its action to the subspace $\boldsymbol{X}$. To determine the operation of $g \in \mathcal{G}$ on $\mathfrak{T}$, we let $Z \in \mathfrak{T}$ and, parallel to what was done in earlier sections, proceed to solve the equation

$$p'_Z \cdot g = p_A k p'_{Z_1}, \tag{87}$$

for $Z_1 \in \mathfrak{T}$, $A \in \mathfrak{J}_C$, and $k \in \mathcal{S}_C$. If we put

$$Z \cdot g = Z_1, \tag{88}$$

then this will determine a biholomorphic transformation (of $\mathfrak{T}$) assigned to each $g \in \mathcal{G}$ and hence an action of $\mathcal{G}$ on $\mathfrak{T}$, once we have shown that (87) has a unique solution for each element g of a set of generators of $\mathcal{G}$. As remarked above, the group $\mathcal{G}$ is connected and is generated by $\boldsymbol{P}^+ = P_R^+$ and ι. If $g = p'_B$, $B \in \mathfrak{J}$, then $p'_Z \cdot p'_B = p'_{Z+B}$; hence, $Z \cdot g = Z + B$. To determine the action of ι, we use the formula for p'_B, $B \in \mathfrak{J}_C$, that for p_A, indicated above, and the easily verified formula

$$k \cdot (X, \cdots, \xi') = (k \cdot X, \cdots, \nu(k)\xi'), \quad k \in \mathcal{S}_C, \tag{89}$$

to find the conditions for (87) to hold. The result is

$$Z \cdot \iota = -(\det Z)^{-1} \cdot (Z \times Z) = -Z^{-1} \tag{90}$$

(where Z^{-1} is the Jordan algebra inverse of $Z \in T$, which *exists* because $Z = X + iY$ with Y positive-definite, i.e., $Y \in \mathfrak{R}^+$). It follows by direct calculation that $\mathcal{S}_C$ normalizes P^+ and P^-, and that the product mapping

$$P^- \times \mathcal{S}_C \times P^+ \to P^- \cdot \mathcal{S}_C \cdot P^+$$

is a homeomorphism. Therefore, the solution to (87) is unique when it is known to exist, and since the group law in $\mathcal{G}_C$ is associative, the action defined by means of (88) is legitimate. Moreover, the operation of $\mathcal{G}$ is transitive on $\mathfrak{T}$: if $Z \in \mathfrak{T}$, there exists $g \in P_R^+$ such that $Z \cdot g = iY \in i\mathfrak{R}^+$, and by well-known facts about $\mathfrak{J}$ (e.g., the principal axis theorem [19a]), there exists $g_1 \in \mathcal{S}$ such that $g_1 Y = (\delta_{ij}) = E$, so that there exists $g_2 \in \mathcal{S} \subset \mathcal{G}$ such that $iY \cdot g_2 = iE$. This proves the transitivity of $\mathcal{G}$ on $\mathfrak{T}$.

Finally, we define an arithmetic subgroup Γ of $\mathcal{G}$. First let $\boldsymbol{W}_o$

be the lattice in $\boldsymbol{W}_R$ given by

(91) $$\boldsymbol{W}_o = \boldsymbol{X}_o \oplus \boldsymbol{Z} \oplus \boldsymbol{X}'_o \oplus \boldsymbol{Z},$$

where $\boldsymbol{X}_o$ and $\boldsymbol{X}'_o$ are identified with the lattice $\mathfrak{J}_o$, $\mathfrak{J} \cong \boldsymbol{X} \cong \boldsymbol{X}'$. Then let

(92) $$\Gamma = \{g \in \mathcal{G} \mid g \cdot \boldsymbol{W}_o = \boldsymbol{W}_o\}.$$

Clearly Γ is an arithmetic subgroup of $\mathcal{G}$. We define $P_o^+ = \{p'_B \mid B \in \mathfrak{J}_o\}$, and $P_o^- = \{p_B \mid B \in \mathfrak{J}_o\}$. It is seen that $P_o^{\pm} = \Gamma \cap P^{\pm}$. The arithmetic group Γ has several interesting properties [2i], some of which will be used later. One that may be remarked now is that Γ is generated by P_o^+ and P_o^- or, alternatively, by P_o^+ and ι. Thus its image under the canonical mapping of $\mathcal{G}$ onto Hol($\mathfrak{T}$) is generated by the translations $t_B : Z \mapsto Z + B$, $B \in \mathfrak{J}_o$, and by $\iota : Z \mapsto -Z^{-1}$.

It should be noted here that $\mathcal{G}$ is not isomorphic to Hol($\mathfrak{T}$) $=$ Hol$(\mathfrak{T})^0$. In fact, $\mathcal{G}$ has a center $Z_2 = \{\pm id.\}$, and we actually have $\mathcal{G}/Z_2 \cong$ Hol($\mathfrak{T}$).

§ 7. Remarks

In sections 3 and 5, for a given group G, we have constructed subgroups $P_1^{\pm}$ and K_{1C} of G_C, and for a certain element c of exp $\mathfrak{p}$ we have shown that

(93) $$c \cdot G_R^0 \subset P_1^- K_{1C} P_1^+.$$

Things in section 6 were arranged a little differently, but still with an eye to the same objectives. It may be checked without great difficulty in this case, too, that one has a relation similar to (93). In fact, (93) expresses a general result of Harish-Chandra [26c, 27] for our special cases. As it happens, all our examples are of so-called tube-type domains, i.e., domains equivalent to domains of the form

$$D(n, \mathfrak{R}) = \{z \in \boldsymbol{C}^n \mid z = x + iy,\ y \in \mathfrak{R}\},$$

where $\mathfrak{R}$ is a homogeneous convex open cone in $\boldsymbol{R}^n$. One might also consider as an exercise the case of the unitary group of a Hermitian form of signature $(1, n)$ which operates on the unit sphere

(94) $$\sum_{j=1}^{n} |z_j|^2 < 1$$

in $\boldsymbol{C}^n$, which is not equivalent to a tube domain.

CHAPTER 7
ALGEBRAIC GROUPS

§1. Basic definitions and theorems [6a, c; 9; 53]

As in Chapter 6, section 1, we define an algebraic linear group as a subgroup G of $GL(V)$ for some finite-dimensional vector space V such that $G=GL(V)\cap W$, where W is an algebraic subset of End(V). If K is a field of definition for W, then K is called a field of definition for G. We set down here a number of results for the proofs of which we refer the reader to [6a]. Unless stated otherwise, we assume all fields considered are of characteristic zero. When we need to speak of an algebraic group variety (possibly not connected) in the sense of [60b], we shall use the phrase "abstract algebraic group".

If G is a group and if A and B are two subsets of G, we define the commutator $[A, B]$ as the smallest subgroup of G containing all the commutators $aba^{-1}b^{-1}$, $a\in A$, $b\in B$. We define two series $\{C^iG\}_{i=0,1,2,\cdots}$ and $\{\mathscr{D}^iG\}_{i=0,1,2,\cdots}$ of subgroups of G by: $C^0G=\mathscr{D}^0G=G$, $\mathscr{D}^iG=[\mathscr{D}^{i-1}G, \mathscr{D}^{i-1}G]$, $C^iG=[G, C^{i-1}G]$, $i>0$; and we define $\mathscr{D}^*G=\bigcap_{i\geqq 0}\mathscr{D}^iG$, $C^*G=\bigcap_{i\geqq 0}C^iG$. Then, if G is an algebraic group, all $\mathscr{D}^iG$ and C^iG, as well as $\mathscr{D}^*G$ and C^*G, are algebraic subgroups of G and are connected if G is. An algebraic group G is called solvable (resp. nilpotent) if $\mathscr{D}^*G=\{e\}$ (resp. if $C^*G=\{e\}$) which amounts to saying that $\mathscr{D}^iG=\{e\}$ (resp. $C^iG=\{e\}$) for some sufficiently large i because of the descending chain condition on algebraic subvarieties. It is clear that a nilpotent algebraic group is solvable.

If G and G' are linear algebraic groups, a mapping $f: G\to G'$ is called a rational homomorphism if its coordinates are rational functions on G, all defined (in the sense of algebraic geometry) at some point of every component of G, and if $f(x)\cdot f(y)=f(xy)$ whenever f is defined at x, at y, and at xy; when that is so, then f is defined at every point of G. The homomorphism f is said to be defined over the field K if its graph, which is an algebraic subgroup of $G\times G'$, is defined over K. It is called an (algebraic) isomorphism of G onto G' if $f(G)=G'$ and $\ker(f)=\{e\}$ (recalling that K is of characteristic zero).

PROPOSITION 1. *Let* $f\colon G \to G'$ *be a rational homomorphism. Then* $\ker(f)=f^{-1}(e')$ *is an algebraic subgroup of* G, $f(G)$ *is an algebraic subgroup of* G', *and*

$$\dim(\ker(f))+\dim(f(G))=\dim G.$$

If f *is defined over* K, *then so are* $\ker(f)$ *and* $f(G)$.

Let G be an abstract algebraic group, H an algebraic subgroup of G, both defined over K. An abstract variety V defined over K is called a quotient variety of G by H with respect to K if there exists an everywhere defined rational mapping π, defined over K, of G onto V such that (a) $\pi(g)=\pi(g')$ if and only if $g \in g'H$; (b) a rational mapping f of G into some variety W right-invariant under H (in other words, $f(gh)=f(g)$ whenever $g \in G$, $h \in H$, and f is defined at g and at gh) may be factored by way of V:

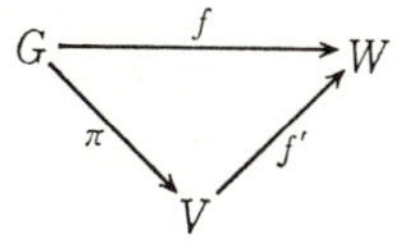

where $f=f'\circ\pi$ generically; and (c) if H is normal in G, then V is an abstract algebraic group and π, a rational homomorphism.

THEOREM 13. *If* G *is an algebraic group and* H, *a closed subgroup, both defined over* K, *then a quotient variety of* G *by* H *with respect to* K *exists.*

The proof may be found in [60c]. The quotient variety V, which is characterized up to a biregular birational isomorphism, is denoted by G/H.

We use G_m (as before) to denote the algebraic group $GL(1)$ (so $\dim V=1$) and G_a to denote the algebraic group consisting of a one-dimensional vector space V together with the additive structure on V as group law; G_a may be identified with the group of all the 2×2 matrices $\begin{pmatrix}1 & x\\ 0 & 1\end{pmatrix}$ in order to make it into a linear algebraic group in our definition. Then if f is a rational homomorphism of G_a into a linear algebraic group G, it follows from [6a : § 9.3] that $f(G_a)$ is a unipotent subgroup of G. One may show that every one-dimensional, connected, linear, algebraic group is isomorphic to G_a or to G_m and that a solvable, connected, linear, algebraic group has a composition series

$$G = G_0 \supset G_1 \supset \cdots \supset G_l = \{e\}$$

such that each of the factor groups G_i/G_{i+1} is isomorphic either to G_a or to G_m. And in fact one has

THEOREM 14. [6a : p. 53]. *Let G be a connected, solvable, algebraic subgroup of $GL(V)$ defined over an algebraically closed field k and let B be the group of upper triangular matrices in $GL(V)$ with respect to some pre-assigned basis of V. Then there exists $g \in GL(V)_k$ such that $gGg^{-1} \subset B$.*

An element $g \in GL(V)$ is called unipotent if all the roots of its characteristic equation are 1. A linear algebraic group is called unipotent if all its elements are unipotent; it is then automatically connected. In the above theorem, if G is unipotent and if $G \subset B$, then it is clear that in fact $G \subset B_u$, the group of all unipotent, upper triangular matrices in B.

If G is a connected, linear algebraic group, then the family of connected, solvable, algebraic subgroups of G, ordered by inclusion, contains, obviously, maximal elements. They are called the Borel subgroups of G, and any two Borel subgroups of G are conjugate by an inner automorphism [6a : p. 66]. If H is an algebraic subgroup of G, then the quotient variety G/H is a complete projective variety if and only if H contains a Borel subgroup of G; such a subgroup H is called a parabolic subgroup of G. The groups that we have designated by H or H_1 in Chapter 6, sections 2 and 3, are parabolic, for in all cases they contain Borel subgroups of G. In fact, in our examples, the system of coordinates in $\boldsymbol{C}^n$ has been chosen so that, luckily, the intersection of G with the group of upper triangular matrices of $GL(V)$ is a Borel subgroup of G, an event which may fail to occur (and would have in our examples if the matrix I had been replaced by the identity matrix E of the same dimension); that G/Borel is complete in our examples may be seen by observing that the flags discussed there, supplied with Plücker coordinates, constitute the points of an algebraic subvariety of projective space; this is a special case of the proof used in general [6a : pp. 65–66].

If G is a linear algebraic group, there exists a unique, maximal, normal, connected, solvable subgroup $R(G)$ of G and a unique, maximal, normal, unipotent subgroup $R_u(G)$; we have $R_u(R(G))=R_u(G)$. These are called, respectively, the radical and the unipotent radical of G and are defined over the same field as G. If $R(G)=\{e\}$, then G is called semi-simple, while if $R_u(G)=\{e\}$, then G is called reductive. We observe that $R(G)$ is contained in every Borel subgroup B of G and $R_u(G)\subset R_u(B)$. The quotient groups $G/R(G)$ and $G/R_u(G)$ are respectively semi-simple and reductive. The quotient mapping π: $G\to G/R_u(G)$ (in characteristic zero) has a cross-section L which is a reductive subgroup of G defined over the same field k as G and $G=L\cdot R_u(G)$ (semi-direct product); any two such L are conjugate by an element of $R_u(G)_k$ if both are defined over k. L is called a Levi complement of $R_u(G)$. For example, in section 2 of Chapter 6, we may take $G=H$, $R(H)=R_H$, $R_u(H)=U_H$, and $L=Z(T_H)$. We recall the definition of torus from section 2 of Chapter 6. Then if G is reductive, $R(G)=T$ is a torus consisting of central elements of G and there is a semi-simple subgroup L' of G such that $L'\cap T$ is finite and $G=L'\cdot T$. The latter is not a direct product (in general); however, to preserve related notions of some usefulness we say that G is the almost semi-direct (resp. the almost direct) product of two subgroups A and B if B is normal, $A\cap B$ is finite, and $G=A\cdot B$ (resp. if these conditions hold and in addition $ab=ba$ for all $a\in A$, $b\in B$).

If G is connected and T is a torus in G, the centralizer $Z_G(T)=Z(T)$ is a connected algebraic subgroup of G [6a : p. 60] and the normalizer $\boldsymbol{N}_G(T)=\boldsymbol{N}(T)$ of T in G is an algebraic subgroup of G. One knows that if G is reductive (resp. semi-simple), then $Z(T)=T\cdot Z'$, where $T\cap Z'$ is finite and Z' is reductive (resp. semi-simple). By a character of an algebraic group G, we mean a rational homomorphism of G into G_m. Over an algebraically closed field, a torus T is isomorphic to $G_m{}^l$, $l\geqq 0$, and l is called the rank of T. It is elementary to see that any character of G_m is of the form $\zeta\to\zeta^k$, $k\in\boldsymbol{Z}$, hence the character group $\boldsymbol{X}^*(T)$ of T is a free Abelian group of rank $l=\dim T$, and if $t_1, t_2\in T$, $t_1\neq t_2$, there exists a character χ of T such that $\chi(t_1)\neq\chi(t_2)$. If T is a torus contained in the connected, linear, algebraic group G, then $\boldsymbol{N}(T)$ operates on $\boldsymbol{X}^*(T)$ by inner automorphisms and the kernel is exactly $Z(T)$; thus $W(T)=\boldsymbol{N}(T)/Z(T)$ has a faithful representation as a subgroup of $GL(l, \boldsymbol{Z})$. It is easy to see that the set of elements of

T having finite orders is a Zariski-dense subset of T; let T_N be the set of these whose orders divide N. Then $\boldsymbol{N}(T)$ normalizes T_N and hence its identity component $\boldsymbol{N}(T)^0$ centralizes T_N; this being true for all N, it follows that $\boldsymbol{N}(T)^0 \subset Z(T)$. Since $Z(T)$ is connected, it follows that $\boldsymbol{N}(T)^0 = Z(T)$ [6a : § 7.9]. Therefore, $W(T)$ is a finite group, called the Weyl group of T.

If T is a torus defined over k, not necessarily algebraically closed, then T is said to split over k or to be k-split if T is isomorphic over k to $G_m{}^l$; this is equivalent to saying that every $\chi \in \boldsymbol{X}^*(T)$ is defined over k.

Let G be a connected, linear, algebraic group defined over a field k, not necessarily algebraically closed. Then [9 : § 4.21] all maximal k-split tori of G are conjugate by inner automorphisms coming from elements of G_k. If ${}_kT$ is one of these, its dimension is called the k-rank of G. Let ${}_kW = W({}_kT)$; one may prove that every $w \in {}_kW$ has a representative in $\boldsymbol{N}({}_kT)_k$.

EXAMPLE. The group $G = SO(Q_{r,s})$ of Chapter 6 has an $\boldsymbol{R}$-split torus T of rank r and $Z(T)_{\boldsymbol{R}}$ is the almost direct product of $T_{\boldsymbol{R}}$ and of a compact group; it follows that T is a maximal $\boldsymbol{R}$-split torus in G (in our examples, we have dealt with the cases $r = 0, 1, 2$, and $[\dim V/2]$, where $[x]$ denotes the biggest integer in x). If S is any $\boldsymbol{R}$-split torus in G, the eigenspace in V corresponding to any non-trivial character of S is totally isotropic, and certain direct sums of such eigenspaces which are $\neq \{0\}$ may be expressed as a flag of totally isotropic subspaces of V; and that flag may be imbedded in a maximal flag. This together with Witt's theorem implies the statements of the preceding paragraphs in the special case $G = SO(Q_{r,s})$ with $k = \boldsymbol{R}$.

In the case when k is algebraically closed, one customarily omits the subscripts k and refers to $W(T) = \boldsymbol{N}(T)/Z(T)$ (T a maximal torus of G) as the Weyl group of G.

§ 2. Representations and root systems

Let G be a connected, semi-simple, linear, algebraic group defined over the field k and let $S = {}_kT$ be a maximal k-split torus of G.

Let ρ be a rational representation of G on a vector space $\boldsymbol{W}$; in other words, ρ is a rational homomorphism of G into $GL(\boldsymbol{W})$. If $\chi \in \boldsymbol{X}^*(S)$, let

$$\boldsymbol{W}_\chi = \{w \in \boldsymbol{W} \mid \rho(s) \cdot w = \chi(s)w, \ s \in S\}.$$

Since $\rho(S)$ is completely reducible, $\boldsymbol{W}$ is the direct sum of the subspaces $\boldsymbol{W}_\chi \neq 0$. The χ such that $\boldsymbol{W}_\chi \neq 0$ are called the (S-)weights of the representation ρ.

In particular, if $\rho = Ad$ is the adjoint representation of G on the Lie algebra of G, then ρ is a rational representation of G defined over k and its non-trivial weights will be called the k-roots of G. The set of k-roots is denoted by ${}_k\Sigma$ and is viewed as a subset of $\boldsymbol{X}^*(S)_Q = \boldsymbol{X}^*(S) \otimes_Z \boldsymbol{Q}$ on which a linear ordering $\succ$ is taken to have been defined (if $\alpha \succ 0$, $\beta \succ 0$, $r \in Q$, $r > 0$, then $\alpha + \beta \succ 0$, $r\alpha \succ 0$). A k-root $\alpha \succ 0$ is called simple if α cannot be written as $\beta + \gamma$ where β and γ are also positive k-roots. The set of simple, positive k-roots is to be denoted by ${}_k\Delta$. Let (,) be a positive inner product on $\boldsymbol{X}^*(S)_Q$ invariant under $W(S)$ (which operates on $\boldsymbol{X}^*(S)$ as indicated before). If $\alpha \in {}_k\Sigma$, let r_α be the reflection in the plane perpendicular to α. Then r_α is an element of $W(S)$ and $W(S)$ is generated by the set $\{r_\alpha \mid \alpha \in {}_k\Delta\}$. If ${}_k\Sigma$ cannot be written as the union of mutually perpendicular subsets in non-trivial fashion, it is called irreducible. In any case, ${}_k\Sigma$ can be written as the union of mutually perpendicular irreducible subsets. If ${}_k\Sigma$ is irreducible, so is ${}_k\Delta$, and the possibilities for irreducible ${}_k\Sigma$ are those of the Cartan-Killing classification, in addition to one denoted by BC_l which may occur when k is not algebraically closed, and which is obtained from a system ${}_k\Sigma$ of type C_l by adding the halves of the longest roots. In general, the subspace of the Lie algebra of G corresponding to an element of ${}_k\Sigma$ will have dimension greater than one.

An alternative method of characterizing the roots of G is the following. Let $\chi \in \boldsymbol{X}^*(S)$, $\chi \neq 1$, and let φ_χ be a rational homomorphism of G_a into G; necessarily $\varphi_\chi(G_a)$ is unipotent; we assume that for every $u \in G_a$, $s \in S$, we have $s\varphi_\chi(u)s^{-1} = \varphi_\chi(\chi(s) \cdot u)$, where $\chi(s) \cdot u$ means scalar multiplication of u by $\chi(s)$. Then χ is called a root of G if there exists such a φ_χ different from the trivial homomorphism sending all of G_a into e; such a non-trivial φ_χ is necessarily an injection and for fixed χ we denote the union of all the sets $\varphi_\chi(G_a)$ for non-trivial φ_χ by U^χ; the latter is a closed subset of G but may not be a subgroup of G, and in general the multiplicity dim U_χ of χ is equal to or greater than one. If η is another root $\succ 0$, then $[U_\chi, U_\eta]$ is contained in the product of the U_ξ for those $\xi \in {}_k\Sigma$ which are positive linear combinations of χ and η.

§ 3. Parabolic subgroups of G [9]

We let G and $S={}_kT$ have the same meanings as in section 2. If θ is a subset of ${}_k\Delta$, let

$$S_\theta = (\bigcap_{\alpha \in \theta} \ker_S \alpha)^0,$$

so S_θ is a torus and $\dim S_\theta + \operatorname{card} \theta = l = \operatorname{rank} S$. Let U be the subgroup of G generated by all U_η for $\eta \in {}_k\Sigma$, $\eta \succ 0$. Then U and $Z_G(S_\theta)$ generate a subgroup ${}_kP_\theta$ of G. The main result [9 : § 4] is that ${}_kP_\theta$ is a parabolic subgroup of G defined over k and every k-parabolic subgroup of G is conjugate by an element of G_k to one of the "standard" parabolic groups ${}_kP_\theta$ for precisely one subset θ of ${}_k\Delta$. If $\theta \supset \theta'$, then ${}_kP_\theta \supset {}_kP_{\theta'}$, and ${}_kP_\emptyset = {}_kP$ is a minimal k-parabolic subgroup of G. The reader is invited to compare this with the examples of parabolic subgroups considered in Chapter 6 for $GL(n, \boldsymbol{C})$ and for $SO(Q_{r,s})$.

If k is a subfield of K, let ${}_KT$ be a maximal K-split torus of G containing $S={}_kT$. If $T={}_KT$ is a maximal torus of G, then K is called a splitting field for G. Every character of ${}_KT$ becomes one of ${}_kT$ under the restriction mapping $r: \boldsymbol{X}^*({}_KT) \to \boldsymbol{X}^*({}_kT)$. Orderings $\succ$ on $\boldsymbol{X}^*({}_kT)$ and $\succ'$ on $\boldsymbol{X}^*({}_KT)$ are called compatible if $\alpha \in \boldsymbol{X}^*({}_KT)$ and $r(\alpha) \succ 0$ imply $\alpha \succ' 0$; then we (can and do) choose compatible orderings on $\boldsymbol{X}^*({}_kT)$ and $\boldsymbol{X}^*({}_KT)$ and denote them by the same symbol $\succ$ (omitting the $'$). If $\theta \subset {}_k\Delta$ and if $\bar\theta = \{\alpha \in {}_K\Delta \mid r(\alpha) \in \theta \cup \{0\}\}$, we have ${}_kP_\theta = {}_KP_{\bar\theta}$. In particular, if K is a splitting field for G, then ${}_K\Delta^0 = \{\alpha \in {}_K\Delta \mid r(\alpha) = 0\}$ is the (absolute) simple root system for $Z({}_kT)$; more precisely, as remarked earlier, $Z({}_kT) = {}_kT \cdot Z'$, where Z' is semi-simple, and then ${}_K\Delta^0$ is the root system for Z'.

Let ρ be an irreducible rational representation of G in $GL(\boldsymbol{W})$. Let T be a maximal torus of G contained in a Borel subgroup B of G, let Σ be the root system of G with respect to T, and let Δ be the set of positive simple roots in Σ with respect to the ordering determined by B; we let $\boldsymbol{X}^*(T)$, be supplied with an inner product (,) invariant under the Weyl group of T. It is known that a connected, solvable, linear group acting on a complete projective variety has a fixed point there [6a : § 15]. Hence $\rho(B)$ stabilizes a line D_ρ in $\boldsymbol{W}$, which is unique because ρ is irreducible [53], and the representation given by ρ of T on D_ρ is a character $\chi_\rho \in \boldsymbol{X}^*(T)$, which is the highest weight of ρ and, as such, characterizes ρ up to equivalence. Denote the equivalence class of ρ by $\varepsilon(\rho)$. Let $P=\{g\in G \mid \rho(g)\cdot D_\rho = D_\rho\}$. Then P is a parabolic subgroup of G whose conjugacy class is determined by $\varepsilon(\rho)$; denote that conjugacy class by $\mathcal{P}_{\varepsilon(\rho)}$.

It is known that we have

$$(1) \qquad 2(\chi_\rho, \alpha) = (\alpha, \alpha) m_\alpha$$

for all $\alpha \in \Delta$, where m_α are certain non-negative integers. Conversely, if G is simply-connected, then for every set of non-negative integers m_α there exists an irreducible representation ρ whose highest weight χ_ρ satisfies the equations (1) for all $\alpha\in\Delta$. We let l_α be the element of $\boldsymbol{X}^*(T)$ determined by the equations

$$(2) \qquad 2(l_\alpha, \alpha) = (\alpha, \alpha), \quad (l_\alpha, \beta) = 0, \quad \beta \neq \alpha, \ \alpha, \beta \in \Delta.$$

The elements l_α are called the fundamental highest weights of G; they form a basis of $\boldsymbol{X}^*(T)_{\mathbf{Q}} = \boldsymbol{X}^*(T)\otimes_{\mathbf{Z}}\boldsymbol{Q}$.

Assume now that G is defined over the field k, that ${}_kT \subset T = {}_KT$, that we have given compatible orderings on $\boldsymbol{X}^*({}_kT)$ and on $\boldsymbol{X}^*(T)$, and that $r : \boldsymbol{X}^*(T) \to \boldsymbol{X}^*({}_kT)$ is the restriction mapping. The equivalence class $\varepsilon = \varepsilon(\rho)$ is called strongly rational if there exists $\rho_1 \in \varepsilon$ such that ρ_1 is defined over k and $P_1 \in \mathcal{P}_\varepsilon$ such that P_1 is defined over k. If $\beta \in {}_k\Delta$, let $l_{(\beta)}$ be the sum of the elements l_α, $\alpha\in\Delta$, such that $r(\alpha)=\beta$, and let $\boldsymbol{m}_\beta = r(l_{(\beta)})$. Let $\boldsymbol{m}_\varepsilon$ be the restriction to ${}_kT$ of the highest weight of some $\rho\in\varepsilon$. Then one has [9 : § 12.13]

PROPOSITION 2. *If ε is strongly k-rational, then $\boldsymbol{m}_\varepsilon$ is a linear combination with non-negative integral coefficients of the weights $\boldsymbol{m}_\beta$, $\beta\in{}_k\Delta$. Conversely, for each $\beta \in {}_k\Delta$ there exists a positive integer d_β such that $n\cdot d_\beta \cdot \boldsymbol{m}_\beta$ is the highest k-weight of an irreducible strongly k-rational representation of G for every positive integer n. Moreover, if n_β are non-negative integers not all zero, then $\sum_{\beta\in{}_k\Delta} n_\beta d_\beta \boldsymbol{m}_\beta$ is the highest k-weight of a strongly k-rational, irreducible representation ρ of G.*

We need this proposition in treating the properties of automorphic forms constructed from special series.

§4. The Bruhat decomposition

Let G be a reductive, connected, linear algebraic group defined over the field k, let $S={}_kT$ be a maximal k-split torus in G, let ${}_kP$ be a minimal k-parabolic subgroup of G containing S, and let ${}_kW=W(S)$ be the Weyl group of S. Then [9 : § 5.15] one has

(3)
$$G_k=\bigcup_{w\in{}_kW_k}{}_kP_kw_kP_k \quad \text{(disjoint union)},$$

where $w\in{}_kW_k$ means that w runs over a set of representatives in $N(S)_k$ of the elements of ${}_kW$. The reflections r_α, $\alpha\in{}_k\Delta$, are called fundamental reflections. If r is a fundamental reflection and if $w\in{}_kW_k$, then one has [9 : § 5.16]

(4)
$$r_kP_k\{w, rw\}_kP_k={}_kP_k\{w, rw\}_kP_k, \quad \text{and}$$

(5)
$$r_kP_kr\neq{}_kP_k.$$

The decomposition (3) may be rewritten in the following way. Let ${}_kU$ be the unipotent radical of ${}_kP$, so that ${}_kP=Z({}_kT)\cdot{}_kU$ (semi-direct product); if $w\in{}_kW$ is represented by $n\in N({}_kT)_k$, then $Z({}_kT)$ is normalized by n, so that we have

(6)
$${}_kP_k\cdot n\cdot{}_kP_k={}_kU_k\cdot n\cdot{}_kP_k.$$

Let

(7)
$${}_k\Sigma_+(n)=\{\alpha\in{}_k\Sigma_+(=\text{positive roots in }{}_k\Sigma)\,|\,n^{-1}U_\alpha n=U_\beta,\ \beta>0\},$$

and

(8)
$${}_k\Sigma_-(n)=\{\alpha\in{}_k\Sigma_+\,|\,n^{-1}U_\alpha n=U_\beta,\ \beta<0\}.$$

Then one can go further to write

(9)
$${}_kU_k\cdot n\cdot{}_kP_k={}_kU^-(n)_k\cdot n\cdot{}_kP_k,$$

where ${}_kU^-(n)_k$ is the subgroup of ${}_kU$ generated by all U_α with $\alpha\in{}_k\Sigma_-(n)$.

The decomposition in (3) and the rules (4) for multiplication of its cells together with the restriction (5) may be abstractly formulated as follows [12a, c; 57a]: Let W be a Coxeter group, that is, a group with a finite minimal set R of involutive generators; let G be a group; and let B be a subgroup of G such that there is a bijection $w \to Bg_wB$ of W onto the double cosets of B in G with the following properties:

(10) $$Bg_rB \cdot Bg_wB \subset Bg_wB \cup Bg_{rw}B, \quad \text{all } r \in R,\ w \in W,$$

(11) Bg_rB is not a single coset of B.

With these axioms, one may show [57a] that if J is a subset of R and if W_J is the subgroup of W generated by the elements of J, then the union $B\overline{W}_JB$ of all the double cosets Bg_wB, $w \in W_J$, is a subgroup of G, and that every subgroup of G containing B is of this form and is its own normalizer.

§ 5. The Cartan and Iwasawa decompositions

We assemble here for reference some properties of semi-simple Lie algebras and Lie groups that will be needed later. The general reference for these is [27 : especially pp. 152–225].

Let $\mathfrak{g}_0$ be a semi-simple Lie algebra over $\boldsymbol{R}$ and let $\mathfrak{g} = \mathfrak{g}_0 + i\mathfrak{g}_0$ be its complexification. By a real form of $\mathfrak{g}$, we mean an $\boldsymbol{R}$-sub-Lie algebra $\mathfrak{g}_k$ of $\mathfrak{g}$ having the same dimension over $\boldsymbol{R}$ as $\mathfrak{g}_0$ and such that $\mathfrak{g} = \mathfrak{g}_k + i\mathfrak{g}_k$. The real form $\mathfrak{g}_k$ is called compact if the restriction to it of the Killing form B of $\mathfrak{g}$ is negative-definite. This is equivalent to saying that the analytic group associated to $\mathfrak{g}_k$ in the adjoint group of $\mathfrak{g}$ is compact. It is known [27 : p. 155] that a compact real form of a given complex, semi-simple Lie algebra always exists. Let σ be the conjugation of $\mathfrak{g}$ with respect to $\mathfrak{g}_0$: $\sigma(X+iY) = X-iY$ for $X, Y \in \mathfrak{g}_0$.

A direct vector space decomposition $\mathfrak{g}_0 = \mathfrak{p}_0 + \mathfrak{k}_0$ of $\mathfrak{g}_0$ into a subalgebra $\mathfrak{k}_0$ and a subspace $\mathfrak{p}_0$ is called a Cartan decomposition if there is a compact real form $\mathfrak{g}_k$ of $\mathfrak{g}$ such that $\sigma\mathfrak{g}_k = \mathfrak{g}_k$, $\mathfrak{k}_0 = \mathfrak{g}_0 \cap \mathfrak{g}_k$, and $\mathfrak{p}_0 = \mathfrak{g}_0 \cap i\mathfrak{g}_k$. It is known [27 : Chap. III, § 7] that $\mathfrak{g}_0$ always has a Cartan decomposition and that any two Cartan decompositions are conjugate by an inner automorphism of $\mathfrak{g}_0$. If $\mathfrak{g}_0 = \mathfrak{k}_0 + \mathfrak{p}_0$ is a Cartan decomposition, then $B(,)$ is negative-definite on $\mathfrak{k}_0$ and positive-definite on $\mathfrak{p}_0$, and $\theta : T+X \mapsto T-X$, $T \in \mathfrak{k}_0$, $X \in \mathfrak{p}_0$, extends to an automorphism of $\mathfrak{g}$ for which $\theta\mathfrak{g}_0 \subset \mathfrak{g}_0$ and $\theta\mathfrak{g}_k \subset \mathfrak{g}_k$; moreover, $\mathfrak{k}_0$ is a compactly imbedded subalgebra of $\mathfrak{g}_0$ (i.e., the corresponding analytic subgroup of the adjoint group is compact), and is maximal among subalgebras of $\mathfrak{g}_0$ with that property. Hence $\mathfrak{k}_0$ is reductive in $\mathfrak{g}_0$, by which one means that the adjoint representation of $\mathfrak{k}_0$ in $\mathfrak{g}_0$ is fully reducible, and so its complexification $\mathfrak{k}$ is reductive in $\mathfrak{g}$. Since θ is an automorphism of $\mathfrak{g}$ of which $\mathfrak{k}$ and $\mathfrak{p}$ are respectively the $+1$ and -1 eigenspaces, we have $[\mathfrak{k}, \mathfrak{k}] \subset \mathfrak{k}$, $[\mathfrak{k}, \mathfrak{p}] \subset \mathfrak{p}$, and $[\mathfrak{p}, \mathfrak{p}] \subset \mathfrak{k}$.

Let $\mathfrak{h}_{\mathfrak{p}_0}$ be a subalgebra of $\mathfrak{g}_0$ contained in $\mathfrak{p}_0$, of maximum dimension; evidently it is Abelian and therefore we may extend it to a maximal Abelian subalgebra $\mathfrak{h}_0$ of $\mathfrak{g}_0$. If $X \in \mathfrak{h}_0$ and $Y \in \mathfrak{h}_{\mathfrak{p}_0}$, then $[\theta X, Y] = -[\theta X, \theta Y] = -\theta[X, Y] = 0$, hence $X - \theta X$ commutes with all elements of $\mathfrak{h}_{\mathfrak{p}_0}$ and is anti-symmetric with respect to θ, therefore $X - \theta X \in \mathfrak{p}_0$ and consequently $X - \theta X \in \mathfrak{h}_{\mathfrak{p}_0}$ since $\mathfrak{h}_{\mathfrak{p}_0}$ is a maximal subalgebra of $\mathfrak{g}_0$ contained in $\mathfrak{p}_0$. Hence, $\theta\mathfrak{h}_0 = \mathfrak{h}_0$ and so $\mathfrak{h}_0 = \mathfrak{h}_{\mathfrak{p}_0} + \mathfrak{h}_{\mathfrak{k}_0}$, with $\mathfrak{h}_{\mathfrak{k}_0} = \mathfrak{h}_0 \cap \mathfrak{k}_0$. Now $i\mathfrak{p}_0$ and $\mathfrak{k}_0$ are contained in $\mathfrak{g}_k$, so from dimensional considerations $\mathfrak{g}_k = \mathfrak{k}_0 + i\mathfrak{p}_0$ and $\mathfrak{k}_0 + i\mathfrak{p}_0$ is compact, therefore, $ad\, X$ is semi-simple for all $X \in \mathfrak{k}_0 \cup \mathfrak{p}_0$. Thus, $ad\, X$ is semi-simple for all $X \in \mathfrak{h}_0$, and so for all $X \in \mathfrak{h} = \boldsymbol{C}\mathfrak{h}_0$. Since $\mathfrak{h}_0$ is maximal Abelian in $\mathfrak{g}_0$, any element $a + ib \in \mathfrak{g}$, $a, b \in \mathfrak{g}_0$, which centralizes $\mathfrak{h}_0$ is in $\mathfrak{h}$. Hence, $\mathfrak{h}$ is a Cartan subalgebra of $\mathfrak{g}$. Let $\mathfrak{h}_{\mathfrak{p}} = \mathfrak{h} \cap \mathfrak{p}$, $\mathfrak{h}_{\mathfrak{k}} = \mathfrak{h} \cap \mathfrak{k}$, and let $H_1, \cdots, H_l$ be an $\boldsymbol{R}$-base of $\mathfrak{h}_0$ such that $H_1, \cdots, H_m$, $m < l$, is a base for $\mathfrak{h}_{\mathfrak{p}_0}$, $H_{m+1}, \cdots, H_l$, a base for $\mathfrak{k}_0 \cap \mathfrak{h}$. Let $H_j^* = H_j$ if $i \leqq m$ and $H_j^* = iH_j$ if $i > m$, and let $\mathfrak{h}^*$ be the $\boldsymbol{R}$-vector space spanned by all H_j^*. If λ is a linear functional on $\mathfrak{h}$, define $H_\lambda \in \mathfrak{h}$ by requiring

$$B(H_\lambda, H) \equiv \lambda(H) \quad \text{for all} \quad H \in \mathfrak{h}, \tag{12}$$

and call λ real if $H_\lambda \in \mathfrak{h}^*$. We impose the lexicographic ordering on $\mathfrak{h}^*$ with respect to the basis H_j^* (an element is $\succ 0$ if in its expression as a linear combination $\sum c_j H_j^*$, the non-vanishing c_j of lowest index is > 0), and define an ordering among linear functionals by : If $\lambda - \mu$ is real, then $\lambda - \mu \succ 0$ if $H_{\lambda-\mu} \succ 0$. Define $(\theta\lambda)(H) = \lambda(\theta H)$.

One then may show that the root system Σ with respect to $\mathfrak{h}$ consists of real linear functionals and that $\theta\Sigma = \Sigma$. For each $\alpha \in \Sigma$, let $\mathfrak{g}_\alpha$ be the corresponding root space :

$$\mathfrak{g}_\alpha = \{X \in \mathfrak{g} \mid [H, X] = \alpha(H) \cdot X\}, \tag{13}$$

and choose $X_\alpha \in \mathfrak{g}_\alpha$ such that

$$B(X_\alpha, X_{-\alpha}) = 1 \tag{14}$$

and such that with

(15) $$H_\alpha = [X_\alpha, X_{-\alpha}],$$

we have $B(H_\alpha, H) = \alpha(H)$ [27 : p. 142 eq. (3), etc.]. Let Σ_+ (resp. Σ_-) be the set of positive (resp. negative) roots in Σ, and let Δ denote the set of simple positive roots. Let

(16) $$P_+ = \{\alpha \in \Sigma_+ \mid \theta\alpha \neq \alpha\}, \quad \text{and}$$

(17) $$P_- = \{\alpha \in \Sigma_+ \mid \theta\alpha = \alpha\}.$$

Then [27 : p. 222] $\theta\alpha < 0$ for $\alpha \in P_+$ and H_β, X_β, and $X_{-\beta}$ all belong to $\mathfrak{k}$ for $\beta \in P_-$. Moreover (loc. cit.), $\alpha > \beta$ for all $\alpha \in P_+$, $\beta \in P_-$. Let $\mathfrak{n} = \sum_{\alpha \in P_+} \boldsymbol{C}X_\alpha$, $\mathfrak{m} = \mathfrak{h}_t + \sum_{\pm\beta \in P_-} \boldsymbol{C}X_\beta$. Then $\mathfrak{n}$ is a nilpotent subalgebra of $\mathfrak{g}$. If $\mathfrak{n}_0 = \mathfrak{g}_0 \cap \mathfrak{n}$, then (Iwasawa) $\mathfrak{g}_0 = \mathfrak{k}_0 + \mathfrak{h}_{\mathfrak{r}_0} + \mathfrak{n}_0$, and if G is any algebraic Lie group with Lie algebra $\mathfrak{g}_0$ and if K, A_p, and N are analytic subgroups of G with Lie algebras $\mathfrak{k}_0$, $\mathfrak{h}_{\mathfrak{r}_0}$, and $\mathfrak{n}_0$ respectively, such that K is a maximal compact subgroup in G and A_p and N are connected, then the product mapping $(k, a, n) \mapsto kan$ is an analytic diffeomorphism of $K \times A_p \times N$ onto G; A_p and N are simply-connected; and $A_pN = S$ is a solvable, maximal, connected, $\boldsymbol{R}$-trigonalizable (reducible to upper triangular form) subgroup of G [27; 8].

The reader is again referred to the examples of Chapter 6, where in each case the Cartan involution is given on the Lie algebra by $X \mapsto -{}^tX$ and where the maximal compact subgroup K of G_R is the set of $g \in G_R$ satisfying ${}^tg = g^{-1}$. In those cases we explicitly verified the Iwasawa decomposition by finding a homogeneous space of G_R on which A_pN acted in a simply transitive manner and such that K was the stability group of a certain point.

Now assume that G is a connected and *simply-connected* linear algebraic group defined over $\boldsymbol{R}$. One may apply Theorem 7.2, p. 272 of [27] to the maximal compact subgroups K and L appearing respectively in Iwasawa decompositions of $G = G_C$ and of G_R, chosen such that $L = K \cap G_R$ and such that K is stable under complex conjugation [27 : p. 156]. We know that $\pi_1(K) = \pi_1(G) = \{e\}$. Since complex conjugation is an involutive, analytic automorphism of G, it follows that L is connected and hence that G_R is connected.

The following lemmas are taken from [26a].

LEMMA 1. *The centralizer of* $\mathfrak{h}_\mathfrak{p}$ *in* $\mathfrak{g}$ *equals* $\mathfrak{h}_\mathfrak{p} + \mathfrak{m}$, *and* $\mathfrak{m}$ *is the centralizer in* $\mathfrak{k}$ *of* $\mathfrak{h}_\mathfrak{p}$. *Further, there exists* $H \in \mathfrak{h}_\mathfrak{p}$ *such that the centralizer* $Z_\mathfrak{g}(H)$ *of* H *in* $\mathfrak{g}$ *is just* $\mathfrak{h}_\mathfrak{p}+\mathfrak{m}$. *Moreover,* $\mathfrak{m}$ *is a subalgebra of* $\mathfrak{k}$ *and we have* $\dim\mathfrak{k}-\dim\mathfrak{m}=\dim\mathfrak{p}-\dim\mathfrak{h}_\mathfrak{p}=\dim\mathfrak{n}$.

PROOF. The -1 eigenspace of θ is $\mathfrak{p}$, hence for a root α, we have $\alpha|\mathfrak{h}_\mathfrak{p}=0$ if and only if $\alpha=\theta\alpha$, i.e., either α or $-\alpha$ is in P_-. From this, the assertion about centralizers follows from a simple calculation in terms of a basis X_α, $X_{-\alpha}$, of $\mathfrak{g}/\mathfrak{h}$ (as a vector space) and from choosing H in $\mathfrak{h}_\mathfrak{p}$ such that $\alpha(H)\neq 0$ for all $\alpha\in P_-$.

If $\alpha\in P_+$, then $\theta\alpha\prec 0$, hence $\alpha\mapsto-\theta\alpha$ is a permutation of order two of P_+. Moreover, $\theta X_\alpha=cX_{\theta\alpha}$, where c is a non-zero constant. Hence $X_\alpha\pm\theta X_\alpha$, $\alpha\in P_+$, are all linearly independent modulo $\mathfrak{h}_\mathfrak{p}+\mathfrak{m}$. Letting

$$\mathfrak{q}=\sum_{\alpha\ P_+}\boldsymbol{C}(X_\alpha-\theta X_\alpha),\quad \mathfrak{l}=\sum_{\alpha\ P_+}\boldsymbol{C}(X_\alpha+\theta X_\alpha),$$

we have $\mathfrak{q}\subset\mathfrak{p}$, $\mathfrak{l}\subset\mathfrak{k}$, and $\mathfrak{q}+\mathfrak{l}+\mathfrak{h}_\mathfrak{p}+\mathfrak{m}=\mathfrak{g}=\mathfrak{p}+\mathfrak{k}$. Hence, $\mathfrak{q}+\mathfrak{h}_\mathfrak{p}=\mathfrak{p}$, $\mathfrak{l}+\mathfrak{m}=\mathfrak{k}$, and if we put $q=\mathrm{card}(P_+)$, then $\dim\mathfrak{n}=q=\dim\mathfrak{q}=\dim\mathfrak{p}-\dim\mathfrak{h}_\mathfrak{p}$. Likewise, $q=\dim\mathfrak{l}=\dim\mathfrak{k}-\dim\mathfrak{m}$. Since $\mathfrak{m}=\{X\in\mathfrak{k}\,|\,ad\,X|\mathfrak{h}_\mathfrak{p}=0\}$, $\mathfrak{m}$ is a subalgebra of $\mathfrak{k}$.

If $\mathfrak{l}$ is any Lie algebra and $\mathfrak{k}$, a subalgebra, then we say that $\mathfrak{k}$ is reductive in $\mathfrak{l}$ if the adjoint representation of $\mathfrak{k}$ in $\mathfrak{l}$ is fully reducible; $\mathfrak{l}$ is called reductive if it is reductive in itself. It is the same thing as saying that the quotient of $\mathfrak{l}$ by its center is semi-simple. Then [16b: t. III, Chap. IV, § 4, Theorem 4 and Proposition 3] $\mathfrak{l}$ is reductive if and only if it has a faithful semi-simple linear representation; and in order for a representation ρ of a reductive Lie algebra $\mathfrak{l}$ to be semi-simple, it is necessary and sufficient that $\rho(C)$ be semi-simple for every C in the center $\mathfrak{c}$ of $\mathfrak{l}$; moreover, if $\mathfrak{l}$ is reductive, then [loc. cit., Proposition 1] the derived algebra $\mathfrak{l}'$ of $\mathfrak{l}$ is semi-simple and $\mathfrak{l}=\mathfrak{c}+\mathfrak{l}'$ (direct sum); and [loc. cit., Proposition 5] if $\mathfrak{l}$ is a finite-dimensional Lie algebra, and if $\mathfrak{k}$ is reductive in $\mathfrak{l}$, then any semi-simple representation ρ of $\mathfrak{l}$ induces a semi-simple representation of $\mathfrak{k}$.

Returning to our previous notation where $\mathfrak{g}_0$ is semi-simple, $\mathfrak{g}$, its complexification, etc., we have

LEMMA 2. *The algebra* $\mathfrak{m}$ *is reductive in* $\mathfrak{k}$ *and in* $\mathfrak{g}$, *and* $[\mathfrak{m},\mathfrak{n}]\subset\mathfrak{n}$. *If* σ *is the adjoint representation of the algebra* $\mathfrak{h}_\mathfrak{p}+\mathfrak{n}$ *(note that* $\mathfrak{h}$ *normalizes* $\mathfrak{n}$ *and hence so does* $\mathfrak{h}_\mathfrak{p}$*), then*

$$\mathrm{tr}(\sigma(H))=\sum_{\alpha\in\Sigma_+}\alpha(H),\quad H\in\mathfrak{h}_\mathfrak{p}.$$

PROOF. Let G be the identity component of the adjoint group of $\mathfrak{g}_0$ and K, the analytic subgroup corresponding to $\mathfrak{k}_0$, and put $\mathfrak{m}_0=\mathfrak{m}\cap\mathfrak{g}_0$. If $x\in K$ and $H\in\mathfrak{h}_{\mathfrak{p}_0}$, let $x\cdot H$ denote the adjoint operation of x on H and define

$$M=\{x\in K\,|\,x\cdot H=H \text{ for all } H\in\mathfrak{h}_{\mathfrak{p}_0}\}.$$

Then M is a closed subgroup of K, hence is compact, and by Lemma 1, $\mathfrak{m}_0$ is the Lie algebra of M. Hence $\mathfrak{m}_0$ is reductive in $\mathfrak{k}_0$ and in $\mathfrak{g}_0$, so $\mathfrak{m}$ is reductive in $\mathfrak{k}$ and in $\mathfrak{g}$.

Let $\alpha\in P_+$, $\beta\in P_-$; then $\alpha\pm\beta>0$, and $\theta(\alpha+\beta)=\theta\alpha+\beta\neq\alpha+\beta$ since $\theta\alpha\neq\alpha$, hence $\alpha+\beta\in P_+$, if it is a root. Likewise $\alpha-\beta\in P_+$ if it is a root. Hence, $[X_\alpha, X_\beta]$ and $[X_\alpha, X_{-\beta}]$ belong to $\mathfrak{n}$. Since $[\mathfrak{h}, \mathfrak{n}]\subset\mathfrak{n}$, and $\mathfrak{m}$ is spanned by $\mathfrak{h}_{\mathfrak{t}}$ and X_β, $\pm\beta\in P_-$, we have $[\mathfrak{m}, \mathfrak{n}]\subset\mathfrak{n}$.

Clearly

$$\operatorname{tr}(\sigma(H))=\sum_{\alpha\in P_+}\alpha(H),\quad H\in\mathfrak{h}_{\mathfrak{p}}.$$

Since $\alpha\in P_-$ implies that $\alpha(\mathfrak{h}_{\mathfrak{p}})=0$, we have

$$\operatorname{tr}(\sigma(H))=\sum_{\alpha\in\Sigma_+}\alpha(H),$$

as required.

Let $\mathfrak{a}_1$ and $\mathfrak{a}_2$ be two subalgebras of $\mathfrak{g}_0$ contained in $\mathfrak{p}_0$, of maximum dimension; they are automatically Abelian as remarked before. It follows from [27 : p. 211] that there exists $k\in K^*$, the subgroup of the adjoint group with Lie algebra $\mathfrak{k}_0$, such that $k\cdot\mathfrak{a}_1=\mathfrak{a}_2$, and that every element of $\mathfrak{p}_0$ belongs to such a maximal subalgebra. (To obtain these facts from loc. cit., one must multiply certain vector spaces by $i=\sqrt{-1}$.) Moreover, by [27 : Theorem 1.1, pp. 214-215] the mapping $\varphi:(X, k)\mapsto\exp X\cdot k$ of $\mathfrak{p}_0\times K$ into G_0 is a diffeomorphism of $\mathfrak{p}_0\times K$ onto G_0, where G_0 is a connected, semi-simple, real Lie group of non-compact type with Lie algebra $\mathfrak{g}_0$ and K is the analytic subgroup with Lie algebra $\mathfrak{k}_0$. One has $k\cdot\exp X\cdot k^{-1}=\exp(Ad\, k\cdot X)$. Combining these facts, we obtain

$$G_0=KAK, \tag{18}$$

where A is any analytic subgroup of G_0 whose Lie algebra $\mathfrak{a}$ is a maximal (Abelian) subalgebra of $\mathfrak{g}_0$ contained in $\mathfrak{p}_0$. In fact, in (18) A may be replaced by $A_+=\exp\mathfrak{a}_+$, $\mathfrak{a}_+$ being a Weyl chamber in $\mathfrak{a}$, i.e., a fundamental domain for the adjoint operation of the normalizer of $\mathfrak{a}$ in K. This decomposition of G_0 may be referred to as a Cartan decomposition of G_0 corresponding to the Cartan decomposition $\mathfrak{g}_0=\mathfrak{k}_0+\mathfrak{p}_0$ of its Lie algebra.

The Iwasawa and Cartan decompositions, explained here for semi-simple groups, may be extended to the case of reductive real algebraic groups (connected or not) without difficulty [8 : § 1.11]. Our main concern is with the semi-simple case, but it is useful to record this fact as it will afford a smoother transition to the illustrative examples of the next section.

§ 6. The p-adic Iwasawa and Cartan decompositions

Let k be a $\boldsymbol{p}$-adic number field, by which we mean a finite algebraic extension of the $\boldsymbol{p}$-adic completion $\boldsymbol{Q}_p$ of the rational number field for some finite rational prime $\boldsymbol{p}$. Then k is a locally compact field and we denote its ring of integers by $\mathfrak{O}$ and the maximal ideal of $\mathfrak{O}$ by $\mathfrak{P}$, so that a basis of compact, open neighborhoods of 0 in k is given by the powers $\mathfrak{P}^n$ of $\mathfrak{P}$. Then the Iwasawa and Cartan decompositions of the last section have useful extensions which cover the case in which the locally compact field $\boldsymbol{R}$ is replaced by the locally compact field k.

These can be easily illustrated in the case of the general linear group $GL(n, k)$. First of all, $\mathfrak{O}$ is a principal ideal domain and all its ideals are powers of the maximal ideal $\mathfrak{P}$. Let π be a prime element of $\mathfrak{P}$, and let F be the finite field with $q=N\mathfrak{P}$ elements. Let K be the group $GL(n, \mathfrak{O})$ of integral $n\times n$ matrices with unit determinant, let A be the group of $n\times n$ diagonal, non-singular matrices over k, and let N be the group of unipotent $n\times n$ upper triangular matrices over k. Within A, let A_+ be the subset of diagonal matrices with diagonal entries $d_j=u_j\pi^{m_j}$, where u_j is a unit and $m_1\leqq m_2\leqq\cdots\leqq m_n$. Then it is clear from well-known facts about elementary divisors and elementary row and column operations on matrices that we have

$$G=GL(n, k)=KAN, \tag{19}$$

and

$$G=KA_+K. \tag{20}$$

However, it should be observed that if $g\in G$, the decomposition of g into a product kan implied by (19) is not unique. Of course K is a maximal compact subgroup of G: This follows from (20), because if $a \in A_+$ has entries $u_j \pi^{m_j}$ and not all $m_j = 0$, then the positive and negative powers of a will form an unbounded set in the $\mathfrak{P}$-adic topology.

Let G be an absolutely almost simple, connected, and simply-connected linear algebraic k-group ($G = G_k$). One may generalize what has been said above as follows [12c]. We assume $G \subset GL(V)$. Then there exist a maximal k-split torus S, minimal k-parabolic subgroup P generated by $Z(S)$ and by its unipotent radical U, and a k-lattice Λ [60e] in V such that if $K = \{g \in G \mid g \cdot \Lambda = \Lambda\}$, then K is a maximal compact subgroup of G and we have

$$G = G_k = K \cdot P_k. \tag{21}$$

Moreover, we have

$$G = K \cdot S_k \cdot K. \tag{22}$$

If G is a connected and simply-connected, semi-simple, linear algebraic group contained in $GL(V)$ and defined over $\boldsymbol{Q}$, and if G is almost $\boldsymbol{Q}$-simple (that is, G has no normal, closed subgroups defined over $\boldsymbol{Q}$ of dimension > 0), then we may write $G = R_{k/\boldsymbol{Q}} G'$, where $R_{k/\boldsymbol{Q}}$ is the groundfield reduction functor [60d], and where G' is a connected, absolutely almost simple, and simply-connected linear algebraic group defined over the algebraic number field k. We assume that k is totally real and that $G_{\boldsymbol{R}}$ has no compact simple factors of positive dimension. Then $G_{\boldsymbol{R}}$ is connected (section 5). Let $\mathfrak{P}$ be a prime ideal in k, let $\boldsymbol{p} = \mathfrak{P} \cap \boldsymbol{Q}$, and denote by $k_{\mathfrak{P}}$ and $\boldsymbol{Q}_p$ the respective completions of k and of $\boldsymbol{Q}$. Under our assumptions, strong, and hence weak, approximations hold in G (provided possibly that $G' \neq E_8$)[37a, b], so that $G_{\boldsymbol{Q}}$ is dense in $G_{\boldsymbol{Q}_p}$ and G'_k is dense in $G'_{k_{\mathfrak{P}}}$.

Let $\Gamma \subset G_{\boldsymbol{Q}}$ be an arithmetic subgroup of $G_{\boldsymbol{R}}$ and let P be a $\boldsymbol{Q}$-parabolic subgroup of G; then there exists [6f] a finite subset $\boldsymbol{A}$ of $G_{\boldsymbol{Q}}$ such that $G_{\boldsymbol{Q}} = \Gamma \boldsymbol{A} P_{\boldsymbol{Q}}$. For each finite prime $\boldsymbol{p}$, let Γ_p be the closure of Γ in $G_{\boldsymbol{Q}_p}$. It is easy to see that if Λ is an $\boldsymbol{R}$-lattice [60e] of V contained in $V_{\boldsymbol{Q}}$, then there exists a finite set S of primes of $\boldsymbol{Q}$ with the property that if $\boldsymbol{p}$ is a prime of $\boldsymbol{Q}$ not in S, then for each $a \in \boldsymbol{A}$ we have, in fact, $a \in \Gamma_p$, and Γ_p is the stabilizer in $G_{\boldsymbol{Q}_p}$ of the lattice (open, compact subgroup) $\Lambda_p = \Lambda \otimes_{\boldsymbol{Z}} \boldsymbol{Z}_p$ in $V_{\boldsymbol{Q}_p}$. For each $\boldsymbol{p} \in S$, we apply the results of Bruhat and Tits [12c] (by applying them to $G'_{k_{\mathfrak{P}}}$ for each $\mathfrak{P} \mid \boldsymbol{p}$ and using Theorem 1.3.3 of [60d]) and obtain a special maximal compact subgroup K_p of G_{Q_p} such that

$$G_{Q_p} = K_p \cdot P_{Q_p}$$

(noting by [9 : § 6.19] that $P = R_{k/Q}P'$ for some k-parabolic subgroup P' of G'). Then K_p is the stability group of a lattice Λ''_p in V_{Q_p}. There exists [60e : p. 84] a unique $\boldsymbol{R}$-lattice Λ' in V contained in V_Q such that $\Lambda'_p = \Lambda' \otimes_{\boldsymbol{Z}} \boldsymbol{Z}_p = \Lambda_p$ for all $\boldsymbol{p} \notin S$ and $\Lambda'_p = \Lambda''_p$ for all $\boldsymbol{p} \in S$. If we let $\Gamma' = \{\gamma \in G_R \mid \gamma \cdot \Lambda' = \Lambda'\}$, and if Γ'_p denotes the closure of Γ' in G_{Q_p}, then by strong approximation we have $\Gamma'_p = K_p$ for $\boldsymbol{p} \in S$ and $\Gamma'_p = \Gamma_p$ for $\boldsymbol{p} \notin S$. Therefore, $G_{Q_p} = \Gamma'_p \cdot P_{Q_p}$ for every $\boldsymbol{p}$. Now, in fact, by "adjusting" the lattice Λ' at a finite number of $\boldsymbol{p}$, we may assume that Γ'_p is a *special* maximal compact subgroup of G_{Q_p} for every $\boldsymbol{p}$; a proof of this fact has been communicated to me by H. Hijikata. The resulting arithmetic group Γ' is then "special", and appears to be associated with some kind of natural arithmetic structure on G. It would be interesting to see how close a special arithmetic group comes to being a maximal arithmetic subgroup of G_R.

§ 7. Harish-Chandra's realization of bounded symmetric domains

The original reference for the results to be cited here is [26c], and one may also refer to [27 : Chap. VIII]. Let G be an absolutely almost simple, linear, connected, algebraic group defined over $\boldsymbol{R}$. Let $G_1 = G_R^0$. We assume that G_1 is non-compact, and denoting by K a maximal compact subgroup of G_1, we assume K has non-discrete center Z, so that $\dim Z = 1$. This implies that a Cartan subgroup H of K will also be a Cartan subgroup of G_1. Let $\mathfrak{g}$ be the Lie algebra of G_1, $\mathfrak{k}$ that of K, and denote by $\mathfrak{p}$ the orthogonal complement of $\mathfrak{k}$ with respect to the Killing form; then $\mathfrak{g} = \mathfrak{k} + \mathfrak{p}$ is a Cartan decomposition of $\mathfrak{g}$. Let $\mathfrak{h}$ be the Lie algebra of H; then $\mathfrak{h}$ is a Cartan subalgebra of $\mathfrak{g}$ and contains the one-dimensional center $\mathfrak{z}$ of $\mathfrak{k}$, which is the Lie algebra of $Z \subset H$. Let Σ be the set of roots of $\mathfrak{g}_C$ with respect to $\mathfrak{h}_C$. A root α is called compact if $\alpha(\mathfrak{z}) = 0$ and otherwise is called non-compact. The space $\mathfrak{p}_C$ is the sum of the non-compact root spaces and we denote by π^+ (resp. by π^-) the set of positive (resp. negative) non-compact roots and by $\mathfrak{p}^+$ (resp. $\mathfrak{p}^-$) the span of the corresponding root spaces, so that $\mathfrak{p}_C$ is the direct sum of $\mathfrak{p}^+$ and $\mathfrak{p}^-$

and

$$\mathfrak{g}_C=\mathfrak{k}_C+\mathfrak{p}^+ +\mathfrak{p}^-.$$

Let $E_\mu(\mu\in\Sigma)$ be root vectors and H_μ, elements of $\mathfrak{h}_C$ such that

$$[E_\mu, E_{-\mu}]=H_\mu,\quad \nu(H_\mu)=2(\nu, \mu)\cdot(\mu, \mu)^{-1},\quad \mu, \nu\in\Sigma, \tag{23}$$

where (,) is the Killing form, and such that complex conjugation of $\mathfrak{g}_C$ with respect to $\mathfrak{g}$ interchanges E_μ and $E_{-\mu}$ when μ is non-compact.

Two linearly independent roots μ and ν are called strongly orthogonal if neither $\mu+\nu$ nor $\mu-\nu$ are roots; then μ and ν are orthogonal with respect to (,). One may choose a maximal set $\{\mu_1, \cdots, \mu_m\}$ of strongly orthogonal roots, where m is the real rank of G, and we write $E_{\pm i}$ for $E_{\pm\mu_i}$, $i=1, \cdots, m$.

Let $P^\pm=\exp(\mathfrak{p}^\pm)$ and let K_C be the analytic subgroup of G_C with Lie algebra $\mathfrak{k}_C$. One may see that $P^-K_CP^+$ is a Zariski-open subset of G_C and that the product mapping is a homeomorphism of the Cartesian product $P^-\times K_C\times P^+$ onto it; for example, in the cases considered in Chapter 6, the possibility of solving the equation

$$g=p^-kp^+ \tag{24}$$

depended on an "open" property of some submatrix of g, and when that property was satisfied, then the solution for p^+, k, and p^- was unique. As that property was always satisfied when $g\in G_1$, in the examples referred to, one always had $G_1\subset P^-K_CP^+$; likewise, in the general case one has, in fact, $G_1\subset P^-K_CP^+$. For $g\in P^-K_CP^+$, let $\zeta(g)$ be the unique element of $\mathfrak{p}^+$ such that

$$g\in P^-K_C\exp(\zeta(g)).$$

Then, just as in the examples of Chapter 6, $D=\zeta(G_1)$ is a bounded domain in the complex vector space $\mathfrak{p}^+$ and $g'\in G_1$ operates on it on the right by : $\zeta(g)\cdot g'=\zeta(gg')$. Thus, if $g=p_1^-k_1p_1^+$, then $p_1^+=\exp(\zeta(g))$ and

$$p_1^+\cdot g'=p_2^-k_2p_2^+, \tag{25}$$

where $p_2^+=\exp(\zeta(gg'))$; moreover, one may then prove without difficulty that [3 : § 1.9]

$$j_D(\zeta(g), g')=\det Ad_{\mathfrak{p}^+}(k_2^{-1}). \tag{26}$$

It can be shown that the quotient of G_1 by its center is naturally isomorphic to $\mathrm{Hol}(D)^0$; this follows from the fact that given an irreducible, symmetric, bounded domain D, the group $\mathrm{Hol}(D)^0$ is of the type of real, semi-simple, Lie group G_R^0 considered, and that, but for taking finite coverings of G, no two distinct groups give the same domain (cf. [14; 27 : Chap. VIII]).

Let b be a non-negative integer $\leq m$ and put

$$c_b = \prod_{i=1}^{b} \exp\left(\frac{\pi}{4}(E_{-i} - E_i)\right).$$

We have considered such an element of $\exp(\mathfrak{p}_C)$ in Chapter 6, sections 3 and 5. Then one may prove (as in the special cases, loc. cit.) that $G_1 \cdot c_b \subset P^- K_C P^+$ and that the domain $D_b = \zeta(G_1 \cdot c_b)$ is biholomorphically equivalent to D under the mapping $\zeta(g \cdot c_b) \mapsto \zeta(g)$, $g \in G_1$. Now D_b may be described in terms of three subspaces W, E, and $\mathfrak{p}_b{}^+$ of $\mathfrak{p}^+$, of which $\mathfrak{p}^+$ is the direct sum, as follows: There is a non-degenerate, open, self-adjoint, convex cone $V \subset W' = W_R$, a bounded domain $F = F_b \subset \mathfrak{p}_b{}^+$, and for each $t \in F$, a real bilinear mapping $L_t : E \times E \to W$ with the properties

1) $L_t(v, u) = \overline{L_t(u, v)}, \quad u, v \in E,$

2) $L_t(u, u) \in \bar{V}, \quad$ and $\quad L_t(u, u) = 0 \quad$ only if $\quad u = 0,$

such that

$$D_b = \{(z, u, t) \in W \times E \times F \mid \mathrm{Im}\, z - L_t(u, u) \in V\}.$$

A domain described in the same manner as D_b, in terms of F, V, L_t, etc., is called a Siegel domain of the third kind [46]. One may compare this description of D_b with that of the domain given by formulae (79)–(82) of Chapter 6. We shall develop this subject further in Chapter 11, section 3.

Suppose, for the sake of example, that $D = \sigma_1 \times \cdots \times \sigma_m$, the Cartesian product of m copies $\sigma_1, \cdots, \sigma_m$ of the unit disc, and let $c_{(j)}$ be the Cayley transform of σ_j onto the ordinary upper half-plane in $\boldsymbol{C}$ which carries 0 into i and three appropriately ordered points α_j, β_j, and γ_j of $\partial\sigma_j$ onto 0, 1, and ∞ respectively. Then c_b is the product of $c_{(1)}, \cdots, c_{(b)}$, and if $F = \gamma_1 \times \cdots \times \gamma_b \times \sigma_{b+1} \times \cdots \times \sigma_m$, then c_b carries F into a subset of a hyperplane at ∞. The general case is in many ways parallel to this, and we refer to c_b in general as a partial Cayley transform of D.

Now assume that G is a connected, linear, semi-simple algebraic group defined over $\boldsymbol{R}$ and, for simplicity, assume that G is a direct product of absolutely simple subgroups G_j, $j=1, \cdots, l$ (in any case, we know that G is isogeneous to such a direct product). We assume that each G_j is defined over $\boldsymbol{R}$ and assume that if K_j is a maximal compact subgroup of G_{jR}^0, then K_j has a non-discrete center. The earlier considerations of this section may be applied and one sees that the quotient of G_R^0 by a maximal compact subgroup K is a bounded domain $D=D_1\times\cdots\times D_l$, where $D_j=K_j\backslash G_{jR}^0$. Moreover, one has a set of partial Cayley transforms of D which arise by taking the products of partial Cayley transforms of $D_1, \cdots, D_l$ in the sense described above.

§ 8. Discrete groups acting on D

Let G be a connected, semi-simple, linear algebraic group defined over $\boldsymbol{Q}$. We assume that G satisfies the assumptions made in the last paragraph of section 7. Thus, if K is a maximal compact subgroup of G_R^0, then $X=K\backslash G_R^0$ is (biholomorphically equivalent to) a bounded symmetric domain D. We assume that G is given as a subgroup defined over $\boldsymbol{Q}$ of $GL(V)$ for some vector space V of finite dimension defined over $\boldsymbol{Q}$, and let Λ be a lattice of V_R (i.e., a discrete subgroup of V_R such that V_R/Λ is compact), $\Lambda\subset V_Q$. Let $G_\Lambda=\{g\in G \mid g\Lambda=\Lambda\}$. Then any subgroup of G_R commensurable with G_Λ is called arithmetic (Chapter 6, section 1). If Γ is an arithmetic subgroup of G_R^0, then its image in $\mathrm{Hol}(D)$ is discrete, and the quotient space D/Γ has a natural structure as a complex analytic space (Chapter 5, section 1). We have given examples of such domains D with arithmetic groups acting on them in Chapter 6.

If f is a holomorphic function on D, then f is called an automorphic form of weight l with respect to Γ if we have $f(Z\cdot\gamma)\,j(Z,\gamma)^l \equiv f(Z)$ for all $\gamma\in\Gamma$, $Z\in D$. Succeeding chapters will be devoted to the analysis, construction, and properties of such functions. This will lay a foundation for following a proof given elsewhere [3] of the fact that D/Γ is a Zariski-open set on an algebraic projective variety.

CHAPTER 8
REPRESENTATIONS OF COMPACT GROUPS

§1. Measure theory and convolution on a locally compact group

We continue here with the references, notation, and definitions of Chapter 4, section 1. However, in this chapter, and henceforth, $M(X)$ will be used to denote the *complex* measures with compact support on the space X, hence is the same as what would be denoted by $M_c(X)_C = M_c(X) + iM_c(X)$ in the previous notation. Let $(X_1, \mu_1), \cdots, (X_n, \mu_n)$ be locally compact measure spaces, let X be the Cartesian product $\prod X_i$, and let $\mu = \prod \mu_i$ (product measure); denote by φ a mapping of X into a locally compact space Y. Then $\mu_1, \cdots, \mu_n$ are said to be convolvable with respect to φ if φ is μ-proper, and then we write $\varphi(\mu) = \mu_1 * \cdots * \mu_n$. The case we are interested in is when $X_1 = \cdots = X_n = G$, a locally compact topological group and $\varphi: (x_1, \cdots, x_n) \mapsto x_1 \cdots x_n$ is the product mapping. By [B, Chap. VIII, §1, no. 3], if $\mu_1, \cdots, \mu_n$ are measures on G, then they are convolvable if a) they are all bounded, or if b) all but one have compact support. If all have compact support, then their convolution product has compact support $\mathcal{S}(\mu_1) \cdot \cdots \cdot \mathcal{S}(\mu_n)$.

Hence, the measures $M(G)$ of compact support form an algebra under the convolution product. If $\alpha, \beta \in M(G)$, we let $\alpha\beta = \alpha * \beta$ (i.e., omit the $*$). From the definition of convolution product, we have for $f \in \mathcal{K}(G)$,

$$\begin{aligned} \alpha\beta(f) &= \int_G \int_G f(g'g'') d\alpha(g') d\beta(g'') \\ &= \int_G \int_G \varepsilon_{g'g''}(f) d\alpha(g') d\beta(g''), \end{aligned} \tag{1}$$

so that we may write, in an appropriate sense,

$$\alpha\beta = \int_G \int_G \varepsilon_{g'g''} d\alpha(g') d\beta(g''). \tag{2}$$

If K is a compact subgroup of G, a measure $\alpha \in M(K)$ will be made into a measure on G by: If $\theta \in \mathcal{K}(G)$, then $\alpha(\theta)=\alpha(\theta \mid K)$. From the above formula (2), it is clear that this mapping of $M(K)$ into $M(G)$ is a homomorphism of algebras; clearly $\alpha(\theta \mid K)=0$ for all $\theta \in \mathcal{K}(G)$ if and only if $\alpha=0$ as an element of $M(K)$. Hence, $M(K)$ may be viewed as a subalgebra of $M(G)$. Fix a Haar measure dg on G. Let $f \in L^1(G)$; we assign to f a measure on G, also denoted by f, by

$$(3) \qquad f(h)=\int_G hf\,dg, \quad h \in \mathcal{K}(G).$$

Then [B, Chap. VIII, §4, no. 2] if $\mu \in M(G)$, $f \in \mathcal{K}(G)$, it follows that $\mu * f$ and $f * \mu$ are measures coming from continuous functions of compact support on G; in other words, $\mathcal{K}(G)$ is a two-sided ideal in $M(G)$. We have the inclusions:

$$(4) \qquad \mathcal{K}(G) \subset M(G), \quad \mathcal{K}(K) \subset M(K) \subset M(G).$$

At the same time we note that the assignment of measures on G to elements of $L^1(G)$ makes the latter into an algebra under the convolution product: If f_1 and $f_2 \in L^1(G)$, then $f_1 * f_2$ is the element of $L^1(G)$ given by

$$(5) \qquad f_1 * f_2(g)=\int_G f_1(g') f_2(g'^{-1}g)\,dg',$$

where the fact that the integral on the right side defines an element of $L^1(G)$ follows from Fubini's theorem:

$$(6) \qquad \|f_1 * f_2\|_1 \leqq \int_G \left(\int_G |f_2(g_1^{-1}g)|\,dg\right) |f_1(g_1)|\,dg_1 = \|f_1\|_1 \cdot \|f_2\|_1,$$

the Haar measure dg being left-invariant. Moreover, the associative law $(f*g)*h=f*(g*h)$ may be verified by a trivial calculation using the left invariance of Haar measure. In this way, L^1 becomes a Banach algebra, which is commutative if and only if G is.

The group G is called unimodular if its Haar measure is both left- and right-invariant. In general, if $g' \in G$ one has $d(gg')=\Delta(g')dg$, where Δ is a continuous, positive, real-valued function satisfying $\Delta(g_1g_2)=\Delta(g_1)\Delta(g_2)$, and G is unimodular if and only if $\Delta \equiv 1$. If G is compact, Δ must be bounded, hence $\Delta \equiv 1$; if G is a semi-simple Lie group, it has no proper Abelian factor groups, so that $\Delta \equiv 1$; if G is Abelian, it is clear that a right-invariant measure is also left-invariant; and combining the last two cases one sees that a reductive Lie group is also unimodular. If G is an algebraic Lie group (real or complex) and if $R(G)$ is its radical with a maximal split torus T and unipotent radical $R_u(G)$, then it is a direct computation to show that G is unimodular if and only if T centralizes $R_u(G)$; in fact, if $\mathfrak{u}$ is the Lie algebra of $R_u(G)$, then

(7) $$\Delta(g) = |\det Ad_{\mathfrak{u}}(g)|^r,$$

for some positive real exponent r.

Now let G be unimodular and let p, q, and r be three real numbers satisfying $p \geqq 1$, $q \geqq 1$, $1/r = 1/p + 1/q - 1 \geqq 0$, so that of course $r \geqq 1$, too. Then [60a : p. 55] if $f_1 \in L^p(G)$ and $f_2 \in L^q(G)$, the convolution product $f_1 * f_2$ is defined and belongs to $L^r(G)$ and one has

(8) $$||f_1 * f_2||_r \leqq ||f_1||_p \cdot ||f_2||_q.$$

The particular case $p = r = \infty$, $q = 1$ is independently obvious and gives for $f_1 \in L^\infty(G)$, $f_2 \in L^1(G)$ that $f_1 * f_2 \in L^\infty(G)$ and

(9) $$||f_1 * f_2||_\infty \leqq ||f_1||_\infty ||f_2||_1.$$

One also has

(10) $$||f_1 * f_2||_\infty \leqq ||f_1||_p ||f_2||_{p'},$$

if $1/p + 1/p' = 1$ and $f_1 \in L^p(G)$, $f_2 \in L^{p'}(G)$.

§2. Representations on a locally convex space [22b, 26d]

Let G be a topological group and let V be a Hausdorff, locally convex, real or complex, topological vector space. Let $\mathrm{End}(V)$ be the algebra of continuous linear mappings of V into itself and let $\mathrm{End}(V)^*$ be the group of its invertible elements (having continuous inverses). By a representation of G on V we mean a homomorphism ρ of G into $\mathrm{End}(V)^*$ such that the mapping $(g, v) \mapsto \rho(g) \cdot v$ is a continuous mapping of $G \times V$ into V.

Let $\mathcal{S}$ be the family of all continuous semi-norms on V. Then, by Chapter 3, section 1, and by [64 : Theorem 1, p. 26] these define the topology on V, and a basis of neighborhoods of 0 is given by the family of convex, balanced, and absorbing sets $\mathcal{U}\{\nu_\alpha, \varepsilon_\alpha\}$ given as follows: If $\{\nu_\alpha\}$ is a finite set of continuous semi-norms and if for each α we are given $\varepsilon_\alpha > 0$, define

(11) $$\mathcal{U}\{\nu_\alpha, \varepsilon_\alpha\} = \{v \in V \mid \nu_\alpha(v) \leqq \varepsilon_\alpha, \text{ each } \alpha\}.$$

We now assume that G is locally compact and one has [26d : p. 5]

LEMMA 1. *Suppose that ρ is a homomorphism of G into* End$(V)^*$ *such that for each $v \in V$, the mapping $g \mapsto \rho(g)\cdot v$ of G into V is continuous and that there exists a neighborhood $\mathcal{U}$ of $e \in G$ with the property: Given $\nu_0 \in \mathcal{S}$, one may find $\nu \in \mathcal{S}$ such that*

$$\nu_0(\rho(g)\cdot v) \leqq \nu(v) \tag{12}$$

for all $v \in V$, $g \in \mathcal{U}$.

Then ρ is a representation of G on V.

PROOF. Choose $x_0 \in G$ and $v_0 \in V$. Then we have

$$\rho(x_0 u)\cdot v - \rho(x_0)\cdot v_0 = \rho(x_0)\rho(u)\cdot(v - v_0) + \rho(x_0)\cdot(\rho(u)\cdot v_0 - v_0)$$

for $u \in \mathcal{U}$, $v \in V$, and from this, from the properties of semi-norms, and from the hypotheses, it follows that $\rho(x_0 u)\cdot v$ approaches $\rho(x_0)\cdot v_0$ as $u \to e$ and $v \to v_0$. This proves the lemma.

One also easily verifies

LEMMA 2. *Let ρ be a representation of G on V. Let C be a compact subset of G. Then for any neighborhood $\mathcal{U}_0$ of zero in V, there exists another neighborhood $\mathcal{U}$ of zero such that $\rho(C)\cdot\mathcal{U} \subset \mathcal{U}_0$. Moreover, given $\nu_0 \in \mathcal{S}$, there exists $\nu \in \mathcal{S}$ such that $\nu_0(\rho(x)\cdot v) \leqq \nu(v)$ for all $x \in C$, $v \in V$.*

PROOF. The first statement follows from the definition of continuity in u and v, and from the fact that ρ is a homomorphism. The second statement follows from the first and from the way the topology on V is related to the semi-norms.

So now assume V is complex and complete, that G is locally compact and unimodular, and that dg is the (two-sided invariant) Haar measure on G. Let ρ be a representation of G on V and let $f \in \mathcal{K}(G)$. Then the integral

$$\int_G f(g)\rho(g)\cdot v\, dg$$

is defined as an element of V, which we denote by $\rho(f)\cdot v$. The convergence of the integral and its continuity as a function of v follow from [B, Chap. VI, § 1, no. 7] and from Lemma 2. One defines as usual the convolution of f and $g \in \mathcal{K}(G)$, and then one has $\rho(f * g) = \rho(f)\rho(g)$, of which the verification is a formality based on Fubini's theorem.

For purposes of the present chapter, we now concentrate on the case when $V = \mathcal{H}$, a Banach space. If T is a linear mapping of $\mathcal{H}$ into itself, then $||T||$ is defined as

$$\sup_{||v||=1} ||Tv||,$$

$||\ ||$ being the Banach space norm, and T is called bounded if this is finite. If T is bounded and T^{-1} exists and is bounded, then T is called non-singular. If ρ is a representation of G on $\mathcal{H}$, it follows from Lemma 2 that ρ is a homomorphism of G into the group of non-singular bounded operators on $\mathcal{H}$. Let $n(g) = ||\rho(g)||$, $g \in G$. Then by [B, Chap. VIII, § 2, no. 1, Lemmas 1 and 2], n is lower semi-continuous on G (thus, for every $g \in G$, $\underline{\lim}_{g' \to g} n(g') = n(g)$), $n(st) \leqq n(s)n(t)$ for all $s, t \in G$, and n is bounded on every compact set. If n is bounded on all of G, then ρ is called bounded. A lower semi-continuous function n on G which is everywhere $\geqq 0$ and which satisfies

$$n(st) \leqq n(s)n(t), \quad s, t \in G, \tag{13}$$

is called a semi-norm on G.

If n is a semi-norm, not identically zero, on G and if $f \in \mathcal{K}(G)$, we define

$$||f||_n = \int_G |f(g)| n(g) dg.$$

Then (13) implies

$$||f_1 * f_2||_n \leqq ||f_1||_n ||f_2||_n. \tag{14}$$

Let $A_n = \{f | f$ absolutely integrable with respect to the measure $n(g)dg\}$. Then $A_n = A_n(G)$ becomes a normed algebra with respect to the convolution product, which is complete because $L^1(G)$ is complete and $n(g) > 0$ everywhere (since it is not identically zero). And $\mathcal{K}(G)$ is dense in $A_n(G)$ because $\mathcal{K}(G)$ is dense in $L^1(G)$ (definition of the integral) and if $f \in \mathcal{K}(G)$ is close to $h \in n \cdot A_n(G)$, then $n^{-1}f$ is close to $n^{-1}h \in A_n(G)$ with respect to the L^1-norm for the measure $n(g)dg$.

Now take $n(g) = ||\rho(g)||$ again. Then [loc. cit., p. 137] if μ is a measure on G such that n is μ-integrable, the integral $\int_G \rho(g) d\mu(g)$ defines a bounded linear transformation of $\mathcal{H}$ into itself by

$$(15)\qquad \left(\int_G \rho(g)d\mu(g)\right)\cdot v=\int_G \rho(g)\cdot v\, d\mu(g), \quad v\in\mathcal{H},$$

having norm $\leqq \int_G n(g)d\mu(g)$; we denote this bounded linear transformation by $\rho(\mu)$. Thus, in particular, we obtain a representation of the (convolution) algebra of measures $M(G)$ with compact support, and of its two-sided ideal $\mathcal{K}(G)$, by an algebra of bounded linear transformations of $\mathcal{H}$ (the verification of $\rho(\alpha\beta)=\rho(\alpha)\rho(\beta)$ is again a formal triviality using Fubini's theorem [loc. cit., p. 145]). Likewise, if ρ is bounded, one obtains a representation of the convolution algebra $L^1(G)$ (identified as before with an algebra of bounded measures) by bounded transformations of $\mathcal{H}$. Thus, if $f\in L^1(G)$, we have

$$\rho(f)\cdot v=\int_G f(g)\rho(g)\cdot v\, dg, \quad v\in\mathcal{H},$$

and $\rho(f*g)=\rho(f)\rho(g)$. Moreover, under suitable hypotheses to insure convergence, we define $\rho*f\colon G\to \mathrm{End}(\mathcal{H})$ by

$$\rho*f(x)=\int_G f(g^{-1})\rho(xg)dg$$

and obtain

$$(16)\qquad \rho*f(x)=\rho(x)\int_G f(g^{-1})\rho(g)dg=\rho(x)\rho(\check{f}),$$

where $\check{f}(x)=f(x^{-1})$.

This is a convenient place to record three notions of irreducibility, distinguished as such by Godement [22b]. *With notation as above, ρ is said to be:*

A. *Algebraically irreducible if, disregarding topology, there exists no proper algebraic invariant subspace of ρ in $\mathcal{H}$.*

B. *Completely irreducible if ρ is a linear representation of a complex associative algebra $\mathfrak{A}$ by an algebra of bounded operators on a Banach space $\mathcal{H}$ and if every bounded operator T on $\mathcal{H}$ is a strong limit of operators $\rho(x)$, $x\in\mathfrak{A}$.* (By T being a strong limit of operators $\{\rho(x)\}_{x\in\mathfrak{A}}$ on $\mathcal{H}$, we mean that for any finite subset E of $\mathcal{H}$ and for any $\varepsilon>0$, there exists $x\in\mathfrak{A}$ such that $\|\rho(x)\cdot v-T\cdot v\|<\varepsilon$ for every $v\in E$.)

C. *Topologically irreducible if there exists no proper, topologically closed, invariant subspace of ρ in $\mathcal{H}$.*

For finite-dimensional representations, the three notions coincide if in B., we understand $\mathfrak{A}$ to be the group algebra of G (Burnside's theorem). For the rest of this chapter we are mainly concerned with the case when G is compact. We shall return to the general case later.

§3. The Peter-Weyl theorem [60a]

Let G be a compact, Hausdorff topological group. Let ρ be a representation of G on a finite-dimensional real or complex vector space V. If V is real (resp. complex), let Q' be any positive-definite quadratic (resp. Hermitian quadratic) form on V; we may define a G-invariant quadratic (resp. Hermitian quadratic) form Q by

$$Q(x)=\int_G Q'(\rho(g)\cdot x)dg,$$

where dg is the Haar measure such that $vol(G)=1$. Then Q defines a Euclidean (resp. Hermitian) inner product (,) on V invariant under G. Suppose W is a subspace of V such that $\rho(g)\cdot W=W$ for every $g\in G$; then $W^{\perp}$, the orthogonal complement of W with respect to (,), is also G-invariant and $V=W\oplus W^{\perp}$. Therefore, the representation ρ of G is completely reducible. Thus, any finite-dimensional real or complex representation of a compact group is fully reducible, and is a unitary representation with respect to a suitably chosen positive-definite inner product on the representation space.

Consequently, the main problem for finite-dimensional representations of G becomes that of studying the irreducible ones. If ρ is an irreducible real representation of G on a real vector space V, then the naturally associated representation of G on V_C is either irreducible (i.e., ρ is absolutely irreducible) or is the sum of an irreducible representation and of its complex conjugate. Henceforth, we consider absolutely irreducible representations of G on a *complex* vector space. We now give a sketch of the treatment of this problem as presented in [60a].

Let $\mathfrak{d}$ and $\mathfrak{d}'$ be two equivalence classes of finite-dimensional, irreducible representations of G and let, in matrix notation, $\rho=(\rho_{ij})\in\mathfrak{d}$, $\rho'=(\rho_{kl}')\in\mathfrak{d}'$, $1\leqq i, j\leqq d=\dim\rho$, $1\leqq k, l\leqq d'=\dim\rho'$; denote by V the representation space of ρ and by V' the representation space of ρ'. Let $\check{\mathfrak{d}}$ be the equivalence class of $\check{\rho}={}^t\rho^{-1}$. Then, by

Schur's lemma, one sees that ρ and ρ' are equivalent if and only if the tensor product $\check{\rho}\otimes\rho'$, which acts on the space of $d\times d'$ rectangular matrices, has a non-zero invariant vector. (See below, Chapter 9, section 2.)

If f is a complex-valued function on G and if $g\in G$, define $l(g)f$ to be the function on G given by $(l(g)f)(x)=f(g^{-1}x)$. In the notation of the preceding section, one has

$$\rho(l(g)f)=\int_G f(g^{-1}g')\rho(g')dg'=\rho(g)\int_G f(g')\rho(g')dg'=\rho(g)\rho(f). \tag{17}$$

If f is the constant function $\equiv 1$ on G, then $l(g)\cdot f=f$ and, denoting f simply by 1,

$$\rho(1)=\rho(g)\rho(1), \quad g\in G, \tag{18}$$

so that every column of the matrix $\rho(1)$ is an invariant vector of $\rho(g)$ for all $g\in G$. And conversely, if v is an invariant vector of $\rho(g)$ for all $g\in G$, it will be an invariant vector of $\rho(1)$, hence a linear combination of the columns of $\rho(1)$. Therefore, the rank of $\rho(1)$ is equal to the dimension of the space of invariant vectors of the representation ρ. Replacing ρ by the representation $\check{\rho}\otimes\rho'$ above and calculating $(\check{\rho}\otimes\rho')(1)$ from its definition, one obtains the orthogonality relations

$$\int_G \rho_{ij}(g)\rho_{kl}'(g^{-1})dg=\begin{cases}0 & \text{if } \mathfrak{d}\neq\mathfrak{d}',\\ 0 & \text{if } \rho=\rho',\ j\neq k \text{ or } i\neq l,\\ d^{-1} & \text{if } \rho=\rho',\ j=k,\ i=l.\end{cases} \tag{19}$$

These allow one to calculate easily

$$\rho_{ij}*\rho_{kl}'=\begin{cases}0 \text{ if } \mathfrak{d}\neq\mathfrak{d}',\\ 0 \text{ if } \rho=\rho',\ j\neq k,\\ d^{-1}\rho_{il} \text{ if } \rho=\rho',\ j=k.\end{cases} \tag{20}$$

If f is a function on G, define $f^{\sim}(x)=\overline{f(x^{-1})}$. Let Φ be an element of $L^1(G)$ such that $\Phi=\Phi^{\sim}$ and define $T=T_\Phi$ to be the linear operator on $L^2(G)$ defined by $T\varphi=\Phi*\varphi$ (recalling (8) with $q=r=2$, $p=1$). Then T is a self-adjoint, bounded operator on L^2 and, since G is compact, is completely continuous (or compact, i.e., if S denotes the closed unit sphere in L^2, then $T(S)$ is relatively compact); if Φ is essentially bounded, and this case is sufficient for our purposes, this is obvious, since then T is of Hilbert-Schmidt type [64 : p. 277], and in general the result follows from a criterion in [60a : p. 53]. Then one knows [64 : p. 284] that the set of eigenvalues of T is a bounded countable subset of $\boldsymbol{C}$ with no limit point $\neq 0$, and each eigenvalue $\lambda \neq 0$ has finite multiplicity. Since T is self-adjoint, all the eigenvalues are real, and if $T \neq 0$, there exists at least one eigenvalue different from 0; in fact, it is known (for example, see [16a : p. 207]) that if $\lambda_0 = \|T\|$, then at least one of $\pm\lambda_0$ is an eigenvalue of T. Much, if not all, of the preceding may also be found in [48]. Fix a non-zero eigenvalue μ of T and let φ_i, $i=1, \cdots, n$, be a basis of the vector space V_μ of solutions φ in $L^2(G)$ of $\Phi * \varphi = \mu\varphi$; one may assume

$$(\varphi_i, \varphi_j) = \int_G \varphi_i(g)\overline{\varphi_j(g)}dg = \delta_{ij}.$$

If one defines

$$\rho_{ij} = \tilde{\varphi}_i * \varphi_j, \quad i, j = 1, \cdots, n, \tag{21}$$

it is easily calculated that $\rho = (\rho_{ij})$ is a unitary representation of G in V and that

$$\varphi_i(g) = \mu^{-1} \sum_k \rho_{ki}(g)\int_G \Phi(g'^{-1})\varphi_k(g')dg'. \tag{22}$$

Therefore $\Phi * \rho \neq 0$, hence $\rho(\bar{\Phi}) \neq 0$, and by decomposing ρ into its irreducible components, one sees that there exists one of them, say σ, such that $\sigma(\bar{\Phi}) \neq 0$.

Now let $\varphi \in L^2(G)$, $\varphi \neq 0$, and put $\Phi = \varphi * \tilde{\varphi}$; then choosing ρ and σ as above, we have $\sigma(\bar{\Phi}) = \sigma(\tilde{\varphi} * \varphi) = \sigma(\bar{\varphi})^t\sigma(\varphi) \neq 0$, which implies $\sigma(\bar{\varphi}) \neq 0$. Note that since $\varphi, \tilde{\varphi} \in L^2(G)$, Φ is essentially bounded by (10), and therefore the Hilbert-Schmidt criterion alone is sufficient for compactness of T.

Since $\sigma(\bar{\varphi}) \neq 0$, it follows that φ cannot be orthogonal to all the coefficients ρ_{ij} of ρ. It follows from this and from (19) that the system of coefficients $\{\rho_{ij}\}_{i,j,\mathfrak{d}}$, as $\mathfrak{d}$ ranges over all the equivalence classes of irreducible (finite-dimensional) unitary representations of G, is a complete orthogonal system for $L^2(G)$, so that any $\varphi \in L^2(G)$ can be expanded in a series

$$\varphi(g) = \sum_{i,j,\mathfrak{d}} m_{ij,\mathfrak{d}}\rho_{ij}(g), \tag{23}$$

which converges in L^2, where

$$m_{ij,\mathfrak{d}} = d(\mathfrak{d}) \int_G \overline{\rho_{ij}(g')} \varphi(g') dg', \quad d(\mathfrak{d}) = \deg(\mathfrak{d}). \tag{24}$$

Since $\sum_j \rho_{ij}(g)\rho_{ji}(g') = \rho_{ii}(gg')$ and since ρ is unitary, one easily proves, substituting (24) in (23), that (23) may be rewritten in the form

$$\varphi = \sum_{\mathfrak{d}} d(\mathfrak{d}) \chi * \varphi, \tag{25}$$

where $\chi(g) = \sum_i \rho_{ii}(g)$ is the character of $\mathfrak{d}$.

A continuous function $\psi \geqq 0$ on G such that $\psi = 0$ outside a small neighborhood $\mathcal{N}$ of the identity and such that $\int_G \psi(g)dg = 1$ is called an approximate identity, and a sequence $\{\psi_m\}$ of such functions corresponding to neighborhoods $\{\mathcal{N}_m\}$ of e such that $\mathcal{N}_{m+1} \subset \mathcal{N}_m$, $\bigcap_m \mathcal{N}_m = \{e\}$, and $\mathcal{S}(\psi_m) \subset \mathcal{N}_m$ is called a Dirac sequence.

Let φ be a continuous function on G. Then for any $\varepsilon > 0$ we can find an approximate identity ψ such that $\|\varphi - \varphi * \psi\|_\infty < \varepsilon$. On the other hand, convoluting (25) on the right with ψ, we obtain

$$\varphi * \psi = \sum_{\mathfrak{d}} d(\mathfrak{d}) \chi * \varphi * \psi, \tag{26}$$

which converges in L^∞, i.e., essentially uniformly on G. Any finite partial sum of this series is a finite linear combination of representative functions ρ_{ij} which are *continuous;* thus, any continuous function on G can be uniformly approximated by a finite linear combination of the representative functions. This is the Peter-Weyl theorem.

A function φ on G is called a central function on G if $\varphi(gg') = \varphi(g'g)$ for $g, g' \in G$ in an almost everywhere sense. Of course, this is equivalent to the a.e. invariance of φ under inner automorphisms of G. If χ is the character $\sum \rho_{ii}$ of a representation of G, then it is clear that χ is a central function because of the well-known relation trace(AB) = trace(BA). It is a trivial calculation to see that the central functions in $L^1(G)$ constitute just the center of the convolution algebra $L^1(G)$. Then, using Schur's lemma and the orthogonality relations (19), one sees that for φ a central function in $L^2(G)$, the series development (23) becomes simply

$$\varphi(g) = \sum_{\mathfrak{d}} (\varphi, \chi_{\mathfrak{d}}) \cdot \chi_{\mathfrak{d}}(g), \tag{27}$$

and by using the same argument as before, with a central approximate identity, one sees that any continuous central function φ can be uniformly approximated by linear combinations of characters of irreducible, finite-dimensional representations of G. And, as one sees from (19), we have for classes $\mathfrak{d}$ and $\mathfrak{d}'$ of irreducible representations,

$$(\chi_{\mathfrak{d}}, \chi_{\mathfrak{d}'}) = \begin{cases} 0 \text{ if } \mathfrak{d} \neq \mathfrak{d}', \\ 1 \text{ if } \mathfrak{d} = \mathfrak{d}', \end{cases} \tag{28}$$

so that if ρ is any finite-dimensional representation of G with character $\chi = \chi_\rho$, we have

$$\rho = \oplus_{\mathfrak{d}} (\chi, \chi_{\mathfrak{d}}) \rho_{\mathfrak{d}}, \tag{29}$$

where $\rho_{\mathfrak{d}} \in \mathfrak{d}$.

§ 4. Some applications

If σ and τ are two finite-dimensional representations of G with characters χ_σ and χ_τ, then the character $\chi_{\sigma\otimes\tau}$ of their tensor product is simply $\chi_\sigma \chi_\tau$, as direct calculation with matrices shows. If ρ is a third representation, which we assume to be irreducible with character χ_ρ, then the number of times that ρ occurs in $\sigma \otimes \tau$ is given very simply as

$$\int_G \chi_\sigma(g) \chi_\tau(g) \overline{\chi_\rho(g)} dg, \tag{30}$$

which we denote by $m(\sigma, \tau; \rho)$. If σ and τ are also irreducible, then evidently m has certain simple properties which may be expressed by

$$m(\sigma, \tau; \rho) = m(\bar{\rho}, \tau; \bar{\sigma}) = m(\tau, \sigma; \rho) = m(\bar{\rho}, \sigma; \bar{\tau}). \tag{31}$$

If $\varphi \in L^2(G)$, $x \in G$, define $r(x)\varphi(y) = \varphi(yx)$ for all $y \in G$. This clearly gives us a unitary representation of G on $L^2(G)$. The Hilbert space $L^2 = L^2(G)$ can be written as a Hilbert space direct sum

$$L^2 = \oplus_{\mathfrak{d}} L_{\mathfrak{d}}, \tag{32}$$

where $L_{\mathfrak{d}}$ is the space of dimension $d(\mathfrak{d})^2$ spanned by the coefficients ρ_{ij} of a unitary representation $\rho \in \mathfrak{d}$, and where the term on the left is understood to be the closure of the restricted direct sum of the terms on the right. Since $\rho_{ij}(yx) = \sum_l \rho_{il}(y)\rho_{lj}(x)$, it is clear that $L_{\mathfrak{d}}$ is an invariant subspace of the representation r and that the restriction of r to this subspace is equivalent to the direct sum of $d(\mathfrak{d})$ copies of ρ. Therefore, under r, L^2 decomposes into a (Hilbert space) direct sum of finite-dimensional subspaces, and the equivalence class $\mathfrak{d}$ occurs $d(\mathfrak{d})$ times in this direct sum for every equivalence class $\mathfrak{d}$ of finite-dimensional, irreducible representations of G.

§ 5. The Frobenius reciprocity theorem [60a : § 23]

Let G be a compact group and H, a closed subgroup. Let $\mathfrak{d}$ be an equivalence class of finite-dimensional, irreducible representations of G and let δ be an equivalence class of finite-dimensional, irreducible representations of H. Fix a unitary $\rho\in\mathfrak{d}$ and $\mu\in\delta$, let $d=\dim(\mathfrak{d})$, $d'=\dim(\delta)$, let V be the representation space of ρ and W, that of μ. We assume that μ (resp. ρ) is unitary with respect to an inner product $(\ ,\ )$ (resp. $<\ ,\ >$) on W (resp. on V).

We consider the linear space of all W-valued measurable functions w on G which satisfy

$$w(g\xi)=w(g)\mu(\xi), \quad g\in G,\ \xi\in H, \tag{33}$$

(for convenience representing the elements of W by $1\times d'$ row vectors) and denote by $\mathcal{H}$ the subspace of those satisfying

$$\|w\|_2=\int_G |w(g)|^2 dg=\int_{G/H}|w|^2(x)dx<+\infty, \tag{34}$$

where $|w|^2(x)$ means the (well-defined by (33)) square of the length of $w(g)\in W$ for any $g\in x=gH$, and dx is the G-invariant measure on the homogeneous space $X=G/H$ coming from the Haar measure dg on G that gives $vol(G)=1$; we also take the Haar measure dh on H such that $vol(H)=1$. We cause G to operate on $\mathcal{H}$ on the *right* by the representation i_μ given by

$$w(g)i_\mu(g')=w(g'g), \quad g'\in G,\ w\in\mathcal{H}. \tag{35}$$

The representation i_μ is called the representation of G induced by μ.

Each $w\in\mathcal{H}$ has an expansion

$$w=\sum_{\mathfrak{d}_1}\dim(\mathfrak{d}_1)\chi_{\mathfrak{d}_1}*w, \tag{36}$$

converging in the L^2 sense, where $\mathfrak{d}_1$ runs over the equivalence classes of (finite-dimensional) irreducible representations of G and

$\chi_{\mathfrak{d}_1}$ is the character of $\mathfrak{d}_1$. It is easily seen that $\chi_{\mathfrak{d}_1} * w \in \mathscr{H}$. The linear space of all the elements of the form $\chi_{\mathfrak{d}_1} * w$, $w \in \mathscr{H}$, is denoted by $\mathscr{H}_{\mathfrak{d}_1}$, and these are just the $1 \times d'$ row vectors satisfying (33) whose vector components lie in the span of the coefficients ρ_{1ij} of ρ_1, $\rho_1 \in \mathfrak{d}_1$, hence $\mathscr{H}_{\mathfrak{d}_1}$ is stable under $i_\mu(G)$. We now proceed to decompose $\mathscr{H}_{\mathfrak{d}}$ for our given $\mathfrak{d}$.

Let $w = (w_1, \cdots, w_{d'}) \in \mathscr{H}_{\mathfrak{d}}$, so that each $w_k = \sum_{i,j} c_{jik}\rho_{ij}$; then from (33) it follows that the matrix $C_i = (c_{jik})_{1 \leqq j \leqq d,\ 1 \leqq k \leqq d'}$ satisfies

$$\rho(\xi)C_i = C_i\mu(\xi). \tag{37}$$

Let $(\mathfrak{d} : \delta)$ be the number of times $\geqq 0$ the class δ is represented in the restriction of ρ to H. Then the dimension of the space $\mathrm{Hom}_H(W, V)$ of intertwining operators A satisfying

$$\rho(\xi)A = A\mu(\xi) \tag{38}$$

is easily seen from Schur's lemma to be $(\mathfrak{d} : \delta)$. If we put

$$\Lambda(x) = \rho(x)A, \quad x \in G, \tag{39}$$

for such an intertwining operator $A \neq 0$, the rows of Λ span an $i_\mu(G)$-stable subspace of dimension d in $\mathscr{H}_{\mathfrak{d}}$ and the restriction of i_μ to that subspace is a representation in $\mathfrak{d}$. Moreover, $\Lambda(g'xh) = \rho(g')\Lambda(x)\mu(h)$, $g', x \in G$, $h \in H$, and every $d \times d'$ matrix-valued function Λ on G satisfying this condition is of the form $\rho(x)A$, $A \in \mathrm{Hom}_H(W, V)$. Since the functions ρ_{ij} are mutually orthogonal and linearly independent by (19), while $\rho(x)$ has d rows and $\dim \mathrm{Hom}_H(W, V) = (\mathfrak{d} : \delta)$, it follows that the span of all rows of all such $\Lambda(x)$ has dimension $d(\mathfrak{d} : \delta)$ and decomposes into the direct sum of $(\mathfrak{d} : \delta)$ subspaces of dimension d, on each of which the restriction of i_μ is in $\mathfrak{d}$; from (37) and from the equation for w_k, it is clear that that span is just $\mathscr{H}_{\mathfrak{d}}$. Thus one proves

THEOREM 15 (*Frobenius reciprocity*). *The number of times $\mathfrak{d}$ occurs in i_μ is just the number of times δ is represented in the restriction of $\rho \in \mathfrak{d}$ to H.*

One notes that if v is any continuous W-valued function on G, then

$$w(x) = \int_H v(x\xi)\mu(\xi^{-1})d\xi \tag{40}$$

is in $\mathcal{H}$. It is easy to find v such that the integral in (40) is not identically zero, hence, for any μ, the space $\mathcal{H}$ of i_μ is $\neq \{0\}$, hence $\mathcal{H}_\mathfrak{d} \neq 0$ for some $\mathfrak{d}$. This proves most of

COROLLARY. *If δ is any class of finite-dimensional irreducible representations of H, then δ is contained in the restriction to H of some class $\mathfrak{d}$ of irreducible representations of G. If H is a proper subgroup of G, then we may take $\mathfrak{d}$ to be non-trivial.*

Everything is obvious from what has gone before, except the very last assertion about the non-triviality of $\mathfrak{d}$. As for that, it is sufficient to choose v in (40) such that w is non-constant; this is easily done by choosing the values for v on H such that $\int_H v(\xi)\mu(\xi^{-1})d\xi \neq 0$, choosing $v=0$ on some coset of H other than H itself, and applying Tietze's extension theorem.

The theorem and its corollary are, of course, generalizations of well-known [13; 29] results for finite groups.

As a special case, let $H=\{e\}$, and let μ be the trivial (one-dimensional) representation of H. Then $\mathcal{H}=L^2(G)$ and i_μ is just the regular representation of G on $L^2(G)$. Theorem 15 implies that $\mathfrak{d}$ is contained in i_μ just $\dim(\mathfrak{d})$ times, which we already know from the preceding section.

§ 6. A compact Lie group is algebraic [16b]

Again let G be a compact group. Let $\mathcal{O}$ be the complex algebra of continuous functions on G generated by all the complex-valued continuous functions ρ_{ij} as $\rho=(\rho_{ij})$ runs over all real or complex finite-dimensional representations of G. From our previous discussion, it is obvious that this ring is generated by ρ_{ij} for ρ running through irreducible unitary representations, and since a unitary representation may be viewed as an orthogonal representation of twice the dimension, orthogonal representative functions ρ_{ij} will suffice. Suppose that ρ is a real, faithful, orthogonal representation of G (which is known to exist from the Peter-Weyl theorem), and let $\mathcal{O}'$ be the complex subalgebra of $\mathcal{O}$ generated by the functions ρ_{ij}. Since ρ is faithful, the functions of $\mathcal{O}'$ separate the points of G, and since ρ is real, if $f \in \mathcal{O}'$, then $\bar{f} \in \mathcal{O}'$. Clearly, if $g \in G$, there exists

$f \in \mathcal{O}'$ such that $f(g) \neq 0$. Therefore, by the Stone-Weierstrass approximation theorem, every continuous function on G may be uniformly approximated by elements of $\mathcal{O}'$. Using the orthogonality relations (19), it follows that $\mathcal{O}'$ must contain all representative functions ρ'_{ij} for all orthogonal representations ρ' of G, because in any event the product of two representative functions is again a representative function (coming from a tensor product) so that $\mathcal{O}'$ is the linear span of the representative functions which it contains. Therefore, $\mathcal{O}' = \mathcal{O}$.

THEOREM 16. *Let V be a finite-dimensional real vector space and let K be a compact subgroup of $GL(V)$. Then there exists an algebraic subgroup G of $GL(V_C)$ such that G is defined over $\boldsymbol{R}$ and such that $K = G_R = G \cap GL(V)$. If K is connected, so is G.*

PROOF. [16b]. We have $V \subset V_C$ and $GL(V) = GL(V_C)_R$. Let G be the Zariski-closure of K in $GL(V_C)$. If $g \in G$, then $\bar{g} \in G$, so that G is defined over $\boldsymbol{R}$. Let $G_1 = G_R$. Of course, $K \subset G_1$. By appropriate choice of a basis of V over $\boldsymbol{R}$ (which will then be a basis of V_C over $\boldsymbol{C}$), we may assume that K is contained in the orthogonal group $O(V) = \{g \in GL(V) \mid {}^t g g = id.\}$; since the latter is algebraic, we have $G_1 \subset O(V)$, and clearly G_1 is closed in $O(V)$ since it is determined by real polynomial equations. Therefore G_1 is compact. Suppose K were a proper subgroup of G_1. Then by the corollary of Theorem 15, there exists an *irreducible* non-trivial representation σ of G_1 of which the restriction to K contains the trivial representation of K. Thus, by a suitable choice of coordinates in the representation space V_σ of σ, we may assume $\sigma(K)$ to be contained in the group L of matrices (a_{kl}) such that $a_{11} = 1$ and $a_{1k} = a_{k1} = 0$ for $k \neq 1$. The functions $\sigma_{kl} \in \mathcal{O}$, hence are polynomial functions of the coordinate functions g_{ij} of g in the faithful representation of G_1 by the inclusion mapping as a subgroup of $GL(V)$. But since G is the Zariski-closure of K, every polynomial function on $GL(V)$ that vanishes on K must vanish on G_1, hence $\sigma(G_1) \subset L$, which contradicts the irreducibility of σ. Thus we must have $K = G_1 = G_R$. If K is connected, then $K = G^0 \cap GL(V)$, whence the last assertion.

§7. Compact and algebraic Lie groups

Let G be a connected and compact Lie group. It follows from the Peter-Weyl theorem that G has at least one faithful representation as a closed subgroup of the orthogonal group $O(V)\subset GL(V)$ of a finite-dimensional, *real* vector space V. Identify G with its image in $O(V)$ under this representation and let G_C be its Zariski-closure in $GL(V_C)$. Then $\dim_R G = \dim_C G_C$ and the Lie algebra $\mathfrak{g}_C$ of G_C is the complexification of the Lie algebra $\mathfrak{g}$ of G. By Theorem 16, we have $(G_C)_R = G$. Let $\boldsymbol{H}$ be a maximal, connected Abelian subgroup of G with Zariski-closure $\boldsymbol{H}_C \subset G_C$ and let $\mathfrak{h}$ (resp. $\mathfrak{h}_C$) be the Lie algebra of $\boldsymbol{H}$ (resp. of $\boldsymbol{H}_C$). Since $\boldsymbol{H}$ is commutative and compact, hence fully reducible, it is diagonalizable (over $\boldsymbol{C}$) and consists of semi-simple elements, and so the same must be true of its Zariski-closure $\boldsymbol{H}_C$ which is therefore a torus (Theorems 12 (p. 57) and 16). Then $\boldsymbol{H}$ and $\boldsymbol{H}_C$ (resp. $\mathfrak{h}$ and $\mathfrak{h}_C$) are Cartan subgroups (resp. Cartan subalgebras) of G and of G_C respectively (resp. of $\mathfrak{g}$ and of $\mathfrak{g}_C$ respectively) and $(\boldsymbol{H}_C)_R = \boldsymbol{H}$. In general, a connected, compact, commutative Lie group $\boldsymbol{H}$ will be called a compact torus, and, when we are given a faithful representation of $\boldsymbol{H}$ in $GL(V)$, V being a real vector space, its Zariski-closure in $GL(V_C)$ is an algebraic torus $\boldsymbol{H}_C$. If χ is a character of $\boldsymbol{H}$ of absolute value 1, there corresponds naturally to it a rational character of $\boldsymbol{H}_C$. In fact, $\boldsymbol{H}$ is a product of copies of the unit circle C_1 and $\boldsymbol{H}_C$ is the product of the same number of copies of $\boldsymbol{C}^*$; and the character group of C_1 is naturally isomorphic to $\boldsymbol{X}^*(\boldsymbol{C}^*)$ (each character group consisting of the m-th power homomorphisms, $m \in \boldsymbol{Z}$).

Let exp be the exponential mapping and let Λ be the kernel of the restriction $\exp|\mathfrak{h}$ of exp to $\mathfrak{h}$; then Λ is a lattice in the real vector space $\mathfrak{h}$. Let e be the mapping $z \to e^z$ from $\boldsymbol{C}$ into $\boldsymbol{C}^*$; $\ker(e) = 2\pi i\boldsymbol{Z}$. If λ is a linear mapping from $\mathfrak{h}$ into $\boldsymbol{C}$, we see that it can be transferred to a character e^λ from $\boldsymbol{H}$ into $\boldsymbol{C}^*$ if and only if $\lambda^{-1}(2\pi i\boldsymbol{Z})$ contains Λ; $e^\lambda(h)$ is defined as $e^{\lambda(x)}$ if $\exp(x) = h$, which is well-defined under the condition

$$\lambda(\Lambda) \subset 2\pi i\boldsymbol{Z}. \tag{41}$$

Let $\mathfrak{h}_C{}^*$ be the dual space of $\mathfrak{h}_C$ and let Λ^* be the set of $\lambda \in \mathfrak{h}_C{}^*$ satisfying (41); denote by $\mathfrak{h}^*$ the real vector subspace of $\mathfrak{h}_C{}^*$ spanned by Λ^*.

Let ρ be a linear representation of G in a complex vector space $\boldsymbol{W}$ and let $d\rho$ be the associated representation of $\mathfrak{g}$ in $\mathfrak{gl}(\boldsymbol{W})$. Then $\rho(\boldsymbol{H})$ and $d\rho(\mathfrak{h})$ are diagonalizable; specifically $\boldsymbol{W}=\bigoplus \boldsymbol{W}_\lambda$ (direct sum), where

$$\boldsymbol{W}_\lambda=\{\boldsymbol{w}\in \boldsymbol{W}\,|\, d\rho(x)\cdot \boldsymbol{w}=\lambda(x)\boldsymbol{w},\ x\in\mathfrak{h}\}$$
$$=\{\boldsymbol{w}\in \boldsymbol{W}\,|\,\rho(h)\cdot \boldsymbol{w}=e^\lambda(h)\cdot \boldsymbol{w},\ h\in \boldsymbol{H}\},$$

and where λ runs over a finite subset of Λ^*, the "weights" of ρ. In particular, one has the adjoint representation ρ_0 of G on $\mathfrak{g}_c$, and its weights are called the "roots" of G. If we impose a linear ordering on the vector space $\mathfrak{h}^*$, the roots divide into positive and negative roots. If α is a positive root, it is called "simple" if it cannot be written in the form $\alpha=\beta+\gamma$, where β and γ are both positive roots. The number l of simple roots $\alpha_1, \cdots, \alpha_l$ is equal to dim $\boldsymbol{H}$, and these are a basis of $\mathfrak{h}^*$. Let B be the Killing form of $\mathfrak{g}$; B is negative-definite because G is compact, and is used to identify $\mathfrak{h}$ with its dual space $\mathfrak{h}^*$ and $\mathfrak{g}$ with its dual space $\mathfrak{g}^*$. If α is a root, let $\mathfrak{g}_\alpha\subset\mathfrak{g}_C$ be defined as above and put $\mathfrak{g}_0=\mathfrak{h}_C$. Then dim $\mathfrak{g}_\alpha=1$ for $\alpha\neq 0$. If $(\rho, \boldsymbol{W})$ is any finite-dimensional complex representation of G, λ, a weight, then $\rho(\mathfrak{g}_\alpha)\cdot \boldsymbol{W}_\lambda\subset \boldsymbol{W}_{\lambda+\alpha}$ $(=\{0\}$ if $\lambda+\alpha$ is not a weight); in particular, $[\mathfrak{g}_\alpha, \mathfrak{g}_\beta]\subset\mathfrak{g}_{\alpha+\beta}$. For each root $\alpha\in\mathfrak{h}^*\cong\mathfrak{h}$, we can and do choose $E_\alpha\in\mathfrak{g}_\alpha-\{0\}$ and $H_\alpha'\in\mathfrak{h}_C$ such that:

$$[H, E_\alpha]=\alpha(H)E_\alpha \text{ (by definition)},\quad H\in\mathfrak{h},$$

$$(42)\qquad [E_\alpha, E_{-\alpha}]=H_\alpha',\quad B(E_\alpha, E_{-\alpha})=1,$$

$$B(H, H_\alpha')=\alpha(H),$$

and put

$$H_\alpha=2B(\alpha, \alpha)^{-1}\cdot H_\alpha',$$

$$(43)\qquad B(H_\alpha', H_\beta')=\langle\alpha, \beta\rangle=\alpha(\beta)=\beta(\alpha),$$

$$H_{\alpha_i}=H_i,\quad i=1, \cdots, l.$$

In other words, $\langle\, ,\rangle$ is the inner product on $\mathfrak{h}^*$ induced by B in identifying $\mathfrak{h}^*$ with its dual space $\mathfrak{h}$. We have $\mathfrak{g}_\alpha\perp\mathfrak{g}_\beta$ if $\alpha+\beta\neq 0$ (in this statement, either α or β may be 0 and the result holds for semi-simple $\mathfrak{g}$, compact or not). In fact, if $X\in\mathfrak{g}_\alpha$, $Y\in\mathfrak{g}_\beta$, we have

$$ad\, X\cdot ad\, Y\cdot \mathfrak{g}_\gamma\subset\mathfrak{g}_{\gamma+\alpha+\beta},$$

so $ad\, X\cdot ad\, Y$ has only zeros along its diagonal, when represented as a matrix with respect to a suitable basis, if $\alpha+\beta\neq 0$, hence

$\mathrm{tr}(ad\, X \cdot ad\, Y) = 0$, i.e., $X \perp Y$.

Let $\boldsymbol{N}(\boldsymbol{H})$ (resp. $\boldsymbol{N}(\boldsymbol{H}_C)$) denote the normalizer of $\boldsymbol{H}$ in G (resp. of $\boldsymbol{H}_C$ in G_C). Then $\boldsymbol{H}$ (resp. $\boldsymbol{H}_C$) is its own centralizer in G (resp. in G_C); in the case of $\boldsymbol{H}$ this follows from its construction [27 : p. 247]; as for $\boldsymbol{H}_C$, it is defined over $\boldsymbol{R}$ so the same is true of its centralizer Z in G_C, therefore Z_R is Zariski-dense in Z, which is connected, and $Z_R \subset G$ which is compact, hence Z_R is compact and therefore (Theorem 16) is connected, and because Z_R centralizes $\boldsymbol{H}$, we have $Z_R \subset \boldsymbol{H}$, so that Z is contained in the Zariski-closure $\boldsymbol{H}_C$ of $\boldsymbol{H}$. This is another way of seeing, as asserted previously, that $\boldsymbol{H}_C$ is a maximal torus in G_C. The group $W = W(\boldsymbol{H}_C) = \boldsymbol{N}(\boldsymbol{H}_C)/\boldsymbol{H}_C$ is the finite Weyl group of G_C. Since $\boldsymbol{H}$ is Zariski-dense in $\boldsymbol{H}_C$, we have $\boldsymbol{N}(\boldsymbol{H}) \subset \boldsymbol{N}(\boldsymbol{H}_C)$, where $\boldsymbol{N}(\boldsymbol{H})$ denotes the normalizer of $\boldsymbol{H}$ in G, and $\boldsymbol{N}(\boldsymbol{H}) \cap \boldsymbol{H}_C$ consists of elements which centralize $\boldsymbol{H}$, hence must be $\boldsymbol{H}$ itself. Therefore, $\boldsymbol{N}(\boldsymbol{H})/\boldsymbol{H}$ is canonically isomorphic to a subgroup of W. And now in fact it is not hard to see that that subgroup is all of W, i.e., W is canonically isomorphic to $\boldsymbol{N}(\boldsymbol{H})/\boldsymbol{H}$, so W is also called the Weyl group of G. If $w \in W$, let w operate on $\mathfrak{h}^*$ by $(w \cdot \lambda)(x) = \lambda(Ad(w)^{-1} \cdot x)$, $x \in \mathfrak{h}$, $\lambda \in \mathfrak{h}^*$; then W permutes the weights of any representation of G and leaves the Killing form restricted to $\mathfrak{h}$ invariant, hence consists of orthogonal transformations with respect to the latter. If α is a root, $\alpha \in \Lambda^*$, let r_α denote the reflection in the plane orthogonal to α with respect to the Killing form. Then W is faithfully represented as a permutation group on the roots, and as such, it is generated by $r_{\alpha_1}, \cdots, r_{\alpha_l}$. The above considerations are parallel to the treatment in Chapter 7. In Chapter 7 our discussion was for roots and weights with respect to a maximal algebraic torus, while the present discussion presents analogous considerations with respect to a compact torus. The link between the two is the remark made earlier in this section describing how characters of $\boldsymbol{H}$ of absolute value 1 correspond to rational characters of $\boldsymbol{H}_C$.

Now Λ^* is just the (discrete) character group of the (compact) group $\boldsymbol{H}$, so that for every $\lambda \in \Lambda^*$, e^λ is a one-dimensional representation of $\boldsymbol{H}$. It follows from the corollary of Theorem 15 that this is contained in the restriction to $\boldsymbol{H}$ of some finite-dimensional, irreducible representation of G. Therefore Λ^* may be identified with the set of all weights of all finite-dimensional, irreducible representations of G, and must hence be stable (as a set) under W.

Let σ be an irreducible finite-dimensional (complex) unitary representation of G with representation space V, and let λ_0 be the highest weight with respect to a linear ordering on $\mathfrak{h}^*$. Clearly $\lambda_0 \geqq w \cdot \lambda_0$ for every $w \in W$. Since $d\rho(x_\alpha) \cdot V_{\lambda_0} \subset V_{\lambda_0+\alpha}$ if $x_\alpha \in \mathfrak{g}_\alpha$, and since λ_0 is the highest weight, it follows that $d\rho(\mathfrak{g}_\alpha) V_{\lambda_0} = \{0\}$ for all positive roots α, therefore V_{λ_0} is stabilized by the parabolic subgroup of G_C, whose Lie algebra is $\mathfrak{h}_C + \sum_{\alpha>0} \mathfrak{g}_\alpha$. In fact, it is stabilized by the larger parabolic subgroup of G_C whose Lie algebra is the sum of this and of all $\mathfrak{g}_\alpha$, $\alpha < 0$, such that $\alpha \perp \lambda_0$. In what follows, we essentially reproduce the proof given in [54 : Exps. 19 and 21] of the following facts: Given any $\lambda_0 \in \mathfrak{h}^*$ such that $\lambda_0 \geqq w \cdot \lambda_0$ for all $w \in W$, there exists a unique (up to equivalence) irreducible representation σ with highest weight λ_0 and if V is the representation space, then $\dim V_{\lambda_0} = 1$; thus, in the associated projective representation, V_{λ_0} is represented by a point stabilized by a parabolic subgroup of G_C, and that point therefore has a closed orbit; from the irreducibility of σ it follows that the set of vectors $\sigma(g) \cdot V_{\lambda_0}$, $g \in G$, spans V (over $\boldsymbol{C}$).

§ 8. The Weyl character and dimension formulas [54 : Exps. 19, 21]

In the following discussion, we fix a compact Lie group G, a maximal compact torus $\boldsymbol{H}$ of G, and system $\Delta = \{\alpha_1, \cdots, \alpha_l\}$ of positive simple roots in the root system Σ for G with respect to $\boldsymbol{H}$. Denote by ρ the quantity $\frac{1}{2} \sum_{\alpha \in \Sigma_+} \alpha$, where Σ_+ is the system of positive roots. Let H_i, $i = 1, \cdots, l$, H_α', $E_{\pm\alpha}$, B, $\langle\, , \rangle$, etc., be as in (42) and (43).

LEMMA 3. *Let the simple roots be* $\{\alpha_1, \cdots, \alpha_l\}$ *and let* $r_i = r_{\alpha_i}$ *be the reflection in the plane perpendicular to* α_i. *Then* r_i *permutes the positive roots other than* α_i *among themselves and* $r_i\alpha_i = -\alpha_i$.

PROOF. Let $\alpha = \sum m_j \alpha_j$ be a positive root, so all $m_j \geqq 0$. Then $r_i\alpha = m_i'\alpha_i + \sum_{j \neq i} m_j \alpha_j$ because for any $x \in \mathfrak{h}^*$, $r_i(x) = x - 2(\alpha_i, \alpha_i)^{-1}(\alpha_i, x)\alpha_i$, and for $r_i\alpha$ to be a root, m_i' and all m_j $(j \neq i)$ must have the same sign; so if $\alpha \neq \alpha_i$, then $m_j > 0$ for some $j \neq i$, hence $m_i' \geqq 0$ and $r_i\alpha$ is a positive root, while if $\alpha = \alpha_i$, then obviously $r_i\alpha = -\alpha$.

LEMMA 4. *We have* $\rho(H_i)=1,\ i=1, \cdots, r.$

PROOF. By the preceding lemma, r_i permutes the elements of Σ_+ different from α_i and the latter is sent into $-\alpha_i$, while $r_i(\rho)=\rho-2(\alpha_i, \alpha_i)^{-1}(\alpha_i, \rho)\alpha_i$ and $H_i=H_{\alpha_i}=2(\alpha_i, \alpha_i)^{-1}H_{\alpha_i}'$, so that $2(\alpha_i, \alpha_i)^{-1}(\alpha_i, \rho)=\rho(H_i)$, in other words, $\rho(H_i)=1$.

LEMMA 5. *For* $\lambda\in\mathfrak{h}^*$ *to satisfy* $w\cdot\lambda<\lambda$ *for every* $w\in W-\{e\}$, *it is necessary and sufficient that* $\lambda(H_i)>0$ *for all* $i=1, \cdots, l$.

PROOF. For each $i=1, \cdots, l$, one has

$$r_i\lambda=\lambda-\lambda(H_i)\alpha_i, \tag{44}$$

as we see from the expression for H_i given in the proof of Lemma 4.

From this, the necessity of the condition is clear: If some $\lambda(H_i)\leqq 0$, we have $r_i\lambda\geqq\lambda$.

As for the sufficiency, we know that every $w\in W$ may be written in the form $r_{i_1}\cdots r_{i_p}$; if p is the smallest positive integer for which this is so $(w\neq e)$, we write $p=l(w)$. We have to show that if $\lambda(H_i)>0$, $i=1, \cdots, l$, then $w\cdot\lambda<\lambda$ when $l(w)>0$. We proceed by induction on $l(w)$. If $l(w)=1$, then $w=r_j$ and the result follows at once from (44). Suppose the result has already been established for $0<l(w)<p$, and let $l(w)=p\geqq 2$, $w=r_{i_1}\cdots r_{i_p}=w'r_{i_p}$, $w'=r_{i_1}\cdots r_{i_{p-1}}$. Then $w\cdot\lambda=w'r_{i_p}\cdot\lambda=w'\cdot\lambda-\lambda(H_{i_p})w'\alpha_{i_p}$. If $w'\alpha_{i_p}>0$, then $w\cdot\lambda<w'\cdot\lambda<\lambda$ (by induction hypothesis) and we are done (by the minimality of the word for w, the word for w' is minimal). Suppose $w'\alpha_{i_p}<0$. Since w is a minimal word and each r_i is of period two, we see that $i_{p-1}\neq i_p$, hence by Lemma 1, we have $r_{i_{p-1}}\alpha_{i_p}>0$. On the other hand, $w'\alpha_{i_p}<0$, so that the smallest index k for which

$$\beta_l=r_{i_l}\cdot r_{i_{l+1}}\cdots r_{i_{p-1}}\alpha_{i_p}>0$$

for all $l\geqq k$ satisfies $1<k\leqq p-1$. Then $\beta_k>0$ and $\beta_{k-1}=r_{i_{k-1}}\cdot\beta_k<0$, which by Lemma 1 can be true only if $\beta_k=\alpha_{i_{k-1}}$, therefore, $w'=w_1r_{i_{k-1}}w_2$ with $w_2\alpha_{i_p}=\beta_k=\alpha_{i_{k-1}}$. But $w_2r_\alpha w_2^{-1}=r_{w_2\alpha}$ for any root α, and therefore $w'=w_1w_2r_{\alpha_{i_p}}$ and so $w=w_1w_2$, which is of length $\leqq l(w)-2$, a contradiction of the definition of $l(w)$. Hence only the case $w'\alpha_{i_p}>0$ occurs and we are done.

Now the system of linear inequalities $\lambda(H_i)>0$, $i=1, \cdots, l$, and the system of linear inequalities $\lambda>w\cdot\lambda$ for all $w\neq e$ define the same open convex domain in $\mathfrak{h}^*$, and the closures of these domains are the same, the first being given by $\lambda(H_i) \geqq 0$, $i=1, \cdots, l$, and the second by $\lambda \geqq w \cdot \lambda$ for all $w \in W$. Therefore we have

COROLLARY. *$\lambda \geqq w \cdot \lambda$ for all $w \in W$ if and only if $\lambda(H_i) \geqq 0$ for $i=1, \cdots, l$.*

As remarked earlier, Λ^* is the set of weights of irreducible representations of G. Let σ be such a representation, V the representation space, λ a weight of σ, V_λ the corresponding weight space, and α a root. If $\mu \in \Lambda^*$ but is not a weight, define $V_\mu = \{0\}$, and put $\boldsymbol{W} = \sum_{k=-\infty}^{\infty} V_{\lambda+k\alpha}$. Since $d\sigma(\mathfrak{g}_\alpha) V_\mu \subset V_{\mu+\alpha}$, we see that $\boldsymbol{W}$ is stable under $d\sigma(X)$, where $X = (\sqrt{2}\,|\alpha|)^{-1}\pi(E_\alpha + E_{-\alpha})$, hence is stable under $\sigma(\exp X) = \exp(d\sigma(X))$. By direct calculation, using properties of the sine and cosine, we obtain

$$e^{ad\,X} \cdot Y = Y - 2\langle\alpha, \alpha\rangle^{-1}\alpha(Y) \cdot H_\alpha' = r_\alpha Y \tag{45}$$

for $Y \in \mathfrak{h}$, so if $x = \exp X$, $h \in \boldsymbol{H}$, then $xhx^{-1} = r_\alpha h r_\alpha (r_\alpha \in \boldsymbol{N}(\boldsymbol{H}))$. From this we easily deduce the fact that if $\lambda \in \Lambda^*$, then $2\langle\lambda, \alpha\rangle / \langle\alpha, \alpha\rangle \in \boldsymbol{Z}$.

Let P be the set of all $\lambda \in \mathfrak{h}^*$ such that $\lambda(H_i) \in \boldsymbol{Z}$, $i=1, \cdots, l$. As we have seen, $\Lambda^* \subset P$. Moreover, $r_i\lambda = \lambda - \lambda(H_i)\alpha_i$, so that W operates on P as well as on Λ^*. We note that $\rho = \frac{1}{2} \sum_{\alpha \in \Sigma_+} \alpha \in P$ because $\rho(H_i) = 1$, $i=1, \cdots, l$. Let

$$P_+ = \{\lambda \in P \mid \lambda(H_i) \geqq 0, \quad i=1, \cdots, l\},$$

and let $\{\eta_j\}$ be the dual basis to $\{H_i\}$, i.e., $\eta_j \in \mathfrak{h}^*$ and $\eta_j(H_i) = \delta_{ij}$; obviously $\eta_j \in P_+$, $j=1, \cdots, l$, and $\{\eta_j\}$ is a set of free generators of the semi-group P_+.

Let K be a subring of $\boldsymbol{C}$ and let A be the group ring of P over K. Every element of A may be written symbolically in the form

$$a = \sum_{\lambda \in P}{}' a_\lambda e^\lambda, \quad a_\lambda \in K, \tag{46}$$

$\sum'$ denoting finite sums, so that only a finite number of a_λ are different from zero. If in fact $\lambda \in \Lambda^*$, then e^λ is a character of $\boldsymbol{H}$, but in general e^λ are symbols for which we define $e^\lambda \cdot e^\mu = e^{\lambda+\mu}$. If A_+ is the subring of A such that $a_\lambda = 0$ for $\lambda \notin P_+$, then A_+ is a unique factorization domain of integrity. It follows that the same is true of A since every element of A can be written in the form $e^{-\lambda}a_+$, $\lambda \in P_+$,

$a_+ \in A_+$. The Weyl group W operates on this ring by: $w(\sum' a_\lambda e^\lambda) = \sum' a_\lambda e^{w\cdot\lambda}$. Viewing w as a linear transformation of $\mathfrak{h}^*$, we have $\det w = \pm 1$ since w is an orthogonal transformation. Let Q be the K-linear endomorphism of A defined by

$$Q\cdot a = \sum_{w\in W} (\det w) w\cdot a, \quad a\in A.$$

We say that a is symmetric (resp. anti-symmetric) according as $w\cdot a = a$ (resp. $w\cdot a = (\det w)a$) for all $w\in W$; denote the space of symmetric elements (resp. anti-symmetric elements) by A^+ (resp. by A^-). Then $Q\cdot a\in A^-$ for all $a\in A$, and if $a\in A^-$, we have $Q\cdot a = (\text{card } W)a$, so that Q is a linear mapping of A onto A^-; hence, $A^- = \sum'_{\lambda\in P} K\cdot Qe^\lambda$, and in fact, the sum may be restricted to $\lambda\in P_+$ by the corollary to Lemma 5. If $\lambda(H_i)=0$ for some i, then $r_i\lambda=\lambda$, and if

$$W^+ = \{w\in W \mid \det w = +1\} = \{w \mid l(w)\equiv 0 (\text{mod } 2)\},$$

then $W = W^+ \cup W^+ r_i$ so that

$$Qe^\lambda = \sum_{w\in W^+} (\det w\cdot e^{w\lambda} + \det(wr_i)e^{wr_i\lambda}) = 0,$$

since $\det(r_i) = -1$. Thus, by Lemma 5, if $P_* = \{\lambda\in P \mid \lambda > w\cdot\lambda,\ w\in W - \{e\}\}$, we have

$$A^- = \sum_{\lambda\in P_*}{}' K\cdot Qe^\lambda.$$

Suppose $a\in A$, α a root such that $r_\alpha a = -a$. Then a is a linear combination of the terms

$$e^\lambda - r_\alpha e^\lambda = e^\lambda(1 - e^{-\lambda(H_\alpha)\cdot\alpha}), \tag{47}$$

and $\lambda(H_\alpha)\in \boldsymbol{Z}$. Hence, in A, $e^\lambda - r_\alpha e^\lambda$ is divisible by $1-e^{-\alpha}$, and so a is divisible by $1-e^{-\alpha}$. So if $a\in A^-$, then $r_\alpha a = -a$ for all roots α, and $\prod_{\alpha\in\Sigma_+} (1-e^{-\alpha})$ divides a, while $\rho\in P_*$ so that e^ρ is a unit of A; therefore, a is divisible by the quantity

$$D = e^\rho \prod_{\alpha\in\Sigma_+} (1-e^{-\alpha}) = \prod_{\alpha\in\Sigma_+} (e^{\alpha/2} - e^{-\alpha/2}). \tag{48}$$

It follows from Lemma 3 that $r_iD = -D$, $i = 1, \cdots, l$, so $D = \sum_{-\rho\leqq\lambda\leqq\rho} c_\lambda Qe^\lambda$, and each term is divisible by D since it belongs to A^-, i.e., $Qe^\lambda = b_\lambda D$, and comparing terms one obtains $b_\lambda = 0$ if $\lambda \neq \pm\rho$, so that $D = \text{const.}\, Qe^\rho$, and comparing highest terms here gives that the constant is one, or $D = Qe^\rho$. If $a\in A^+$, then $a\cdot D\in A^-$ and so

$$a \cdot D = \sum_{\lambda \in P_*}' a_\lambda Q e^\lambda = \sum_{\lambda \in P_+}' a_\lambda Q e^{\lambda+\rho}, \tag{49}$$

since $\lambda(H_i) > 0$, $\lambda(H_i) \in \boldsymbol{Z}$, and $\rho(H_i) = 1$ imply $(\lambda - \rho)(H_i) \geqq 0$. Hence

$$a = \sum_{\lambda \in P_+}' a_\lambda (Q e^{\lambda+\rho} / Q e^\rho), \tag{50}$$

because $Qe^{\lambda+\rho}$, being anti-symmetric, is divisible by D.

One knows [16b: pp. 410–414; 27: pp. 211–213] that if $\boldsymbol{H}$ is a maximal compact torus in G, then every element of G is conjugate to an element of $\boldsymbol{H}$; and denoting by $\boldsymbol{H}_r$ the set $\{h \in \boldsymbol{H} \mid Z_G(h) = \boldsymbol{H}\}$ and by G_r the set of $g \in G$ conjugate to an element of $\boldsymbol{H}_r$ by an inner automorphism, the set $\boldsymbol{H} - \boldsymbol{H}_r$ (resp. $G - G_r$) consists of a finite number of proper analytic subsets of H (resp. of G). Thus, the mapping $(g, h) \to ghg^{-1}$ sends $G \times \boldsymbol{H}$ onto G, and since $\boldsymbol{H}$ is commutative, we have a naturally associated mapping φ of $(G/\boldsymbol{H}) \times \boldsymbol{H}$ onto G such that for $g \in G_r$, $\varphi^{-1}(g)$ consists of $\omega = \text{card } W$ points of $(G/\boldsymbol{H}) \times \boldsymbol{H}_r$. So if $P = G/\boldsymbol{H}$, then up to a set of measure zero, $P \times \boldsymbol{H}$ is an ω-fold covering of G. If one computes [54: Exp. 21; cf. also 27: Chap. VII, §§ 3, 4 and Chap. X, § 1] the Jacobian determinant of φ at points of $P \times \boldsymbol{H}_r$, one obtains the integral formula:

$$\begin{aligned} \int_G f(g) dg &= \omega^{-1} \int_{\boldsymbol{H}} \prod_{\alpha \in \Sigma} (e^\alpha(h) - 1) \int_P f(ghg^{-1}) d(g \cdot p) dh \\ &= \omega^{-1} \int_{\boldsymbol{H}} \prod_{\alpha \in \Sigma} (e^\alpha(h) - 1) \int_G f(ghg^{-1}) dg\, dh, \quad f \in L(G), \end{aligned} \tag{51}$$

where dg and dh are the Haar measures on G and on $\boldsymbol{H}$, normalized as usual, and $d(g \cdot p)$, the corresponding measure at the point $g\boldsymbol{H}$ on P. In particular, let f be a central function on G; then (51) takes the form

$$\int_G f(g) dg = \omega^{-1} \int_{\boldsymbol{H}} D(h) \overline{D(h)} f(h) dh, \tag{52}$$

where $D(h) = \prod_{\alpha \in \Sigma_+} (e^{\alpha/2}(h) - e^{-\alpha/2}(h))$ (noting that a central function is determined by its values on $\boldsymbol{H}$; the given form of $D(h)$ involves expressions that may be meaningless, in which case its value is to be taken from the values of hyperbolic functions on the Lie algebra $\mathfrak{h}$ (cf. p. 117)).

Letting (σ, V) be a finite-dimensional representation of G, put $\chi(g) = \text{tr}(\sigma(g))$, $m_\lambda = \dim V_\lambda$ for every weight λ. Then $\chi(h) = \sum_\lambda m_\lambda e^\lambda(h)$,

$h \in \boldsymbol{H}$. By obvious properties of the action of W, we see that χ is a finite trigonometric series in e^{λ} with non-negative, integral coefficients, and this expansion (as a function on $\boldsymbol{H}$) is *unique* and *invariant under* W and $m_{\lambda}=m_{w\cdot\lambda}$, $w \in W$. Then $D\cdot\chi$ is anti-symmetric and equals $\sum_{\mu\in P_*} a_{\mu} Q\cdot e^{\mu}$; and since the distinct expressions $Q\cdot e^{\mu}$ have no terms in common, a_{μ}, as the coefficient of e^{μ}, is an integer, because the coefficients of χ and of D are integers. Moreover, since σ is irreducible, we have from (19)

$$1=\int_G |\chi(g)|^2 dg=\omega^{-1}\int_H |D\cdot\chi|^2(h)dh=\omega^{-1}\int_H |\sum a_{\mu} Qe^{\mu}(h)|^2 dh.$$

Since $\mu\neq\lambda$ implies $e^{\mu}\perp e^{\lambda}$ and since Qe^{λ} and Qe^{μ} have no terms in common, it follows that

$$\int_H |Qe^{\mu}|^2 dh=\omega \quad \text{and} \quad \sum |a_{\mu}|^2=1.$$

Since $a_{\mu}\in \boldsymbol{Z}$, it follows that $a_{\mu}=0$ for all but one μ, say of the form $\Lambda+\rho$, $\Lambda\in P_+$, and $a_{\Lambda+\rho}=\pm 1$. Therefore $\chi=\pm Qe^{\Lambda+\rho}/Qe^{\rho}$, and in fact the sign must be $+$ since the coefficients of χ are $\geqq 0$ and the coefficient of e^{Λ} in $Qe^{\Lambda+\rho}/Qe^{\rho}$ is $+1$; this also shows that $m_{\Lambda}=+1$. Moreover, since $\Lambda\in P_+$, we have $\Lambda\geqq w\cdot\Lambda$ for all $w\in W$, and therefore Λ (or e^{Λ} as a character on $\boldsymbol{H}$) is the highest weight of σ. Since inequivalent irreducible representations have orthogonal characters, we see that σ is characterized equivalently by its highest weight Λ or by its character χ.

Suppose, conversely, that for some $\Lambda\in P_+\cap\Lambda^*$, which we denote by Λ_+, there did not exist a representation σ with it as highest weight. Then $\phi=Qe^{\Lambda+\rho}/Qe^{\rho}$ is a W-invariant function on $\boldsymbol{H}$, so it can be extended uniquely to a continuous central function on G, and is orthogonal to all $Qe^{\Lambda'+\rho}/Qe^{\rho}$, $\Lambda'\neq\Lambda$, therefore is orthogonal to *all* characters of representations. But as a central function, it has an approximation by linear combinations of characters, and this leads at once to a contradiction. Hence Λ must be the character of some irreducible representation.

If $\Lambda\in\Lambda_+$, define

$$d_{\Lambda}=\prod_{\alpha\in\Sigma_+}\langle\Lambda+\rho,\alpha\rangle/\langle\rho,\alpha\rangle \tag{53}$$

and

(54) $$\chi_{\Lambda}=Qe^{\Lambda+\rho}/Qe^{\rho}.$$

As G is a compact Lie group, it follows from section 6 that every finite-dimensional continuous representation of G is (real) analytic, if we provide G with any of the (equivalent) analytic structures it inherits from a faithful, finite-dimensional, linear representation. Hence, each χ_{Λ} is the restriction to H of the analytic (and hence all the more C^{∞}) function $\mathrm{tr}(\sigma(g))$ on G.

THEOREM 17. *Let* $f\in C^{\infty}(G)$. *Then*

$$f(e)=\sum_{\Lambda\in\Lambda_{+}} d_{\Lambda}\int_{G} f(g)\chi_{\Lambda}(g)dg \quad (\textit{Plancherel formula}).$$

COROLLARY 1. *If* $\chi_{\Lambda'}$ *is the character of the irreducible representation* (σ, V), *then* $\dim V=\chi_{\Lambda'}(e)=d_{\Lambda'}$.

Proof that Corollary 1 follows from Theorem 17: Let $f=\overline{\chi_{\Lambda'}}$. Clearly

$$\dim V=\chi_{\Lambda'}(e)=\overline{\chi_{\Lambda'}(e)}=\sum_{\Lambda} d_{\Lambda}\int_{G} \overline{\chi_{\Lambda'}(g)}\chi_{\Lambda}(g)dg=d_{\Lambda'}$$

by the orthogonality relations.

COROLLARY 2. $\dim V=O(|\Lambda'|^{M})$, *where* M *is the number of (non-zero) positive roots and* $|\Lambda'|$ *is the length of* Λ' *with respect to the Killing form.*

(The $O(\ )$ notation is the standard one from analysis.)

PROOF. This follows immediately from Corollary 1 and (53).

PROOF OF THEOREM 17. [54: Exp. 21]. The functions χ_{Λ} are central functions, hence if f' is defined by

$$f'(g')=\int_{G} f(gg'g^{-1})dg,$$

it follows that $\int_{G} f'(g)\chi_{\Lambda}(g)dg=\int_{G} f(g)\chi_{\Lambda}(g)dg$; therefore, it is sufficient to treat the case of f being a central function in $C^{\infty}(G)$.

To avoid interrupting the discussion later, we introduce here the symmetric algebras $S(\mathfrak{h})$ and $S(\mathfrak{h}^{*})$ of $\mathfrak{h}$ and of $\mathfrak{h}^{*}$ respectively (q.v. infra, Chapter 9, section 1). There is a canonical pairing $\langle\,,\rangle$ of these which puts the space of elements of degree m of one into duality with the space of elements of degree m of the other by defining

$$\langle x_1\cdots x_m, x_1'\cdots x_m'\rangle=(m!)^{-1}\sum_{\pi}\prod_{j=1}^{m}\langle x_j, x'_{\pi(j)}\rangle, \quad x_j\in\mathfrak{h},\ x_j'\in\mathfrak{h}^*,$$

where π runs over all permutations of $\{1,\cdots,m\}$. The Weyl group W is made to act linearly on $S(\mathfrak{h}^*)$ by defining for a monomial $x_1'\cdots x_m'$, $w(x_1'\cdots x_m')=(wx_1')\cdots(wx_m')$. We need

LEMMA 6. *If p is an anti-symmetric polynomial of $S(\mathfrak{h}^*)$ with respect to W, then p is divisible by $\boldsymbol{D}=\prod_{\alpha\in\Sigma+}\alpha$. In particular, if $p\neq 0$, then* $\deg p\geqq \operatorname{card}\Sigma_+$.

PROOF. Clearly $\boldsymbol{D}$ is anti-symmetric itself (Lemma 3). Let p be anti-symmetric and let α be any positive root; then $p-r_\alpha(p)=2p$. But if M is any monomial, $M=\beta_1\cdots\beta_k$, where $\beta_1,\cdots,\beta_k$ are positive roots, we have

$$r_\alpha(\beta_j)=\beta_j-2(\alpha,\alpha)^{-1}(\alpha,\beta_j)\alpha,$$

so that $M-r_\alpha M$ is immediately seen to be divisible by α. Since no positive root is a multiple of any other, the positive roots are mutually prime linear forms as elements of $\mathfrak{h}^*\subset S(\mathfrak{h}^*)$. Hence, p is divisible by their product $\boldsymbol{D}$.

Henceforth, using the Killing form to identify $\mathfrak{h}$ and $\mathfrak{h}^*$, we shall write elements of either $\mathfrak{h}$ or $\mathfrak{h}^*$ as though they were elements of $\mathfrak{h}^*$.

Now we return to the proof of Theorem 17. Let f be a central function of class C^∞ on G. By (52), one has

$$\int_G f(g)\chi_\Lambda(g)dg=\omega^{-1}\int_H f(h)\overline{D(h)}\sum_{w\in W}(\det w)e^{w(\Lambda+\rho)}(h)dh. \tag{55}$$

Since $\boldsymbol{D}$ is anti-symmetric, we have for $\sigma\in W$

$$\begin{aligned}(\det\sigma)\prod_{\alpha\in\Sigma+}\langle\Lambda+\rho,\alpha\rangle&=\prod_{\alpha\in\Sigma+}\langle\Lambda+\rho,\sigma^{-1}\alpha\rangle\\ &=\prod_{\alpha\in\Sigma+}\langle\sigma(\Lambda+\rho),\alpha\rangle.\end{aligned}\tag{56}$$

If $\lambda\in\Lambda_*$ is such that $\langle\lambda,\alpha\rangle\neq 0$ for all roots $\alpha\neq 0$, simple transitivity of W on the Weyl chambers implies that λ is conjugate under W to just one μ such that $\mu(H_i)>0$ for all i, and $\mu\in\Lambda^*\cap P_+=\Lambda_+$ satisfies $\mu(H_i)\in\boldsymbol{Z}$ for all i, hence $(\mu-\rho)(H_i)\geqq 0$ for all i and therefore $\mu-\rho\in\Lambda_+$.

Therefore, each $\lambda \in \Lambda^*$ such that $\langle \lambda, \alpha \rangle \neq 0$ for all $\alpha \in \Sigma$ can be written *uniquely* in the form $\sigma(\Lambda+\rho)$, $\Lambda \in \Lambda_+$, $\sigma \in W$. Therefore, we have

$$\sum_{\Lambda \in \Lambda_+} d_\Lambda \int_G \chi_\Lambda(g) f(g) dg$$

$$(57) \qquad = \omega^{-1} \prod_{\alpha \in \Sigma_+} \langle \rho, \alpha \rangle^{-1} \sum_{\Lambda \in \Lambda_+, \sigma \in W} \det \sigma \int_H f(h) D(h) q(\Lambda+\rho, \sigma, h) dh$$

$$= (\omega \prod_{\alpha \in \Sigma_+} \langle \rho, \alpha \rangle)^{-1} \sum_{\lambda \in \Lambda^*} \int_H f(h) \overline{D(h)} \prod_{\alpha \in \Sigma_+} \langle \lambda, \alpha \rangle e^\lambda(h) dh,$$

where $q(\Lambda+\rho, \sigma, h) = \prod_{\alpha \in \Sigma_+} \langle \Lambda+\rho, \alpha \rangle e^{\sigma(\Lambda+\rho)}(h)$, and where the term for λ is zero if λ is perpendicular to any root. As functions of λ, the coefficients $\prod_{\alpha \in \Sigma_+} \langle \lambda, \alpha \rangle$ increase like a polynomial function of λ, but $f \cdot \bar{D}$, being of class C^∞ on $\boldsymbol{H} \approx (\boldsymbol{R}/\boldsymbol{Z})^r$, has Fourier coefficients

$$(58) \qquad \int_H f(h) \overline{D(h)} e^\lambda(h) dh$$

that go to zero at ∞ as a function of λ at least as fast as $|\lambda|^{-N}$ for each $N > 0$, as one sees from repeated integration by parts (see below, Chapter 9, section 6), hence their multiples by any fixed polynomial function of λ have the same property (but for an immaterial shift in the exponent by the degree of the fixed polynomial). Therefore, the series in (57) converges absolutely, rearrangement of the terms is legitimate, and we may consider as a distribution (i.e., continuous linear functional on $C^\infty(\boldsymbol{H})$) the sum (convergent *as a series of distributions*)

$$T = \overline{D(h)} dh \sum_{\lambda \in \Lambda^*} e^\lambda(h) \prod_{\alpha \in \Sigma_+} \langle \lambda, \alpha \rangle$$

$$(59) \qquad = \sum_{\lambda \in \Lambda^*, \sigma \in W} e^{\lambda - \sigma \cdot \rho}(h) \det \sigma \cdot \prod_{\alpha \in \Sigma_+} \langle \lambda, \alpha \rangle \, dh$$

$$= \sum_{\mu \in \Lambda^*} (\sum_{\sigma \in W} \det \sigma \cdot \prod_{\alpha \in \Sigma_+} \langle \mu + \sigma \cdot \rho, \alpha \rangle) e^\mu(h) dh.$$

If n is the number of positive roots, one has

$$(60) \qquad \prod_{\alpha \in \Sigma_+} \langle \mu + \sigma \cdot \rho, \alpha \rangle = (n!)^{-1} \langle (\mu + \sigma \cdot \rho)^n, \prod_{\alpha \in \Sigma_+} \alpha \rangle,$$

the last $\langle \, , \rangle$ being the dual pairing of $S(\mathfrak{h})$ and $S(\mathfrak{h}^*)$. In $S(\mathfrak{h}^*)_Q$, one has

$$(n!)^{-1} \sum_{\sigma \in W} \det \sigma \cdot (\mu + \sigma \cdot \rho)^n = \sum_p (p!(n-p)!)^{-1} \mu^{n-p} \sum_{\sigma \in W} \det(\sigma) \sigma(\rho^p),$$

and the last appearing sum is anti-symmetric and of degree $p \leqq n$, and so is zero except for $p=n$ and is divisible by $\boldsymbol{D}$ by Lemma 6; therefore, (60) is independent of μ and thus the sum of the expressions (60), each multiplied by $\det \sigma$, as σ runs over W, is independent of μ and is equal to

$$(n!)^{-1} \sum_{\sigma \in W} (\det \sigma) \prod_{\alpha \in \Sigma_+} \langle \sigma \cdot \rho, \alpha \rangle = (n!)^{-1} \prod_{\alpha \in \Sigma_+} \langle \rho, \alpha \rangle,$$

where $\omega = \operatorname{card} W$. Moreover, the distribution $\sum_{\mu \in \Lambda^*} e^{\mu}(h) dh$ (where we multiply by $f(h)$ and integrate) is, in view of the Fourier (=Peter-Weyl) expansion of C^∞-functions on $\boldsymbol{H}$, just ε_e, the distribution that assigns to $f \in C^\infty(G)$ its value $f(e)$ at the identity e. Hence

$$T = (\omega \prod_{\alpha \in \Sigma_+} \langle \rho, \alpha \rangle) \varepsilon_e,$$

so that

$$\sum_{\Lambda \in \Lambda_+} d_\Lambda \int_G \chi_\Lambda(g) f(g) dg = \omega^{-1} \prod_{\alpha \in \Sigma_+} \langle \rho, \alpha \rangle^{-1} \int_H f(h) dT(h) = \varepsilon_e(f) = f(e),$$

(in the symbolism of distributions) which is what we had to prove.

CHAPTER 9
SOME WORK OF HARISH-CHANDRA

§1. The universal enveloping algebra

Let V be a finite-dimensional vector space over a field k of characteristic zero. Let $T(V)$ be the tensor algebra of V,

$$T(V)=\sum_{n\geqq 0}' T_n(V), \tag{1}$$

where $T_n=T_n(V)$ is spanned by the products $v_1\otimes\cdots\otimes v_n$ of degree n, $T_0(V)=k$, and $T(V)$ is supplied with the obvious non-commutative multiplication such that $T_m\cdot T_n\subset T_m\otimes T_n=T_{m+n}$. Let I be the two-sided ideal in $T(V)$ generated by all elements of the form $X\otimes Y-Y\otimes X$, $X, Y\in T$. The associative, commutative factor algebra $S=S(V)=T(V)/I$ is called the symmetric algebra of V. The image in S of T_n is denoted by $S_n=S_n(V)$ and $S=\sum_{n\geqq 0}' S_n$. Then S is the polynomial algebra $k[x_1, \cdots, x_l]$, where $x_1, \cdots, x_l$ is a basis of V, and the dual space S^k of S may be identified with the space of polynomial functions on V.

If $\mathfrak{A}$ is any algebra over k with a law of composition $\circ$, then a derivation of $\mathfrak{A}$ is a k-linear mapping δ of $\mathfrak{A}$ into itself such that $\delta(k)=0$ and $\delta(f\circ g)=\delta f\circ g+f\circ\delta g$. Thus, if $\mathfrak{A}=\sum_{n\geqq 0}\mathfrak{A}_n$ is a graded algebra generated over $k=\mathfrak{A}_0$ by its elements of degree 1, then δ is determined by its effect on $\mathfrak{A}_1$.

Let $\mathfrak{g}$ be a finite-dimensional Lie algebra over k and let J be the two-sided ideal generated in $T(\mathfrak{g})$ by all elements of the form $X\otimes Y-Y\otimes X-[X, Y]$, $X, Y\in\mathfrak{g}=T_1$. The associative algebra $U=U(\mathfrak{g})=T(\mathfrak{g})/J$ is called the universal enveloping algebra, because if σ is any representation of $\mathfrak{g}$ in an associative algebra $\mathfrak{A}$, then (by the definition of such a representation) $\sigma([a, b])=\sigma(a)\sigma(b)-\sigma(b)\sigma(a)$, so that J is contained in the kernel of the natural extension of σ to $T(\mathfrak{g})$; hence, σ may be factored uniquely by way of U:

$$\begin{array}{ccc} \mathfrak{g} & \xrightarrow{\sigma} & \mathfrak{A} \\ & {\scriptstyle i}\searrow \quad \nearrow{\scriptstyle \sigma^U} & \\ & U(\mathfrak{g}) & \end{array} \tag{2}$$

where $i: \mathfrak{g} \to U(\mathfrak{g})$ is the natural mapping, σ^U is an algebra homomorphism, and $\sigma^U \circ i = \sigma$. If U_n is the natural image of T_n in U, then $U = \sum'_{n \geqq 0} U_n$. We shall see later that $\ker(i) = 0$. If $\mathfrak{g}$ is Abelian, then $U(\mathfrak{g})$ is isomorphic as an algebra to $S(\mathfrak{g})$, but not generally otherwise.

LEMMA 1. *Let ρ be a representation of the Lie algebra $\mathfrak{g}$ in the Lie algebra $\mathfrak{gl}(V)$ of linear endomorphisms of some vector space V over k. Then there exists a unique representation ρ^U of $U(\mathfrak{g})$ as an associative algebra in $\mathfrak{gl}(V)$ such that for all $X \in \mathfrak{g}$ we have $\rho(X) = \rho^U \circ i(X)$. Conversely, given a representation σ of U in $\mathfrak{gl}(V)$, there exists a unique representation $\rho_{\mathfrak{g}}$ of $\mathfrak{g}$ in $\mathfrak{gl}(V)$ such that $(\rho_{\mathfrak{g}})^U = \sigma$.*

PROOF. The first part is a special case of the universality property formulated above.

Conversely, given a representation

$$\sigma : U \to \mathfrak{gl}(V),$$

define $\rho(X) = \sigma \circ i(X)$ for $X \in \mathfrak{g}$. Then it is easily verified that $\rho = \sigma | \mathfrak{g}$ is the representation we want.

It is well-known [16b: p. 333] that any finite-dimensional Lie algebra $\mathfrak{g}$ over a field of characteristic zero has a faithful representation as a Lie subalgebra of $\mathfrak{gl}(V)$ for some finite-dimensional vector space V. In fact, in the cases we consider this will be obvious without appealing to a general result. Hence, we may always assume that $\mathfrak{g}$ is a Lie subalgebra of some $\mathfrak{gl}(V)$.

Now let $\mathfrak{g}$ be the Lie algebra of the Lie group G. If $X \in \mathfrak{g}$, then X determines a tangent vector $\tilde{X}$ at e defined for a C^∞-function f in a neighborhood of e by $\tilde{X}f(e) = \lim_{t \to 0} t^{-1}(f(\exp tX) - f(e))$, which in turn determines a left-invariant vector field, also denoted by $\tilde{X}$, on G whose value at $g \in G$ is given for a function f of class C^∞ by $\tilde{X}_g f(g) = \lim_{t \to 0} t^{-1}(f(g \exp tX) - f(g))$. By using Taylor's expansion [27 : pp. 96–97], one verifies that for the Lie bracket, one has $\widetilde{[X, Y]} = [\tilde{X}, \tilde{Y}]$. Hence, $X \xrightarrow{\tau} \tilde{X}$ is a Lie algebra representation of $\mathfrak{g}$ in the associative algebra $\mathscr{F}$ of left-invariant differential operators on G. Thus, there exists, by Lemma 1, a unique homomorphism (of associative algebras) $\alpha: U(\mathfrak{g}) \to \mathscr{F}$ such that $\alpha \circ i(X) = \tau(X) = \tilde{X}$ for all $X \in \mathfrak{g}$. Now the differential operators at a point are the elements of the symmetric algebra of the tangent space and [27: p. 98 and pp. 391–392] if $X_1, \cdots, X_n$ is a basis of $\mathfrak{g}$ and if $P \in S(\mathfrak{g})$, then the assignment $P \mapsto D_P$ defined by

$$D_P f(g) = [P(\partial_1, \cdots, \partial_n) f(g \exp(x_1 X_1 + \cdots + x_n X_n))](0),$$

where $\partial_i = \partial / \partial x_i$, is a linear mapping of $S(\mathfrak{g})$ onto $\mathscr{F}$; its kernel is $\{0\}$ since, obviously, the differential operators

$$(\partial/\partial x_1)^{e_1} \cdots (\partial/\partial x_n)^{e_n}, \quad e_1, \cdots, e_n \in \boldsymbol{Z},$$

for different sets of $e_i \geqq 0$, are linearly independent (as one may see, for instance, by considering their restrictions to polynomials in the coordinates).

Let $M = x_{i_1} \cdots x_{i_m}$ and $M' = x_{i_1'} \cdots x_{i_m'}$ be two monomials in T of the same degree, where, because of associativity, we always agree to write a power of x_i as $x_{i_1} \cdots x_{i_r}$, $i_1 = \cdots = i_r = i$. Suppose the sets of indices $i_1, \cdots, i_m$ and $i_1', \cdots, i_m'$ are the same. Since $x_{i_j} x_{i_{j+1}} \equiv x_{i_{j+1}} x_{i_j} + [x_{i_j}, x_{i_{j+1}}] \bmod J$, it follows that, viewed as elements of U, i.e., as representatives of their residue classes modulo J, M and M' differ by terms of lower degree in U. An element in U of the form

$$\sum_{0 \leqq r \leqq s} \sum_{(i_1, \cdots, i_r)} a_{i_1 \cdots i_r} x_{i_1} \cdots x_{i_r}$$

is called "normal" if for every r, and set of indices $(i_1, \cdots, i_r)$ we have $i_1 \leqq \cdots \leqq i_r$, and is called "canonical" if the coefficients $a_{i_1 \cdots i_r}$ are symmetric in the indices $i_1, \cdots, i_r$. From our previous remarks, it is clear that, modulo terms of lower degree, any homogeneous element of degree s is congruent to at least one normal element of degree s and to at least one canonical element of degree s. The normal elements are a vector space spanned by all monomials $x_1^{e_1} \cdots x_n^{e_n}$, and the canonical elements, a vector space spanned by all the expressions

$$(3) \qquad (s!)^{-1} \sum_{\pi\{i\}} x_{i_1} \cdots x_{i_s} \quad \text{(non-commutative multiplication)},$$

where $i_1, \cdots, i_s$ run over all permutations $\pi\{i\}$ of a given set of s indices. The critical fact is

LEMMA 2. *J contains no canonical element (or normal element) different from zero.*

There are several proofs of this [27 : p. 98; 16b : p. 343]. Most use the existence of a faithful representation of $\mathfrak{g}$ in some $\mathfrak{gl}(V)$, $\dim V < +\infty$, identifying U with the algebra $\mathfrak{F}$ of left-invariant differential operators on a Lie group whose Lie algebra is $\mathfrak{g}$ and using the fact that $P \mapsto D_P$ is a vector space isomorphism of $S(\mathfrak{g})$ onto $\mathfrak{F}$. (Cf. [27 : pp. 90–99; 54 : Exp. 1].) For a proof not employing the existence of such a faithful representation of $\mathfrak{g}$, see [32b : pp. 151–160].

As a consequence, there is exactly one canonical and one normal element in each residue class of T modulo J, and that canonical (or normal) element is called the canonical (or normal) expression of the elements of that residue class. Therefore, we have a vector space isomorphism λ, called the canonical mapping, of $S(\mathfrak{g})$, spanned by the elements

(4) $$(s!)^{-1} \sum_{\pi\{i\}} x_{i_1} \cdots x_{i_s} \quad (\textit{commutative} \text{ multiplication}),$$

onto $U(\mathfrak{g})$, where λ maps the element (4) onto the similarly denoted element (3). In particular, every linear element (i.e., element of $\mathfrak{g}$) is both canonical and normal, and i is a vector space injection of $\mathfrak{g}$ into $U(\mathfrak{g})$. Furthermore, it follows easily from the remarks preceding Lemma 2 that we have

LEMMA 3. *If F and F' are homogeneous elements of degrees k and k' respectively, then*

$$\lambda(F)\lambda(F') \equiv \lambda(FF')$$

modulo terms of degrees $\leq k + k' - 1$.

We note, finally, that the triple of homomorphisms relating $\mathfrak{F}$, $S(\mathfrak{g})$, and $U(\mathfrak{g})$ gives a commutative diagram; in other words, we have that $\alpha\lambda$ is the same as the mapping $P \mapsto D_P$.

Now we have an automorphism $Ad\, g$ of $\mathfrak{g}$ for every $g \in G$. By Lemma 1, this extends to an automorphism, also denoted by $Ad\, g$, of $U(\mathfrak{g})$. It also extends to an automorphism, to be denoted by $Ad_S g$ of $S(\mathfrak{g})$. Likewise for $X \in \mathfrak{g}$, $ad\, X$ (operating on $\mathfrak{g}$) may be extended to a derivation, denoted by $\sigma(X)$, of $S(\mathfrak{g})$, and to a derivation $ad\,X$ of $U(\mathfrak{g})$. Let $I(\mathfrak{g})$ denote the set of elements of $S(\mathfrak{g})$ invariant under (i.e., annihilated by) $\sigma(X)$ for all $X \in \mathfrak{g}$.

LEMMA 4. *If* $F \in S(\mathfrak{g})$ *and* $X \in \mathfrak{g}$, *then*

$$\lambda(\sigma(X)\cdot F) = [X, \lambda(F)] = X\lambda(F) - \lambda(F)X.$$

PROOF. [26a : Lemma 11]. (Cf. also [16b : p. 345], where this and the fact that λ is a vector space isomorphism are proved together.) If $Y \in \mathfrak{g}$, denote its image in $S(\mathfrak{g})$ by $\bar{Y}$. It is sufficient to prove the lemma when F is of the form $\bar{Y}_1 \cdots \bar{Y}_k$ (product in $S(\mathfrak{g})$), $Y_i \in \mathfrak{g}$. Let $Y_i' = [X, Y_i]$. Now $\sigma(X)$ is a derivation, so $\sigma(X)F = \sum_i \bar{Y}_1 \cdots \bar{Y}_i' \cdots \bar{Y}_k$. As $ad\,X$ is a derivation of U, then for any permutation π of $\{1, 2, \cdots, k\}$,

$$[X, Y_{\pi(1)} \cdots Y_{\pi(k)}] = \sum_i Y_{\pi(1)} \cdots [X, Y_{\pi(i)}] \cdots Y_{\pi(k)}.$$

Thus

$$\lambda(\sigma(X)\cdot F) = (k!)^{-1} \sum_{\pi, i} Y_{\pi(1)} \cdots Y_{\pi(i)}' \cdots Y_{\pi(k)} = [X, \lambda(F)]. \tag{5}$$

LEMMA 5. *The center* $\mathfrak{Z}(\mathfrak{g})$ *of* $U(\mathfrak{g})$ *is the image under* λ *of the linear space of all* $F \in S(\mathfrak{g})$ *for which we have* $\sigma(X)\cdot F = 0$ *for every* $X \in \mathfrak{g}$.

This is a corollary of Lemma 4 [26a]. If G is connected, then $\sigma(X)\cdot F = 0$ for all $X \in \mathfrak{g}$ if and only if $Ad_S g \cdot F = F$ for all $g \in G$. From this [27 : p. 393] one has

LEMMA 6. *Let* G *be connected. Then* $\mathfrak{Z}(\mathfrak{g})$ *may be identified with the space of left-invariant differential operators* D *satisfying* $Ad(g)\cdot D = D$ *for all* $g \in G$, *i.e., may be identified with the space of left-and-right-invariant differential operators on* G.

A second-order two-sided-invariant differential operator $\mathfrak{C}$ on G is obtained in the following way: Let $\{X_i\}$ be a base of $\mathfrak{g}$ and let $\{X^i\}$ be the dual base with respect to the Killing form. Put

$$\mathfrak{C} = \sum_i X_i X^i \in U(\mathfrak{g}).$$

Then $\mathfrak{C} \in \mathfrak{Z}(\mathfrak{g})$ because the Killing form is $Ad\,G$-invariant. The invariant differential operator $\mathfrak{C}$ is called the Casimir operator. If ρ is a finite-dimensional representation of $\mathfrak{g}$, then clearly $\rho(\mathfrak{C})\rho(X)=\rho(X)\rho(\mathfrak{C})$ for all $X\in\mathfrak{g}$. If ρ is irreducible, it follows from Schur's lemma that $\rho(\mathfrak{C})=\omega_\rho I_\rho$, where I_ρ is the identity transformation of the representation space of ρ. It will be proved later that if $\mathfrak{g}$ is semi-simple, then ω_ρ is a positive rational number that increases as the squared length of the highest weight of ρ.

Most of the above can be found in papers of Harish-Chandra; the following two lemmas are from a paper of the same author [26a], and are in preparation for the proof of a result of that paper, from which most of sections 1, 2, and 3 are directly adapted.

If V is a vector space and is the direct sum of two subspaces V' and V'', then $S(V')$ and $S(V'')$ are naturally contained in $S(V)$, and $F'\otimes F''\mapsto F'F''$ determines a vector space isomorphism of $S(V')\otimes S(V'')$ onto $S(V)$. The proof of this is straightforward.

LEMMA 7. *Let the Lie algebra* $\mathfrak{g}$ *be the direct sum of two vector subspaces* $\mathfrak{g}'$ *and* $\mathfrak{g}''$. *Then the linear mapping determined by the assignment* $F'\otimes F''\mapsto\lambda(F')\lambda(F'')$, $F'\in S(\mathfrak{g}')$, $F''\in S(\mathfrak{g}'')$, *is a vector space isomorphism of* $S(\mathfrak{g}')\otimes S(\mathfrak{g}'')$ *onto* $U(\mathfrak{g})$. *Moreover, for* $d\geqq 0$, *the sum of the vector spaces* $\lambda(S_e(\mathfrak{g}))$ *for* $e\leqq d$ *is identical with the sum of the spaces* $\lambda(S_{d'}(\mathfrak{g}'))\lambda(S_{d''}(\mathfrak{g}''))$, $d'+d''\leqq d$.

(N.B. If A and B are subsets of the algebra $\mathfrak{A}$, then AB is the subspace of $\mathfrak{A}$ spanned by the products ab, $a\in A$, $b\in B$, and if k is a positive integer, then A^k will denote the subspace of $\mathfrak{A}$ spanned by the k-fold products of elements of A.)

PROOF. Choosing bases β, β', and β'' of $\mathfrak{g}$, of $\mathfrak{g}'$, and of $\mathfrak{g}''$ respectively such that $\beta=\beta'\cup\beta''$, let M, M', and M'' denote respectively the sets of monomials in the elements of β, of β', and of β''. In the various graded algebras, let the subscript d denote the elements of degree d, and so on.

On the one hand, it is obvious that the vector space sum of the spaces $\lambda(S_e(\mathfrak{g}))$, $e\leqq d$, is spanned by the elements $\lambda(\mu)$, $\mu\in M_e$, $e\leqq d$. On the other hand, it follows at once from Lemma 3 that if $\mu'\in S_{d'}(\mathfrak{g}')$ and $\mu''\in S_{d''}(\mathfrak{g}'')$, then $\lambda(\mu'\mu'')-\lambda(\mu')\lambda(\mu'')$ belongs to the sum of the spaces $\lambda(S_m(\mathfrak{g}))$, $m<d'+d''$. In particular, if $\mu\in M_{d+1}$, then μ may be

written in one and only one way in the form $\mu'\mu''$ with $\mu' \in M'_{d'}$, $\mu'' \in M''_{d''}$, where $d+1=d'+d''$, and therefore $\lambda(\mu)$ lies in the sum of the spaces

$$\lambda(S_{d'}(\mathfrak{g}'))\lambda(S_{d''}(\mathfrak{g}'')), \quad d'+d'' \leqq d+1. \tag{6}$$

It follows thus in straightforward fashion that for any d, the elements $\lambda(\mu')\lambda(\mu'')$ with $\mu' \in M'_{d'}$, $\mu'' \in M''_{d''}$, $d'+d'' \leqq d$, form a base of either of the two vector space sums mentioned in the lemma. From these considerations, the lemma follows easily.

Let $\mathfrak{k}$ be a subalgebra of $\mathfrak{g}$. Assume the Killing form of $\mathfrak{g}$ to be non-degenerate and to have non-degenerate restriction to $\mathfrak{k}$, and let $\mathfrak{p}$ be the orthogonal complement of $\mathfrak{k}$ in $\mathfrak{g}$ with respect to the Killing form. Moreover, assume that $ad\, k(\mathfrak{p}) \subset \mathfrak{p}$ for every $k \in \mathfrak{k}$. Let $\mathfrak{X}$ be the subalgebra of $U=U(\mathfrak{g})$ generated by 1 and $\mathfrak{k}$. It is easily seen that $\lambda(S(\mathfrak{k}))=\mathfrak{X}$ and that $\mathfrak{X}$ may be canonically identified with $U(\mathfrak{k})$. Let $\mathfrak{P}=\lambda(S(\mathfrak{p}))$, $\mathfrak{P}_d=\lambda(S_d(\mathfrak{p}))$.

LEMMA 8. $\lambda(S_d(\mathfrak{p})S(\mathfrak{k})) \subset \sum_{e \leqq d} \mathfrak{P}_e\mathfrak{X}$ *and* $\lambda(S_d(\mathfrak{p})S(\mathfrak{k}))\lambda(S_{d'}(\mathfrak{p})S(\mathfrak{k}))$ *is a vector subspace of* $\sum_{e \leqq d+d'} \mathfrak{P}_e\mathfrak{X}$.

PROOF. The first assertion is implied by Lemma 7. For the second, it is enough to prove that

$$(\mathfrak{P}_d\mathfrak{X})(\mathfrak{P}_{d'}\mathfrak{X}) \subset \sum_{e \leqq d+d'} \mathfrak{P}_e\mathfrak{X}. \tag{7}$$

With $\mathfrak{p}^d$ as determined by the N.B. of Lemma 7, one shows by induction on d that $\mathfrak{p}^d\mathfrak{X} \supset \mathfrak{X}\mathfrak{p}^d$, $d \geqq 0$. This is obvious for $d=0$, follows for $d=1$ from the facts that $ad\, \mathfrak{k}$ normalizes $\mathfrak{p}$ and that $\mathfrak{X}'=\{u \in U \mid u\mathfrak{p} \subset \mathfrak{p}\mathfrak{X}\}$ is a subalgebra of U containing 1 and $\mathfrak{k}$, and follows for general d by an easy induction. Thus $\mathfrak{X}\mathfrak{p}^d \subset \mathfrak{p}^d\mathfrak{X}$. Clearly $\mathfrak{P}_d \subset \mathfrak{p}^d$; therefore,

$$(\mathfrak{P}_d\mathfrak{X})(\mathfrak{P}_{d'}\mathfrak{X}) \subset \mathfrak{p}^{d+d'}\mathfrak{X},$$

so that the proof will be complete if we show $\mathfrak{p}^d\mathfrak{X}$ is contained in the sum of the spaces $\mathfrak{P}_e\mathfrak{X}$, $e \leqq d$. This is true for $d=0$ and 1, and follows for all d by induction on d, using Lemma 7 and (6).

Let $\mathfrak{g}^*$ be the algebraic dual space of $\mathfrak{g}$. The Killing form B, being non-degenerate, determines an isomorphism I of $\mathfrak{g}$ onto $\mathfrak{g}^*$: $I(X)(Y)=B(X, Y)$. Extend I to an isomorphism of $S(\mathfrak{g})$ with $S(\mathfrak{g}^*)$. By the canonical pairing of pp. 126–127, we may identify

$S(\mathfrak{g}^*)$ with the dual space $S(\mathfrak{g})^*$ of $S(\mathfrak{g})$ and $S(\mathfrak{g})^*$ with the polynomial functions on $\mathfrak{g}$ (p. 130). If $F \in S(\mathfrak{g}^*)$, denote its image in $S(\mathfrak{g})^*$ by F'; then $F \mapsto F'$ is a vector space isomorphism independent of the choice of basis of $\mathfrak{g}^*$ (the polynomial changes "contravariantly" with a change of basis and so continues to represent the same function). F' is a polynomial function on $\mathfrak{g}$ (by definition). Thus we have isomorphisms

$$S(\mathfrak{g}) \cong S(\mathfrak{g}^*) \cong (\text{polynomial functions on } \mathfrak{g}).$$

As $B(\mathfrak{k}, \mathfrak{p}) = 0$, $\mathfrak{p}$ is naturally isomorphic with the space of linear functionals on $\mathfrak{g}$ that vanish on $\mathfrak{k}$, so that $S(\mathfrak{p})$ is isomorphic to the algebra of polynomial functions on $\mathfrak{p}$. A polynomial in n variables, representing an element of $S(\mathfrak{g})$ (with respect to a basis of $\mathfrak{g}$ composed of a basis of $\mathfrak{p}$ and of a basis of $\mathfrak{k}$) determines by restriction an element of $S(\mathfrak{p})$ by setting the last $n-p$ variables equal to zero ($p = \dim \mathfrak{p}$, $n = \dim \mathfrak{g}$).

§ 2. Quasi-semi-simple modules

Let $\mathfrak{k}$ be a Lie algebra and let ρ be a representation of $\mathfrak{k}$ on an arbitrary vector space V, and let W be a subspace of V such that $\rho(k) \cdot W \subset W$ for every $k \in \mathfrak{k}$. Denote by ρ_W the representation so obtained of $\mathfrak{k}$ on W. Then ρ_W is called simple (resp. semi-simple) if $\dim W < +\infty$ and if W is a simple (resp. semi-simple) $\mathfrak{k}$-module.

Let Ω be the set of all equivalence classes of simple, finite-dimensional representations of $\mathfrak{k}$. If $\mathfrak{d} \in \Omega$, let $V_\mathfrak{d}$ be the algebraic sum (no convergence) of all $\rho(\mathfrak{k})$-stable, simple $\mathfrak{k}$-submodules W of V such that $\rho_W \in \mathfrak{d}$ (and put $V_\mathfrak{d} = \{0\}$ if no such W exists). Then $V_\mathfrak{d}$ is called the *isotypic* subspace of type $\mathfrak{d}$ in V.

LEMMA 9. *Let W be a subspace of V satisfying $\rho(\mathfrak{k}) W \subset W$. Then*

$$W \cap \left(\sum_{\mathfrak{d} \in \Omega} V_\mathfrak{d}\right) = \sum_{\mathfrak{d} \in \Omega} (W \cap V_\mathfrak{d}),$$

and the (algebraic) sum $\sum_{\mathfrak{d} \in \Omega} V_\mathfrak{d}$ is direct.

PROOF. The elements of $\sum_{\mathfrak{d} \in \Omega} V_\mathfrak{d}$ are finite sums of elements from simple components. Hence, for any $x \in \sum_{\mathfrak{d} \in \Omega} V_\mathfrak{d}$, the smallest ρ-stable subspace V_x containing x is contained in the sum of a finite number of simple $\mathfrak{k}$-submodules of V. Therefore, V_x is semi-simple [16a: Chap. VI], is thus the direct sum of simple $\mathfrak{k}$-modules, and we obtain, consequently, $V_x = \sum_{\mathfrak{d}} (V_x \cap V_{\mathfrak{d}})$. Utilizing this for each element x of $W \cap (\sum_{\mathfrak{d}} V_{\mathfrak{d}})$, one obtains the first part of the lemma. Let $\Omega_1 \subset \Omega$; if $x \in \sum_{\mathfrak{d} \in \Omega_1} V_{\mathfrak{d}}$, then V_x is a subspace of the sum of a finite number of simple $\mathfrak{k}$-modules W' such that the class of $\rho_{W'}$ is in Ω_1, and so every simple summand of ρ_{V_x} is an element of a class in Ω_1, and consequently $V_x \cap V_{\mathfrak{d}} = \{0\}$ if $\mathfrak{d} \notin \Omega_1$. Hence, the sum $\sum_{\mathfrak{d} \in \Omega} V_{\mathfrak{d}}$ is direct.

The representation ρ is called quasi-semi-simple if $V = \sum_{\mathfrak{d} \in \Omega} V_{\mathfrak{d}}$; for this to be so, it is necessary and sufficient that every $x \in V$ belong to some semi-simple $\mathfrak{k}$-submodule of V.

It follows at once from Schur's lemma that if V and W are semi-simple $\mathfrak{k}$-modules and if $\lambda: V \to W$ is a surjective $\mathfrak{k}$-homomorphism, then every simple $\mathfrak{k}$-submodule of W is isomorphic to one in V.

LEMMA 10. *Let ρ be quasi-semi-simple, let $V_0 = \{x \in V \mid \rho(k)x = 0$ for all $k \in \mathfrak{k}\}$ and let V_1 be the span of the set $\{\rho(k) \cdot v \mid k \in \mathfrak{k},\ v \in V\}$. Then $V = V_0 \oplus V_1$ (direct sum).*

PROOF. Let $\mathfrak{d}_0$ be the class of the one-dimensional zero representation, so that $V_0 = V_{\mathfrak{d}_0}$; clearly $V_1 \subset \sum_{\mathfrak{d} \neq \mathfrak{d}_0} V_{\mathfrak{d}}$, because if $X \in \mathfrak{k}$ and $v \in V$, which equals $\sum_{\mathfrak{d}} V_{\mathfrak{d}}$ (because ρ is quasi-semi-simple), then $\rho(X)v$ is a finite sum $\sum \rho(X)v_i$, each v_i belonging to a simple $\mathfrak{k}$-submodule, and $\rho(X)v_i \neq 0$ implies $v_i \in V_{\mathfrak{d}}$ for some $\mathfrak{d} \neq \mathfrak{d}_0$. Hence $V_0 + V_1$ is a direct sum. Suppose W is a $\mathfrak{k}$-simple submodule of V and $\rho_W \in \mathfrak{d} \neq \mathfrak{d}_0$; then $W \subset V_1$. Moreover, $W \supset W'$, where W' is the span of the set $\{\rho(X) \cdot w \mid X \in \mathfrak{k},\ w \in W\}$, and the latter is $\rho(\mathfrak{k})$-stable and is different from $\{0\}$ since $\rho_W \notin \mathfrak{d}_0$; hence, $W = W' \subset V_1$, from which the lemma follows.

Henceforth, we shall assume $\mathfrak{k}$ to be reductive.

Let (ρ, V) and (ρ', V') be representations of $\mathfrak{k}$. Then $V \otimes V'$ is the space of a representation denoted by $\rho \dotplus \rho'$ (tensor sum) and defined by

$$(\rho \dotplus \rho')(X)(v \otimes v') = (\rho(X)v) \otimes v' + v \otimes (\rho'(X)v'), \quad X \in \mathfrak{k}. \tag{8}$$

LEMMA 11. *If ρ and ρ' are quasi-semi-simple, then $\rho \dotplus \rho'$ is* *quasi-semi-simple.*

PROOF. It follows at once from the definitions of quasi-semi-simple and of tensor product that it is enough to consider only the *finite-dimensional case.* By section 5 of Chapter 7, one recognizes that the only problem is to see that the elements of the center $\mathfrak{c}$ of $\mathfrak{k}$ are represented by semi-simple linear transformations under $\rho \dotplus \rho'$. Since ρ and ρ' are semi-simple, we may choose bases of V and of V' with respect to which $\rho(\mathfrak{c})$ and $\rho'(\mathfrak{c})$ respectively consist of diagonal matrices; then from these we obtain a basis of $V \otimes V'$ with respect to which all elements of $(\rho \dotplus \rho')(\mathfrak{c})$ are diagonal. Q.E.D.

DEFINITION. *Let V be a finite-dimensional vector space (over a field k of characteristic zero, as always here) and let V^* be its dual space. If ρ is a representation of a Lie algebra $\mathfrak{g}$ on V, there is a naturally induced representation ρ^* of $\mathfrak{g}$ on V^* determined by the identity*

$$\langle \rho(X)v,\ v^* \rangle + \langle v,\ \rho^*(X)v^* \rangle \equiv 0, \quad X \in \mathfrak{g},\ v \in V,\ v^* \in V^*,$$

where $\langle\ ,\ \rangle$ is the natural pairing of V and V^*. The representation ρ^* is called the representation contragredient to ρ.

If V is identified with V^* by means of a non-degenerate, symmetric, bilinear form B on V, then ρ^* is identified with a representation of $\mathfrak{g}$ on V, called the representation contragredient to ρ with respect to B. Over an algebraically closed field, the equivalence class of the contragredient representation of ρ with respect to such a bilinear form is uniquely determined; however, if k is not algebraically closed, two representations of $\mathfrak{g}$ defined over k and contragredient to ρ in this way may not be equivalent *over k.*

LEMMA 12. *Let (ρ, V) and (ρ', V') be finite-dimensional semi-simple representations of $\mathfrak{k}$ and assume that ρ is irreducible (i.e., that V is a simple $\mathfrak{k}$-module). Then ρ' contains a representation contragredient to ρ if and only if $\rho \dotplus \rho'$ contains the class of $\mathfrak{d}_0$ with positive multiplicity.*

PROOF. This is the Lie algebra version of a result stated in Chapter 8, section 3. Let (σ, W) be any finite-dimensional representation of a Lie algebra (or Lie group) and let (σ^*, W^*) be the contragredient representation; then σ is fully reducible if and only if σ^* is. In particular, σ has an invariant vector if and only if σ^* does.

Now there is a canonical identification of $V^* \otimes V'$ with $\mathrm{Hom}(V, V')$, and the representation ρ of $\mathfrak{k}$ on V becomes a representation ρ_1 of $\mathfrak{k}$ on V^* when we select a vector space isomorphism of V with V^*, given, for example, by choice of a non-degenerate bilinear form on V. If $\rho + \rho'$ contains $\mathfrak{d}_0$, that is the same thing as saying that there exists a non-zero $\phi \in V \otimes V'$ such that $(\rho + \rho')(X)(\phi) = 0$ for all $X \in \mathfrak{k}$ or, equivalently, that there is a non-zero $\phi_1 \in V^* \otimes V'$ such that $(\rho_1 + \rho')(X)(\phi_1) = 0$ for all $X \in \mathfrak{k}$. But ϕ_1 may be viewed as a $\mathfrak{k}$-invariant element of $\mathrm{Hom}(V, V')$ when V is made into a $\mathfrak{k}$-module by means of ρ_1^*, thus the representation (ρ', V') must contain an irreducible $\mathfrak{k}$-submodule isomorphic to (ρ_1^*, V), which is equivalent to (ρ^*, V) since $\rho \sim \rho_1$. Since the steps in the proof may be reversed in evident fashion, the proof is complete.

§3. **The main result** [26a]

Let $\mathfrak{g}_0$ be a semi-simple Lie algebra over $\boldsymbol{R}$ and let $\mathfrak{k}_0$, $\mathfrak{p}_0$, $\mathfrak{g}$, $\mathfrak{k}$, etc. be as in Chapter 7, section 5. Then $\mathfrak{k}$ is reductive. Moreover, let $U = U(\mathfrak{g})$ and let $\mathfrak{X}$ be the subalgebra of U generated by 1 and $\mathfrak{k}$, so that $\mathfrak{X}$ may be identified with $U(\mathfrak{k})$. Then the representations of $\mathfrak{k}$ and of $\mathfrak{X}$ are in a natural one-to-one correspondence, so we agree to denote corresponding representations of $\mathfrak{X}$ and of $\mathfrak{k}$ by the same symbol. Let (π, V) be a representation of U; then π gives rise to a representation of $\mathfrak{k}$ which we denote by π_1. If $\mathfrak{d} \in \Omega$, define $V_{\mathfrak{d}}$ as before to be the $\mathfrak{d}$-isotypic subspace of V with respect to π_1.

LEMMA 13. *The subspace* $V_* = \sum_{\mathfrak{d} \in \Omega} V_{\mathfrak{d}}$ *of* V *satisfies* $\pi(U) \cdot V_* \subset V_*$.

PROOF. Since 1 and $\mathfrak{g}$ generate U, it is sufficient to prove that $\pi(\mathfrak{g}) \cdot V_* \subset V_*$. Since $\mathfrak{g} = \mathfrak{k} + \mathfrak{p}$ and obviously (from the definitions) $\pi(\mathfrak{k}) \cdot V_* \subset V_*$, one need only show that $\pi(\mathfrak{p}) \cdot V_* \subset V_*$. Define σ as the adjoint representation of $\mathfrak{k}$ on $\mathfrak{p}$ and let ρ be the representation of $\mathfrak{k}$ on V_* coming from π. Then $\sigma + \rho$ is a representation of $\mathfrak{k}$ on $\mathfrak{p} \otimes V_*$. Now σ is semi-simple because $\mathfrak{k}$ is reductive in $\mathfrak{g}$. If $x \in V_*$, then x belongs to a semi-simple $\mathfrak{k}$-submodule W of V_*, $\dim W < +\infty$, and

$\mathfrak{p}\otimes W$ is a semi-simple $\mathfrak{k}$-module with respect to the restriction ν of $\sigma+\rho$ to $\mathfrak{k}$. Define the linear mapping $\lambda: \mathfrak{p}\otimes W\rightarrow V$ by $\lambda(Y\otimes w)=\pi(Y)w$. We have

$$\pi(X)\lambda(Y\otimes w)=\pi(X)\pi(Y)w=\pi([X, Y])w+\pi(Y)\pi(X)w$$

and

$$\lambda\nu(X)(Y\otimes w)=\lambda([X, Y]\otimes w+Y\otimes\rho(X)w)$$
$$=\pi([X, Y])w+\pi(Y)\pi(X)w=\pi(X)\lambda(Y\otimes w).$$

Hence, $\pi(X)\lambda=\lambda\nu(X)$, so that λ is a $\mathfrak{k}$-homomorphism of the semi-simple $\mathfrak{k}$-module $\mathfrak{p}\otimes W$ into the $\mathfrak{k}$-module V. It is thus clear that $\lambda(\mathfrak{p}\otimes W)$ is a semi-simple $\mathfrak{k}$-module. As $\pi(\mathfrak{p})\cdot x\subset\lambda(\mathfrak{p}\otimes W)$, and the latter is a semi-simple $\mathfrak{k}$-submodule of V, one has $\pi(\mathfrak{p})\cdot x\subset V_*$, as desired.

If T is a linear endomorphism of the vector space V, T may be extended uniquely to a derivation d_T of the symmetric algebra $S(V)$, mapping S_d into itself for each $d\geqq 0$, such that $T\rightarrow d_T$ is linear and $d_{[T_1, T_2]}=[d_{T_1}, d_{T_2}]$. In section 1, we have defined $\sigma(X)$ as a derivation of $S(\mathfrak{g})$ for $X\in\mathfrak{g}$. Thus σ is a Lie algebra representation of $\mathfrak{g}$ on $S(\mathfrak{g})$; denote its restriction to $\mathfrak{k}$ by $\sigma_{\mathfrak{k}}$.

LEMMA 14. *The representation $\sigma_{\mathfrak{k}}$ of $\mathfrak{k}$ on $S(\mathfrak{g})$ is quasi-semi-simple.*

PROOF. Let $\sigma_d(X)$ be the restriction of $\sigma(X)$ to $S_d(\mathfrak{g})$. Now σ_d is a representation of $\mathfrak{g}$ of finite dimension. As $\mathfrak{k}$ is reductive in $\mathfrak{g}$ and $\mathfrak{g}$ is semi-simple, the representation $\sigma_d|\mathfrak{k}$ of $\mathfrak{k}$ on $S_d(\mathfrak{g})$ is semi-simple (cf. the discussion preceding Lemma 2, Chapter 7, section 5). Since

$$S(\mathfrak{g})=\sum_{d\geqq 0}{}' S_d(\mathfrak{g}) \quad \text{(restricted direct sum)}, \tag{9}$$

the lemma is immediate.

Denote by $\mathfrak{A}$ an associative algebra with unit 1, and let $\mathfrak{J}$ be a left ideal in $\mathfrak{A}$. Then $\mathfrak{A}^*=\mathfrak{A}/\mathfrak{J}$ is an $\mathfrak{A}$-module and we let π_2 be the left-regular representation of $\mathfrak{A}$ on it. Let $\mathfrak{k}$ be a finite-dimensional subspace of $\mathfrak{A}$ such that $\mathfrak{k}$ is a Lie subalgebra of $\mathfrak{A}$ under the bracket defined by $[X, Y]=XY-YX$. The restriction of π_2 determines a natural representation π of $\mathfrak{k}$ on $\mathfrak{A}^*$, as a Lie algebra.

THEOREM 18. [26a: p. 195]. *Denote by $\mathfrak{Y}$ a left ideal in $\mathfrak{X}$, the subalgebra of U generated by* 1 *and* $\mathfrak{k}$. *Assume $\mathfrak{X}/\mathfrak{Y}$ to be finite-dimensional* *and assume the natural representation of* $\mathfrak{k}$ *on* $\mathfrak{X}/\mathfrak{Y}$ *to be semi-simple. Then the natural representation of* $\mathfrak{k}$ *on* $U^* = U/U\mathfrak{Y}$ *is quasi-semi-simple. Let* $\mathfrak{d} \in \Omega$ *and let* $U^*_{\mathfrak{d}}$ *be the* $\mathfrak{d}$*-isotypic subspace of* U^* *with respect to the natural representation* π *of* $\mathfrak{k}$ *on* U^*. *Then* $U^*_{\mathfrak{d}}$ *is a module of finite type over the center* $\mathcal{Z}$ *of* U.

PROOF. We first prove that $U^*_{\mathfrak{d}}$ is in fact a $\mathcal{Z}$-module. Let $x \in U^*_{\mathfrak{d}}$ and let W be the minimal $\pi(\mathfrak{k})$-invariant subspace of U^* containing x. By definition of $U^*_{\mathfrak{d}}$, W is a finite-dimensional, semi-simple $\mathfrak{k}$-module, and every simple summand of W is in $U^*_{\mathfrak{d}}$. If $z \in \mathcal{Z}$, the assignment $y \mapsto zy$ maps W onto a vector subspace zW of U^*. If $X \in \mathfrak{k}$, then $Xzy = zXy \in zW$, so that $\pi(\mathfrak{k})zW \subset zW$ and $\pi(X)zy = z(\pi(X)y)$. Hence zW is a semi-simple $\mathfrak{k}$-module and by Schur's lemma applied to the intertwining operator $y \mapsto zy$ between W and zW, all the simple summands of zW are in $U^*_{\mathfrak{d}}$. Thus $zx \in U^*_{\mathfrak{d}}$, and so $U^*_{\mathfrak{d}}$ is a $\mathcal{Z}$-module.

A. We want now to show that the natural representation of $\mathfrak{k}$ on U^* is quasi-semi-simple. We have

$$\lambda : S(\mathfrak{g}) \to U.$$

Since $\mathfrak{Y} \subset U\mathfrak{Y}$ $(1 \in U)$, one has a canonical linear mapping $f \mapsto f^*$ of $\mathfrak{X}/\mathfrak{Y}$ into U^*. Define Γ to be the linear mapping of $S(\mathfrak{p}) \otimes (\mathfrak{X}/\mathfrak{Y})$ into U^* given by $\Gamma(F \otimes f) = \lambda(F)f^*$ for $F \in S(\mathfrak{p})$ and $f \in \mathfrak{X}/\mathfrak{Y}$.

Assertion (a): Γ *is a linear isomorphism of* $S(\mathfrak{p}) \otimes (\mathfrak{X}/\mathfrak{Y})$ *onto* U^*.

PROOF of (a). The mapping λ gives a vector space isomorphism of $S(\mathfrak{k})$ onto $\mathfrak{X}$. Then for suitable indexing sets I and J, with a distinguished element $0 \in I$, we may choose bases $\{s_i\}$, $i \in I$, of $S(\mathfrak{p})$ with $s_0 = 1$, and $\{t_j\}$, $j \in J$, of $S(\mathfrak{k})$, such that for some subset J' of J, $\{\lambda(t_j)\}_{j \in J'}$ is a basis of $\mathfrak{Y}$. Put $J'' = J - J'$. Lemma 7 implies that $\{\lambda(s_i)\lambda(t_j)\}$, $i \in I$, $j \in J$, is a basis of U, and since $\mathfrak{Y}$ is an ideal in $\mathfrak{X}$, it is easy to see that $\{\lambda(s_i)\lambda(t_j)\}$, $i \in I$, $j \in J'$, is a basis of $U\mathfrak{Y}$ and that $\{\lambda(s_i)\lambda(t_j)\}$, $i \in I$, $j \in J''$, is a basis of U modulo $U\mathfrak{Y}$. As $\mathfrak{X}$ is spanned by $\lambda(s_0)\lambda(t_j)$ $(= \lambda(t_j))$, $j \in J$, we have $U\mathfrak{Y} \cap \mathfrak{X} = \mathfrak{Y}$, so that $\mathfrak{X}/\mathfrak{Y}$ is canonically isomorphic (as a vector space) to its image in U^*. For $j \in J''$, let $\lambda^*(t_j)$ be the residue class of $\lambda(t_j)$ modulo $\mathfrak{Y}$ or modulo $U\mathfrak{Y}$. The elements $\lambda(s_i)\lambda^*(t_j)$, $i \in I$, $j \in J''$, being a basis of U^*, those with $i = 0$

being a basis of $\mathfrak{X}/\mathfrak{Y}$, and as $\Gamma(s_i \otimes \lambda^*(t_j)) = \lambda(s_i)\lambda^*(t_j)$, assertion (a) now follows easily.

Let σ be defined as it was just before Lemma 14. According to the latter lemma, σ is a quasi-semi-simple representation of $\mathfrak{k}$, and since $ad(\mathfrak{k}) \cdot \mathfrak{p} \subset \mathfrak{p}$, it follows that $\sigma(\mathfrak{k}) \cdot S(\mathfrak{p}) \subset S(\mathfrak{p})$. For $X \in \mathfrak{k}$, let $\rho(X) = \sigma(X) \mid S(\mathfrak{p})$. By hypothesis, the natural representation μ of $\mathfrak{k}$ on $\mathfrak{X}/\mathfrak{Y}$ is of finite degree and semi-simple, hence, by Lemmas 9 and 11, the tensor sum $\rho \dotplus \mu$ is a quasi-semi-simple representation of $\mathfrak{k}$ on $S(\mathfrak{p}) \otimes (\mathfrak{X}/\mathfrak{Y})$. From the definition of $\dotplus$, from Lemma 4, which implies that $\lambda(\rho(X)F) = [X, \lambda(F)]$, and from the definition of Γ, one proves easily by direct calculation that

$$((\rho \dotplus \mu)(X))(F \otimes f) = (\rho(X)F) \otimes f + F \otimes Xf, \quad F \in S(\mathfrak{p}),\ f \in \mathfrak{X}/\mathfrak{Y}, \tag{10}$$

and that

$$\Gamma((\rho \dotplus \mu)(X)) = \pi(X)\Gamma, \quad X \in \mathfrak{k}, \tag{11}$$

Γ being a vector space isomorphism. Hence, π is equivalent to $\rho \dotplus \mu$ and is therefore quasi-semi-simple.

B. Let (ν, V) be a simple representation of $\mathfrak{k}$ in the class $\mathfrak{d}^*$ of representations contragredient to those of $\mathfrak{d}$. An element $\phi \in U^* \otimes V$ is, by definition, an invariant of the representation $\pi \dotplus \nu$ of $\mathfrak{k}$ if $((\pi \dotplus \nu)(\mathfrak{k}))\phi = 0$; let $\mathcal{J}$ be the vector space of such invariants. If $\phi \in \mathcal{J}$, $\phi \neq 0$, we may write $\phi = \sum_{i=1}^{m} b_i^* \otimes v_i$, $b_i^* \in U^*$, where $\{v_1, \cdots, v_m\}$ is a base of V; and it follows from Schur's lemma, via the proof of Lemma 12, that if V^* is the span of $b_1^*, \cdots, b_m^*$, then V^* is stable under $\pi(\mathfrak{k})$ and if $\pi_{V^*}(X) = \pi(X) \mid V^*$, then $\dim V^* = \dim V$ and π_{V^*} is a representation contragredient to ν. Thus, $\pi_{V^*} \in \mathfrak{d}$, and $b_i^* \in U^*_{\mathfrak{d}}$, $i = 1, \cdots, m$. Of course, if $\phi = 0$, then all $b_i^* = 0 \in U^*_{\mathfrak{d}}$. On the other hand, if V^* is a $\mathfrak{k}$-submodule of U^* such that π_{V^*} is in $\mathfrak{d}$, then π_{V^*} is contragredient to ν; so again from the proof of Lemma 12, there is seen to exist a base $b_1^*, \cdots, b_m^*$ of V^* with the property that $\phi = \sum_{i=1}^{m} b_i^* \otimes v_i$ is an invariant of the representation $\pi \dotplus \nu$.

As $U^* = U/U\mathfrak{Y}$ is a U-module, we can make $U^* \otimes V$ into a U-module by the formula

$$u(u^* \otimes v) = (uu^*) \otimes v, \quad u \in U,\ u^* \in U^*,\ v \in V. \tag{12}$$

If $z \in \mathcal{Z}$, then left multiplication by z commutes with $(\pi+\nu)(X)$, $X \in \mathfrak{k}$, since only the first component in the tensor product is involved. Hence the module $\mathcal{J}$ of invariants of $\pi+\nu$ is a $\mathcal{Z}$-module. Suppose $\mathcal{J}$ is a $\mathcal{Z}$-module of finite type and let $\mathcal{J}=\sum_{k=1}^{q} \mathcal{Z}\phi_k$, $\phi_k=\sum_{i=1}^{m} b_{k_i}{}^* \otimes v_i$, $b_{k_i}{}^* \in U^*$; then all $b_{k_i}{}^*$ belong to the $\mathfrak{d}$-isotypic subspace of U^*, and every $b^* \in U^*{}_{\mathfrak{d}}$ is either zero or is in a finite-dimensional, $\pi(\mathfrak{k})$-stable subspace in $U^*{}_{\mathfrak{d}}$, hence is a finite sum of elements $b_\alpha{}^*$ such that each $b_\alpha{}^*$ is an element of a basis of some irreducible $\mathfrak{k}$-module of type $\mathfrak{d}$, hence appears in some invariant of the form $\phi=b_\alpha{}^* \otimes v+\cdots$. Therefore, $U^*{}_{\mathfrak{d}}$ is generated as a $\mathcal{Z}$-module by the finite number of elements $b_{k_i}{}^*$. Therefore, it is sufficient to prove that $\mathcal{J}$ is a $\mathcal{Z}$-module of finite type.

One has the (vector space) isomorphism Γ: $S(\mathfrak{p})\otimes(\mathfrak{X}/\mathfrak{Y}) \rightarrow U^*$ defined by $\Gamma(F\otimes f)=\lambda(F)f^*$. From this one has an isomorphism

$$S(\mathfrak{p})\otimes(\mathfrak{X}/\mathfrak{Y})\otimes V \cong U^* \otimes V \tag{13}$$

given by

$$F\otimes f\otimes v \mapsto \Gamma(F\otimes f)\otimes v, \quad F \in S(\mathfrak{p}),\ f \in \mathfrak{X}/\mathfrak{Y},\ v \in V. \tag{14}$$

The linear isomorphism in (13) and (14) is again denoted by Γ. Define $W=(\mathfrak{X}/\mathfrak{Y})\otimes V$ and define the representations ξ, μ', and π' by $\mu'=\mu+\nu$, $\pi'=\pi+\nu$, and $\xi=\rho+\mu'$. If $X \in \mathfrak{k}$ and $A \in S(\mathfrak{p})\otimes W$, we have $\Gamma(\xi(X)A) =\pi'(X)\Gamma(A)$; in fact, if $A=F\otimes f\otimes v$, then we have

$$\begin{aligned}\Gamma(\xi(X)A)&=\Gamma(\sigma(X)F\otimes f\otimes v+F\otimes Xf\otimes v+F\otimes f\otimes \nu(X)\cdot v)\\ &=\lambda(\sigma(X)F)\cdot f^*\otimes v+\lambda(F)\cdot(Xf)^*\otimes v+\lambda(F)\cdot f^*\otimes \nu(X)\cdot v,\end{aligned}$$

and

$$\begin{aligned}\pi'(X)\Gamma(A)&=(\pi+\nu)(X)(\lambda(F)\cdot f^*\otimes v)\\ &=\pi(X)(\lambda(F)\cdot f^*)\otimes v+\lambda(F)\cdot f^*\otimes \nu(X)\cdot v\\ &=X\lambda(F)\cdot f^*\otimes v+\lambda(F)\cdot f^*\otimes\nu(X)\cdot v,\end{aligned}$$

and then we apply the formula of Lemma 4. Let

$$\mathcal{J}'=\{A \in S(\mathfrak{p})\otimes W \mid \xi(X)A=0 \text{ for all } X \in \mathfrak{k}\}.$$

Since Γ is an isomorphism, one has

$$(\pi+\nu)(\mathfrak{k})\Gamma(A)=0$$

if and only if $\xi(\mathfrak{k})A=0$; therefore, Γ maps $\mathcal{J}'$ isomorphically onto $\mathcal{J}$.

Now $S(\mathfrak{p})\otimes W$ becomes an $S(\mathfrak{p})$-module under multiplication on the first factor of the tensor product and if $X\in\mathfrak{k}$, $F\in S(\mathfrak{p})$, and $A\in S(\mathfrak{p})\otimes W$, then direct calculation gives, when $A=s\otimes w$,

$$\xi(X)(FA)=(\rho(X)F)\cdot s\otimes w+F\cdot\xi(X)(s\otimes w),$$

whence by linearization one has

$$\xi(X)(FA)=(\rho(X)F)A+F\cdot(\xi(X)A). \tag{15}$$

If Θ is the space of invariants in $S(\mathfrak{p})$ of the representation ρ, then the fact that $\rho(\mathfrak{k})$ consists of derivations of $S(\mathfrak{p})$ shows Θ to be a subalgebra of $S(\mathfrak{p})$ and (15) shows that $\mathcal{J}'$ is a Θ-module.

The adjoint representation of $\mathfrak{g}$ on itself may be extended to a representation τ of $\mathfrak{g}$ by means of derivations of $S(\mathfrak{g})$ and $\tau|\mathfrak{k}=\sigma$. Let $\mathfrak{J}$ be the set of invariants of τ; it is a subalgebra of the algebra $S(\mathfrak{g})$ of polynomial functions on $\mathfrak{g}$, and according to section 1, if we let $\mathfrak{J}_{\mathfrak{p}}$ be the set of restrictions to $\mathfrak{p}$ of elements of $\mathfrak{J}$, then $\mathfrak{J}_{\mathfrak{p}}$ is a subalgebra of $\mathfrak{J}$. If $J\in\mathfrak{J}$ and $J'=J|\mathfrak{p}$, then viewed as an element of $S(\mathfrak{p})\subset S(\mathfrak{g})$, J' is itself an element of $S(\mathfrak{g})$, obtained from J by setting the variables corresponding to the coordinates in $\mathfrak{k}$ equal to zero. But $J-J'$ is in the ideal $\mathfrak{K}=S(\mathfrak{g})\mathfrak{k}$ of $S(\mathfrak{g})$ and $\mathfrak{K}\cap S(\mathfrak{p})=0$. If $X\in\mathfrak{k}$, then $\sigma(X)J=0$ because $J\in\mathfrak{J}$, while from $[X,\mathfrak{k}]\subset\mathfrak{k}$ and $[X,\mathfrak{p}]\subset\mathfrak{p}$ we conclude that $\sigma(X)\mathfrak{K}\subset\mathfrak{K}$ and $\sigma(X)\cdot S(\mathfrak{p})\subset S(\mathfrak{p})$; therefore, $\rho(X)J'=\tau(X)J'=\sigma(X)(J'-J)$, since $\sigma(X)J=0$, and as $\sigma(X)(J'-J)\in S(\mathfrak{p})\cap\mathfrak{K}$, one has $J'\in\Theta$. Thus $\mathfrak{J}_{\mathfrak{p}}\subset\Theta$, and as $\mathfrak{J}_{\mathfrak{p}}\cdot\mathfrak{J}_{\mathfrak{p}}\subset\mathfrak{J}_{\mathfrak{p}}$, because $\mathfrak{J}$ is an algebra, it follows that $\mathfrak{J}_{\mathfrak{p}}$ is a subalgebra of Θ.

Now assume it to be known that $\mathcal{J}'$ is a $\mathfrak{J}_{\mathfrak{p}}$-module of finite type; we shall later show this to be the case. At this point, we shall show this fact to imply that $\mathcal{J}$ is a $\mathcal{Z}$-module of finite type, as desired.

Letting $S_d(\mathfrak{p})=S(\mathfrak{p})\cap S_d(\mathfrak{g})$, it is easy to see that each of the spaces $S_d(\mathfrak{p})\otimes W$ is stable under the representation ξ of $\mathfrak{k}$. It follows that there are a finite number of *homogeneous* elements $A_i\in S_{d_i}(\mathfrak{p})\otimes W$, $i=1,\cdots,r$, such that $\mathcal{J}'=\sum_{i=1}^{r}\mathfrak{J}_{\mathfrak{p}}A_i$. The main point is to see that one has $\mathcal{J}=\sum_{i=1}^{r}\mathcal{Z}\cdot\Gamma(A_i)$. Define

$$\mathcal{J}'_d=\mathcal{J}'\cap\Big(\sum_{e=0}^{d}S_e(\mathfrak{p})\otimes W\Big),\quad \mathcal{J}_d=\Gamma(\mathcal{J}'_d), \tag{16}$$

and

(17) $$\mathcal{J}_\infty = \sum_{i=1}^{r} \mathcal{Z} \cdot \Gamma(A_i).$$

Of course, $\mathcal{J} = \bigcup_{d\geqq 0} \mathcal{J}_d$, and by examining the degrees of the elements A_i (in the coordinates of $\mathfrak{p}$) and applying (16) with $d=0$, one sees that $\mathcal{J}_0 \subset \mathcal{J}_\infty$; therefore it suffices, since clearly $\mathcal{J}_\infty \subset \mathcal{J}$, to show by induction that

(18) $$\mathcal{J}_{d+1} \subset \mathcal{J}_d + \mathcal{J}_\infty, \quad d \geqq 0.$$

In what follows, we view W as being the subspace $1 \otimes W$ of $S(\mathfrak{p}) \otimes W$.

If $A \in \mathcal{J}' \cap (S_{d+1}(\mathfrak{p}) \otimes W)$, we may write $A = \sum F_i A_i$, where $F_i \in \mathfrak{J}_\mathfrak{p}$. The graded algebra $\mathfrak{J}_\mathfrak{p}$ is closed under projection on homogeneous components since $S(\mathfrak{g})$ is, and each $\tau(Z)$, $Z \in \mathfrak{g}$, is a homogeneous operator. Therefore, we may take $F_i \in S_{d+1-d_i}(\mathfrak{p})$, where $S_k(\mathfrak{g}) = \{0\}$ for negative k. Choose a base $\{w_1, \cdots, w_m\}$ of the finite-dimensional vector space $W = (\mathfrak{X}/\mathfrak{Y}) \otimes V$ and put $A_i = \sum_{j=1}^{m} s_{ij} \otimes w_j$, $s_{ij} \in S_{d_i}(\mathfrak{p})$. Now F_i is the restriction to $\mathfrak{p}$ of an element J_i in $\mathcal{J}$, $i = 1, \cdots, r$, and by throwing away homogeneous terms whose restriction to $\mathfrak{p}$ is zero, we may take J_i to be an element of $S_{d+1-d_i}(\mathfrak{g})$, where $d+1-d_i = \deg(F_i)$. It now follows easily from the definition of Γ and from Lemma 8 that we have

(19) $$\Gamma(A) - \sum_{i=1}^{r} \sum_{j=1}^{m} \lambda(J_i s_{ij}) w_j = \sum_{i,j} \lambda(F_i s_{ij} - J_i s_{ij}) w_j \in \sum_{e \leqq d} (\mathfrak{P}_e \mathfrak{X}) W.$$

Applying Lemmas 3, 7, and 8, one obtains

$$\Gamma(A) - \sum_{i,j} \lambda(J_i) \lambda(s_{ij}) w_j \in \sum_{e \leqq d} (\mathfrak{P}_e \mathfrak{X}) W.$$

Now

$$W = ((\mathfrak{X} + U\mathfrak{Y})/U\mathfrak{Y}) \otimes V \subset U^* \otimes V,$$

so $\mathfrak{X} W \subset W$ and thus we have

(20) $$\begin{aligned} \Gamma(A) - \sum_{i,j} \lambda(J_i)\lambda(s_{ij}) w_j &\in \sum_{e \leqq d} \mathfrak{P}_e W \\ &= \sum_{e \leqq d} \lambda(S_e(\mathfrak{p})) W = \Gamma(\sum_{e \leqq d} S_e(\mathfrak{p}) \otimes W). \end{aligned}$$

But

(21) $$\sum_{i,j} \lambda(J_i) \lambda(s_{ij}) w_j = \sum_i \lambda(J_i) \Gamma(A_i) \in \mathcal{J}_\infty,$$

since $\lambda(J_i) \in \mathcal{Z}$ for all i, because $J_i \in \mathfrak{J}$ and so $\tau(X) J_i = 0$ for all $X \in \mathfrak{g}$.

The left-hand side of (21) is in $\mathcal{J}_\infty$, we have $\Gamma(A)\in\mathcal{J}$, and the set at the right-hand side of (20) intersects $\mathcal{J}$ precisely in $\mathcal{J}_d$. Thus we have

$$\Gamma(A)\in\mathcal{J}_d+\mathcal{J}_\infty, \tag{22}$$

and since $\mathcal{J}_d+\mathcal{J}_\infty=\mathcal{J}_\infty$ by induction hypothesis, we have $\mathcal{J}_{d+1}\subset\mathcal{J}_\infty$, which obviously completes the proof.

PROPOSITION 1. *Θ is a $\mathfrak{J}_\mathfrak{p}$-module of finite type.*

PROOF. We retain the previous notation for $\mathfrak{g}$, for its Cartan subalgebra $\mathfrak{h}$, for $\mathfrak{k}$, $\mathfrak{p}$, etc. Put $h=\dim\mathfrak{h}_\mathfrak{p}$, $p=\dim\mathfrak{p}$, $k=\dim\mathfrak{k}$.

CLAIM: *The degree of transcendency over $\boldsymbol{C}$ of the quotient field of Θ (briefly, "the transcendence degree of Θ") is $\leqq h$.*

PROOF OF CLAIM. If $Y\in\mathfrak{p}$, denote by the corresponding lower case letter y the linear functional on $\mathfrak{g}$ defined by $y(Z)=B(Y, Z)$, B being the Killing form. As B is non-degenerate and $\mathfrak{p}$ is the orthogonal complement of $\mathfrak{k}$ with respect to B, the conditions $y(X)=0$ for all $Y\in\mathfrak{p}$ characterize the elements X of the subspace $\mathfrak{k}$ of $\mathfrak{g}$. Let $Y_1,\cdots,Y_p$ be a base of $\mathfrak{p}$ and let $y_1,\cdots,y_p$ be the corresponding linear functionals. If $\omega\in S(\mathfrak{p})$, then $\omega=F(y_1,\cdots,y_p)$, where F is a polynomial, and the conditions $\rho(X)\omega=0$ for all $X\in\mathfrak{k}$, which are necessary and sufficient for $\omega\in\Theta$, take the form

$$\sum_{i=1}^{p}\frac{\partial F}{\partial y_i}(y_1,\cdots,y_p)\cdot(\rho(X)y_i)=0. \tag{23}$$

Since B is an invariant of degree two of the adjoint representation, we have $B([X, Y], Z)=-B(Y, [X, Z])$, $X, Y, Z\in\mathfrak{g}$. Therefore, if we let $Z\in\mathfrak{p}$, $\zeta_i=y_i(Z)$, $i=1,\cdots,p$, then (23) implies that

$$\sum_{i=1}^{p}\frac{\partial F}{\partial y_i}(\zeta_1,\cdots,\zeta_p)y_i([X, Z])=0,\quad X\in\mathfrak{k}; \tag{24}$$

this follows from the interpretation of the elements of $S(\mathfrak{p})$ as polynomial functions on $\mathfrak{p}$ and from the definition of a derivation. Let $r(Z)$, $Z\in\mathfrak{p}$, be the dimension of the image of $\mathfrak{k}$ in $\mathfrak{p}$ under the linear mapping $ad\,Z$. Let $r=\max_{Z\in\mathfrak{p}} r(Z)$. Then the set $\mathcal{O}=\{Z\in\mathfrak{p}\,|\,r(Z)=r\}$ is a non-empty Zariski-open subset of $\mathfrak{p}$, that is, its complement is determined by a non-trivial set of algebraic equations on $\mathfrak{p}$. The relation (24) implies that if $F_1, \cdots, F_q \in \Theta$, then the rank $r_1(Z)$ of the Jacobian matrix

$$(\partial F_i/\partial y_j)_{i=1,\cdots,q;\ j=1,\cdots,p}$$

for $y_i = \zeta_i$ is equal to or less than $p-r(Z)$ for each $Z \in \mathfrak{p}$. If, say, $F_1, \cdots, F_t$ are algebraically independent, then the mapping $Z \mapsto (F_1(Z), \cdots, F_t(Z))$ of $\mathfrak{p}$ into t-dimensional affine space is generically surjective (in the language of algebraic geometry), and so the rank of its Jacobian matrix must be equal to t at some point of $\mathcal{O}$, thus $t \leqq p-r$. Now $k-r(Z)$ is the dimension of the kernel of $ad\ Z$ in $\mathfrak{k}$ and we know (Chapter 7, Lemma 1) that there exists $Z \in \mathfrak{p}$ such that $r(Z) = p-h$, i.e., whose centralizer in $\mathfrak{p}$ has dimension h. Hence, $h \geqq p-r \geqq t$, which proves the original claim.

We have restriction homomorphisms as indicated by the arrows in the following diagram:

$$\begin{array}{ccccc} & & S(\mathfrak{g}) & & \\ & \swarrow & & \searrow & \\ S(\mathfrak{h}) & & & & S(\mathfrak{p}) \\ & \searrow & & \swarrow & \\ & & S(\mathfrak{h}_\mathfrak{p}) & & \end{array}$$

We have defined $\mathfrak{J}$ as the ring of invariants in $S(\mathfrak{g})$ of the extension τ of the adjoint representation of $\mathfrak{g}$. For each of the subspaces $\mathfrak{h}$, $\mathfrak{p}$, and $\mathfrak{h}_\mathfrak{p}$, we denote by $\mathfrak{J}$ supplied with that subspace as subscript the image of $\mathfrak{J}$ under the corresponding restriction homomorphism (from $S(\mathfrak{g})$ to the symmetric algebra of that subspace) in the above diagram. On the other hand, let W be the Weyl group of $\mathfrak{g}$ with respect to $\mathfrak{h}$ and let $\boldsymbol{I}$ be the subring of W-invariant elements in $S(\mathfrak{h})$. It is not hard to see that $\boldsymbol{I} \supset \mathfrak{J}_\mathfrak{p}$. In fact, Chevalley has shown [cf. 27: pp. 430–434] that $\boldsymbol{I} = \mathfrak{J}_\mathfrak{h}$ and thus in particular that $S(\mathfrak{h})$ is integral over $\mathfrak{J}_\mathfrak{h}$, and $\mathfrak{J}_\mathfrak{h}$ is finitely generated (the latter facts can be derived from $\boldsymbol{I} = \mathfrak{J}_\mathfrak{h}$ using Lemma 2, Chapter 2, section 6; it is the surjectivity of the restriction mapping from $\mathfrak{J}$ that is not trivial). Thus, $\mathfrak{J}_\mathfrak{h}$ is a Noetherian ring. Since restriction from $\mathfrak{h}$ to $\mathfrak{h}_\mathfrak{p}$ maps $S(\mathfrak{h})$ onto $S(\mathfrak{h}_\mathfrak{p})$ (section 1) and $\mathfrak{J}_\mathfrak{h}$ onto $\mathfrak{J}_{\mathfrak{h}_\mathfrak{p}}$ (by definition), it follows that $S(\mathfrak{h}_\mathfrak{p})$ is integral over $\mathfrak{J}_{\mathfrak{h}_\mathfrak{p}}$ and that the latter is a Noetherian ring and must contain h algebraically independent elements. And since $S(\mathfrak{h}_\mathfrak{p})$ is finitely generated as a ring by elements which are, of course, integral over $\mathfrak{J}_{\mathfrak{h}_\mathfrak{p}}$, $S(\mathfrak{h}_\mathfrak{p})$ is a $\mathfrak{J}_{\mathfrak{h}_\mathfrak{p}}$-module of finite type. Let $\Theta_{\mathfrak{h}_\mathfrak{p}}$ be the image of Θ under the restriction homomorphism from $S(\mathfrak{p})$ to $S(\mathfrak{h}_\mathfrak{p})$. The transcendence degree of Θ is $\leqq h$ and $\Theta_{\mathfrak{h}_\mathfrak{p}}$ contains $\mathfrak{J}_{\mathfrak{h}_\mathfrak{p}}$ which has transcendence degree $\geqq h$; thus, since Θ and $\Theta_{\mathfrak{h}_\mathfrak{p}}$ are integral domains, each has transcendence degree h and restriction gives an isomorphism φ of the first onto the second, hence an isomorphism of $\mathfrak{J}_\mathfrak{p}$ onto $\mathfrak{J}_{\mathfrak{h}_\mathfrak{p}}$. We have used the well-known fact from algebraic geometry that restriction of the coordinate ring from an irreducible affine variety to an irreducible subvariety is an isomorphism if and only if the varieties are of the same dimension (and hence coincide). Since $\mathfrak{J}_{\mathfrak{h}_\mathfrak{p}} \subset \Theta_{\mathfrak{h}_\mathfrak{p}} \subset S(\mathfrak{h}_\mathfrak{p})$, and since $\mathfrak{J}_{\mathfrak{h}_\mathfrak{p}}$ is Noetherian, it now follows easily from the above that $\Theta_{\mathfrak{h}_\mathfrak{p}}$ is a $\mathfrak{J}_{\mathfrak{h}_\mathfrak{p}}$-module of finite type. Applying φ^{-1}, it follows that Θ is a $\mathfrak{J}_\mathfrak{p}$-module of finite type.

Now to prove that $\mathcal{J}'$ is a module of finite type over Θ, let $\mathfrak{D}$ be the $S(\mathfrak{p})$-submodule of $S(\mathfrak{p})\otimes W$ generated by $\mathcal{J}'$. As $\dim W < +\infty$, and $S(\mathfrak{p})$ is Noetherian, it follows that we may write $\mathfrak{D} = \sum_{i=1}^{t} S(\mathfrak{p})D_i$, $D_i \in \mathcal{J}'$. Let $J \in \mathcal{J}'$. On the one hand, by Lemma 10, $S(\mathfrak{p})\otimes W$ is the direct sum of $\mathcal{J}'$ and of the module $\mathfrak{W}_1$ spanned by all the elements of the form $\xi(X)a$, $X \in \mathfrak{k}$, $a \in S(\mathfrak{p})\otimes W$, and on the other hand, by the same lemma, $S(\mathfrak{p})$ itself is the direct sum of Θ and of the submodule $\mathfrak{S}_1$ spanned by the elements $\rho(X)\cdot s$, $X \in \mathfrak{k}$, $s \in S(\mathfrak{p})$. Thus we may write $J = \sum_{i=1}^{t} s_i D_i$, $s_i \in S(\mathfrak{p})$, and $s_i = \theta_i + s_i'$, $\theta_i \in \Theta$, $s_i' \in \mathfrak{S}_1$. If we can show that for each i we have $s_i' D_i \in \mathfrak{W}_1$, then, since the sum $\mathcal{J}' + \mathfrak{W}_1$ is direct, it will follow that J, which belongs to $\mathcal{J}'$, is equal to $\sum_{i=1}^{t} \theta_i D_i$ and hence $\mathcal{J}' = \sum_{i=1}^{t} \Theta D_i$, proving what we want. But each s_i' is, by the definition of $\mathfrak{S}_1$, a sum of terms of the form $\rho(X)\cdot s$, $X \in \mathfrak{k}$, $s \in S(\mathfrak{p})$, and from (15) it follows that if $D \in \mathcal{J}'$, then $(\rho(X)\cdot s)D = \xi(X)(s \cdot D)$, which clearly belongs to $\mathfrak{W}_1$, and this completes the proof that $\mathcal{J}'$ is a Θ-module of finite type and hence also the proof of Theorem 18.

§ 4. Representations of a Lie group on a locally convex, complete, linear space [26d]

We return to the notation of Chapter 8, section 2, and now let G be a reductive and hence unimodular Lie group with Haar measure

dg, let V be a Hausdorff, locally convex, complete, complex linear space, and let ρ be a (continuous) representation of G on V. Let $C^\infty(G)$ be the complex-valued functions on G of class C^∞. A vector $v \in V$ is called differentiable if $g \mapsto \rho(g)\cdot v$ is a C^∞-mapping of G into V; this means, by definition, that if $X \in \mathfrak{g}$, the Lie algebra of G, then

$$\rho(X)\cdot v = \lim_{t\to 0} t^{-1}(\rho(\exp tX)\cdot v - v)$$

exists and the mapping $X \mapsto \rho(X)\cdot v$ may be extended uniquely to a linear mapping $X \mapsto \rho(X)\cdot v$ of $U(\mathfrak{g})$ into V such that $\rho(XY)\cdot v = \rho(X)\rho(Y)\cdot v \quad (X, Y \in U(\mathfrak{g}))$. Let V^∞ be the space of differentiable vectors in V. If $f \in C_c^\infty(G) = C^\infty(G) \cap \mathcal{K}(G)$, we define $\rho(f)\cdot v = \int_G f(g)\rho(g)\cdot v\, dg$, so that, as before, one has $\rho(f * g) = \rho(f)\rho(g)$. Then [26d : Lemma 2] one has

LEMMA 15. *If* $f \in C_c^\infty(G)$, *then* $\rho(f)V \subset V^\infty$ *and* $\rho(X)\rho(f)\cdot v = -\rho(\sigma(X)f)\cdot v$, $X \in \mathfrak{g}$, $v \in V$, *where*

$$\sigma(X)f(x) = \lim_{t\to 0} t^{-1}(f(\exp(tX)\cdot x) - f(x)).$$

PROOF. One sees that $\rho(y)\rho(f)\cdot v = \rho({}_y f)\cdot v$, where ${}_y f(x) = f(y^{-1}x)$. If $X \in \mathfrak{g}$, $y_t = \exp(tX)$, and $s, t \in \boldsymbol{R}$, then

$$\frac{d}{ds} f(y_{st}{}^{-1}x) = -t\sigma(X)f(y_{st}{}^{-1}x);$$

integrating from $s=0$ to $s=1$, one obtains

$$t^{-1}(\rho(y_t)-1)\rho(f)\cdot v = -\int_0^1 \rho(y_{st})v'\, ds,$$

where $v' = \rho(\sigma(X)f)\cdot v$. The lemma follows at once from this.

Let $X \mapsto \check{X}$ be the anti-automorphism of $U(\mathfrak{g})$ satisfying $\check{X} = -X$ for $X \in \mathfrak{g}$. Now, σ may be uniquely extended to an anti-isomorphism of $U(\mathfrak{g})$ onto the algebra of right-invariant differential operators on G (as was done in section 1 to obtain an isomorphism onto the left-invariant ones). From the lemma one has, then,

COROLLARY 1. $\rho(X)\rho(f)\cdot v = \rho(\sigma(\check{X})f)\cdot v$ *for* $X \in U(\mathfrak{g})$, $f \in C_c^\infty(G)$.

COROLLARY 2. $\rho(f)\rho(X)\cdot v = \rho(\alpha(\check{X})f)\cdot v$, *for* X *and* f *as in Corollary 1 and where* $\alpha(X)$ *is the left-invariant differential operator associated to X in section 1.*

Recall the definition of Dirac sequence from Chapter 8, section 3.

LEMMA 16. [26d : Lemma 3]. *Denote by* $\{f_m\}$ *a Dirac sequence in* $C_c^\infty(G)$. *We have* $\lim_{m\to\infty} \rho(f_m)\cdot v = v$ *for each* $v \in V$.

PROOF. Let $v \in V$, let ν be a continuous semi-norm, and let $\varepsilon > 0$ be given. Now, by Chapter 8, section 2, there is a neighborhood $\mathcal{U}$ of e in G such that

$$\nu(\rho(x)\cdot v - v) \leqq \varepsilon, \quad x \in \mathcal{U}.$$

Since we have

$$\rho(f_m)\cdot v - v = \int_G f_m(x)(\rho(x)\cdot v - v)dx,$$

it follows that $\nu(\rho(f_m)\cdot v - v) \leqq \varepsilon$ if $\mathcal{S}(f_m) \subset \mathcal{U}$, proving the lemma.

COROLLARY. *The subspace* V^∞ *is dense in* V.

This follows from Lemmas 15 and 16.

We recall the notation $r(g)f(x) = f(xg)$ and $l(g)f(x) = f(g^{-1}x)$, if f is a complex-valued function on G and $g \in G$. If M is a compact subset of G, let $C_M^\infty(G) = \{f \in C_c^\infty(G) \mid \mathcal{S}(f) \subset M\}$; and if X is a left-invariant differential operator on G, define a semi-norm $\nu_{M,X}$ on $C^\infty(G)$ by

$$\nu_{M,X}(f) = \sup_{x \in M} |Xf(x)|.$$

(It is notationally convenient to, and henceforth we do, identify $U(\mathfrak{g})$ with the algebra of left-invariant differential operators on G.) Then the family of semi-norms $\nu_{M,X}$ is used to make $C^\infty(G)$ into a locally convex, Hausdorff, linear space, which is readily seen to be complete. The subspace $C_c^\infty(G)$ is denoted by $D(G)$. Let K be a maximal compact subgroup of G and let $\mathfrak{k}$ be its Lie algebra, let $\mathfrak{X}$ be the universal enveloping algebra of $\mathfrak{k}$, viewed as a subalgebra of $U = U(\mathfrak{g})$, and let $\mathcal{Z}$ be the center of U as before.

LEMMA 17. [26d : § 5]. *The mappings* $g \mapsto l(g)$ *and* $g \mapsto r(g)$ *are representations of* G *on* $V = C^\infty(G)$, *and in this case,* $V = V^\infty$ *for both these representations. Moreover,* $r(X)\cdot f = X\cdot f$ *and* $l(X)\cdot f = \sigma(\check{X})\cdot f$ *for all* $f \in C^\infty(G)$ *and* $X \in U$.

PROOF. First of all, it is clear that

$$\nu_{M,X}(l(g)f)=\nu_{g^{-1}M,X}(f)$$

and secondly $l(g)$ and X commute as operators on $C^\infty(G)$, and from these facts it follows easily by Lemma 1 of Chapter 8, section 2 that l is a representation. Now defining $f^\vee(x)=f(x^{-1})$, one may prove without difficulty that $f\mapsto f^\vee$ is a homeomorphism of $C^\infty(G)$ onto itself, and since

$$l(g)f^\vee(x)=(r(g)f)^\vee(x),$$

it is seen that r is also a representation.

Let $f\in C^\infty(G)$, let $X\in\mathfrak{g}$, and for $t\in\boldsymbol{R}$, define $y_t=\exp tX$. Fix a compact subset M of G and $Y\in U$. Then if $\sigma(X)$ denotes the right-invariant differential operator associated to X (in Lemma 15), one has

$$\begin{aligned}(25)\qquad &\nu_{M,Y}(t^{-1}\{l(y_t)f-f\}+\sigma(X)f)\\ &=\sup_{g\in M}|t^{-1}\{l(y_t)f_1(g)-f_1(g)\}+f_2(g)|,\end{aligned}$$

where $f_2=\sigma(X)f_1$ and $f_1=Yf$. Applying the fundamental theorem of integral calculus as employed in the proof of Lemma 15, one obtains that the right side of (25) is in absolute value equal to or less than

$$\int_0^1\sup_{g\in M}|l(y_{st})f_2(g)-f_2(g)|\,ds,$$

which approaches zero as $t\to 0$. Hence f is differentiable under the representation l and $l(X)f=-\sigma(X)f$. In particular, it follows that for all $X\in U$, we have $l(X)f=\sigma(\check{X})f$. Since this implies $r(X)f(x)=-\sigma(X)f^\vee(x^{-1})=-\alpha(-X)f(x)=\alpha(X)f(x)$ for $X\in\mathfrak{g}$, we see that f is also differentiable with respect to the representation r and obtain that $r(X)f=Xf$, proving the lemma.

Let ρ be a finite-dimensional representation of K on a vector space and let $\mathfrak{U}$ be an ideal of finite codimension in $\mathfrak{Z}$ (i.e., $\dim(\mathfrak{Z}/\mathfrak{U})<+\infty$). A C^∞-mapping $f\colon G\to W$ is said to be of type $(\rho,\mathfrak{U})$ if the conditions i) $f(kx)=\rho(k)f(x)$, $k\in K$, $x\in G$, *and* ii) $l(X)f=0$ for all $X\in\mathfrak{U}$ are satisfied. f is said to be of finite type on the left (resp. on the right) if the set of all the functions $l(k)f$ (resp. $r(k)f$), $k\in K$, spans a finite-dimensional space. Clearly i) implies that f is of finite type on the left. Conversely, suppose f is a complex-valued (scalar) function on G, of finite type on the left with respect to K. Let W' be the span of the set of functions ${}_kf$, $k\in K$, within the vector space of all complex-valued functions on G and let $f_i={}_{k_i}f$, $i=1, \cdots, n=\dim W'$, be a basis of W'. Define h as the (column-)vector-valued function on G given by $h(g)=(f_1(g), \cdots, f_n(g))$. If $k\in K$, then $h(kg)=(f_1(kg), \cdots, f_n(kg))$ and we have $f_i(kg)=\sum_j \rho_{ij}(k)f_j(g)$ for suitable complex numbers $\rho_{ij}(k)$. If we let $\rho(k)$ denote the $n\times n$ matrix $(\rho_{ij}(k))$, then it is easy to see that $h(kg)=\rho(k)h(g)$ for all $k\in K$, and that if k' is any other element of K, then $\rho(kk')=\rho(k)\rho(k')$. Since the functions $f_1, \cdots, f_n$ are linearly independent, it follows that there exist points $g_1, \cdots, g_n\in G$ such that $\det(f_j(g_i))\neq 0$, and solving the n^2 equations $\sum_j \rho_{ij}(k)f_j(g_l)=f_i(kg_l)$, $i, l=1, \cdots, n$, for $\rho_{ij}(k)$, we see that if f is continuous, then ρ is a (continuous) representation of K. All the functions we deal with in this context will be C^∞. Obviously, if f is a function with values in a finite-dimensional complex vector space W, and if f is K-finite on the left, then each coordinate function of f, with respect to some basis of W, will also be K-finite on the left and we can repeat the same arguments. Thus, if f is continuous and K-finite on the left, then there exists a representation (ρ', W') of K such that 1) W is a subspace of W', 2) there is a function h of type ρ' on the left from G into W' (that is, i) is satisfied with h in place of f and ρ' in place of ρ), and 3) there is a projection π of W' onto W such that $f=\pi\circ h$. By abuse of language, we then say that f is of type ρ' on the left. (This is an explanation of a remark at the end of section 5.1 of [3].)

Consider the following hypotheses for a vector-valued function f of differentiability class C^∞:

A) f is K-finite on the left and $l(\mathcal{Z})\cdot f$ has finite dimension (in other words, f is of type $(\rho', \mathfrak{U}')$ for some representation ρ' of K and for some ideal $\mathfrak{U}'$ of finite codimension in $\mathcal{Z}$).

B) $l(\mathcal{Z}\cdot\mathfrak{k})\cdot f$ has finite dimension.

Assume A. Then $\{{}_kf\}_{k\in K}$ has a finite-dimensional span W and for all $X\in\mathfrak{k}$, $f(\exp(tX)\cdot x)\in W$ for every $t\in\boldsymbol{R}$; hence, $\sigma(X)f\in W$, so that $l(X)f\in W$ for every $X\in\mathfrak{k}$, thus, ultimately, for every $X\in\mathfrak{X}$. Therefore, $l(\mathfrak{X})f$ has finite dimension. Let $\mathfrak{U}_1$ be the kernel in $\mathfrak{X}$ of $X\mapsto l(X)f$ and let $\mathfrak{U}_2$ be the kernel in $\mathcal{Z}$ of $X\mapsto l(X)f$. If $X\in\mathfrak{U}_1$, $Y\in\mathfrak{X}$, then $0=l(Y)(l(X)f)=l(XY)f$, so $XY\in\mathfrak{U}_1$, i.e., $\mathfrak{U}_1$ is a right ideal in $\mathfrak{X}$. Moreover, $\mathfrak{U}_2$ is an ideal (two-sided) in $\mathcal{Z}$. Therefore, $\mathfrak{U}_1$ and $\mathfrak{U}_2$

generate a right ideal $\mathfrak{U}'$ of finite codimension in $\mathcal{Z}\mathfrak{X}=\mathfrak{X}\mathcal{Z}$ such that $l(\mathfrak{U}')f=0$, thus B is satisfied. We have shown that A implies B. Though we don't need it, one may also show that B implies A. This is left as an exercise.

LEMMA 18. *Assume that G is semi-simple and that f satisfies* B. *Then f is actually (real) analytic.*

PROOF. We may choose a basis $X_1, \cdots, X_m$ of $\mathfrak{k}$ (on which the Killing form is negative-definite), and a basis $Y_1, \cdots, Y_n$ of the orthogonal complement $\mathfrak{p}$ of $\mathfrak{k}$ (with respect to the Killing form) such that the Casimir operator $\mathfrak{C}\in\mathcal{Z}$ has the form $\sum Y_j^2-\sum X_i^2$. Then $b=\sum Y_j^2+\sum X_i^2\in\mathcal{Z}\cdot\mathfrak{X}$ $(1\in\mathcal{Z}\cap\mathfrak{X})$ and since f is annihilated by an ideal $\mathfrak{U}'$ of finite codimension, in $\mathcal{Z}\cdot\mathfrak{X}$, there exists a positive integer m and $a_0, \cdots, a_m\in\boldsymbol{C}$ such that $\mathcal{E}=b^m+a_{m-1}b^{m-1}+\cdots+a_0\in\mathfrak{U}'$; since $\mathcal{E}$ is an elliptic operator with analytic coefficients which annihilates f, it follows that f is analytic [44b : §§ 3.6–3.8].

Now let $\mathfrak{d}$ be a class of finite-dimensional irreducible representations of K and put $\alpha_{\mathfrak{d}}=\dim(\mathfrak{d})\chi_{\mathfrak{d}}$, where $\chi_{\mathfrak{d}}$ is the character of $\mathfrak{d}$. Let ρ be a representation of G on a locally convex, complete, Hausdorff, complex, topological vector space V. If $v\in V$, then $\rho(\bar{\alpha})$ is the linear mapping of V into itself defined by

$$\rho(\bar{\alpha})\cdot v=\int_K \bar{\alpha}(k)\rho(k)\cdot v\,dk, \quad v\in V. \tag{26}$$

Let $V_{\mathfrak{d}}$ be the $\mathfrak{d}$-isotypic subspace of V with respect to the restriction of ρ to K. Then for any $v\in V$, a simple calculation, based on the properties of the matrix coefficients of a unitary representation of K belonging to $\mathfrak{d}$, reveals that $\rho(\bar{\alpha})v\in V_{\mathfrak{d}}$, and that if $v\in V_{\mathfrak{d}}$, then $\rho(\bar{\alpha})v=v$. Thus, $\rho(\bar{\alpha})$ is a projection of V onto $V_{\mathfrak{d}}$.

One has [26d : Theorem 1]

THEOREM 19. *Let E be a finite-dimensional complex vector space and let f be a C^∞-function from G to E such that the functions Xf, $X\in\mathcal{Z}\mathfrak{X}$, span a finite-dimensional space. Let $\mathcal{N}$ be a neighborhood of the identity e on G and let*

$$J=\{\psi\in C_c^\infty(G)\,|\,\mathcal{S}(\psi)\subset\mathcal{N} \text{ and } \psi(kxk^{-1})=\psi(x), \quad k\in K,\ x\in G\}.$$

Then there exists $\delta\in J$ such that $f*\delta=f$.

PROOF. One may view f as an element of $C^\infty(G)\otimes E$ and extend the representation r to this space by defining $r(g)(a\otimes v)=r(g)a\otimes v$ for $a\in C^\infty(G)$, $v\in E$. It follows from Lemma 17 that every $\psi\in C^\infty(G)\otimes E$ is differentiable with respect to r and $r(X)\psi=X\cdot\psi$, $X\in U$. Let $\mathfrak{U}$ be the kernel in $\mathcal{Z}\mathfrak{X}$ of the mapping $X\mapsto X\cdot f$. By hypothesis, $\mathfrak{U}$ is a left ideal in $\mathcal{Z}\mathfrak{X}$ of finite codimension. Let W be the minimal closed subspace of $C^\infty(G)\otimes E$ such that $f\in W$ and $r(g)\cdot W=W$ for all $g\in G$. If $f_1\in W$, then $r(\exp tX)f_1\in W$, $t\in \boldsymbol{R}$, $X\in\mathfrak{g}$, and since W is closed, we obtain by taking the limit of the difference quotient that $Xf_1\in W$; by repeated application of this we obtain $Xf_1\in W$ for every left-invariant differential operator X on G. Therefore, W contains $r(U)\cdot f$. Let $r(U)\cdot f=W_0$; we shall prove that W is the closure $Cl(W_0)$ of W_0. Suppose not; then by the Hahn-Banach theorem, Chapter 3, there exists a continuous linear functional β on W such that $\beta(W_0)=0$ but $\beta\not\equiv 0$ on W. Let

$$F(x)=\beta(r(x)f), \quad x\in G.$$

If $X\in\mathfrak{g}$, we may use the linearity and continuity of β to obtain

$$\begin{aligned} XF(x)&=\lim_{t\to 0} t^{-1}(F(x\cdot\exp tX)-F(x))\\ &=\lim_{t\to 0}\beta(t^{-1}(r(x\cdot\exp tX)f-r(x)f))\\ &=\beta(\lim_{t\to 0} t^{-1}(r(x\cdot\exp tX)f-r(x)f)), \end{aligned}$$

provided that the function $t^{-1}(r(x\cdot\exp tX)f-r(x)f)$ converges, as $t\to 0$, in the C^∞-topology (i.e., in the topology of the semi-norms $\nu_{M,X}$). But the expression occurring just after t^{-1} is equal to $f(yx\cdot\exp tX)-f(yx)$, y being the "variable" on G, and applying Taylor's formula to this C^∞-function of t and to the similarly defined C^∞-functions of t obtained by replacing f by its mixed partial derivatives of all orders, we obtain the desired C^∞-convergence, so that, in the end, one has

$$XF(x)=\beta(r(x)Xf), \quad X\in\mathfrak{g}.$$

Again, applying this repeatedly, one obtains

$$XF(x)=\beta(r(x)Xf), \quad X\in U=U(\mathfrak{g}). \tag{27}$$

Thus, F is differentiable for r, and (27) tells us that $\mathfrak{U}\cdot F=0$. But by Lemma 18, this implies that F is real analytic. Moreover, $XF(1)=\beta(Xf)=0$, $X\in U$, since $\beta(W_0)=0$. Therefore, all the Taylor's series coefficients of F are zero at a point of G, and hence $F \equiv 0$. Therefore, $\beta \equiv 0$ on W which is the closure of the span of all $r(x)f$, $x \in G$. This is a contradiction; so, we must have $W = Cl(W_0)$.

Let $W_1 = r(\mathfrak{X})f$. Clearly $\dim W_1 < +\infty$ and hence W_1 is closed in W. Moreover, just as before, one sees that $r(K)f \subset W_1$ and thus W_1 is invariant under $r(K)$. This means that W_1 is a fully reducible K-module and $r(\bar{\alpha}_{\mathfrak{d}}) W_1 = W_{1\mathfrak{d}} = \{0\}$ for all but a finite number of classes $\mathfrak{d}_1, \cdots, \mathfrak{d}_m$ of irreducible representations of K. Put $\bar{\alpha} = \sum_{i=1}^{m} \bar{\alpha}_{\mathfrak{d}_i}$, $r(\bar{\alpha}) = \sum_{i=1}^{m} r(\bar{\alpha}_{\mathfrak{d}_i})$. We prove that $W_\alpha = r(\bar{\alpha})W$ is finite-dimensional. It is obvious that $f \in W_1 \subset W_\alpha$ and that W_1 is isomorphic as a K-module to $\mathfrak{X}/(\mathfrak{X} \cap \mathfrak{U})$, and therefore the latter is a semi-simple K-module. Let $\mathfrak{A}$ be the left U-ideal defined by

$$\mathfrak{A} = \{X \in U \mid r(X)f = 0\}.$$

Then W_0 is isomorphic as a K-module to $U/\mathfrak{A}$, and $\mathfrak{A} \supset \mathfrak{U}$. In particular, $\mathfrak{A} \supset \mathfrak{Y} = \mathfrak{X} \cap \mathfrak{U}$, and $\mathfrak{Y}$ is a left $\mathfrak{k}$-module, so that we have a natural $\mathfrak{k}$-module homomorphism φ from $U^* = U/U\mathfrak{Y}$ onto W_0. By Theorem 18, U^* is quasi-semi-simple with respect to the natural representation of $\mathfrak{k}$. Therefore, W_0 is, too, and the restriction $\varphi_{\mathfrak{d}}$ of φ to $U^*_{\mathfrak{d}}$ maps the latter onto $W_{0\mathfrak{d}}$. Now, by Theorem 18, $U^*_{\mathfrak{d}} = \sum_{i=1}^{m} \mathcal{Z} a_i$, so that

$$\varphi_{\mathfrak{d}}(U^*_{\mathfrak{d}}) = r^*(U^*_{\mathfrak{d}})f = \sum_{i=1}^{m} r^*(\mathcal{Z} a_i)f,$$

where r^* is the mapping associated to r of U^* into the set of linear endomorphisms of the space of C^∞-functions annihilated by $r(\mathfrak{A})$, and

$$r^*(\mathcal{Z} a_i)f = r^*(a_i \mathcal{Z})f = r^*(a_i) r(\mathcal{Z})f;$$

since $\dim(r(\mathcal{Z})f) < +\infty$ by hypothesis, we have $\dim(W_{0\mathfrak{d}}) < +\infty$ for every class $\mathfrak{d}$. Moreover, since W_0 is dense in W, we see that $W_{0\mathfrak{d}}$ is dense in $W_{\mathfrak{d}}$, and since $W_{0\mathfrak{d}}$ is finite-dimensional and therefore closed, we have $W_{0\mathfrak{d}} = W_{\mathfrak{d}}$. Consequently, $W_\alpha = r(\bar{\alpha})W$ has finite dimension.

To complete the proof, we need the following

LEMMA 19. *There exists a Dirac sequence* $\{\delta_j\}$ *such that* $\delta_j \in J$ *for every* j.

PROOF. We may replace the neighborhood $\mathcal{N}$ with a smaller neighborhood $\mathcal{N}'$ such that $Ad\, k(\mathcal{N}') = \mathcal{N}'$ for every $k \in K$, because K is compact. If $\{\beta_j\}$ is a Dirac sequence such that $\mathcal{S}(\beta_j) \subset \mathcal{N}'$ for each

j, then define δ_j by

$$\delta_j(x) = \int_K \beta_j(kxk^{-1})dk.$$

Then it is easy to verify that $\{\delta_j\}$ is a Dirac sequence of the type prescribed in the lemma.

To complete the proof of the theorem, we let $\{\delta_j\}$ be a Dirac sequence such that each $\delta_j \in J$. Since W is stable under $r(g)$ for each $g \in G$ and is closed, it is seen that $f * \delta_j \in W$ for each j, and by Lemma 16, $f * \delta_j \to f$ as $j \to \infty$. Let W_2 be the vector space of all functions of the form $f * \delta$, $\delta \in J$; then $W_2 \subset W$, and since $\delta \in J$ is invariant under $Ad\, K$, and since $f \in W_\alpha$, it follows that $W_2 \subset W_\alpha$. Therefore, $\dim(W_2) < +\infty$, and consequently W_2 is closed in W. It follows that $f = \lim_{j \to \infty} f * \delta_j \in W_2$, completing the proof.

COROLLARY 1. *Assume that $f \in C^\infty(G) \otimes E$ as before, but replace the other hypothesis about f by: The span of the functions $l(X)f$, $X \in \mathfrak{Z}\mathfrak{X}$, is finite-dimensional. Then there exists $\delta \in J$ such that $\delta * f = f$.*

PROOF. Merely interchange the roles of "right" and "left" in the proof of Theorem 19.

COROLLARY 2. *Let $f \in C^\infty(G) \otimes E$, assume that f is K-finite on the right (resp. on the left), and assume that $\mathfrak{Z} \cdot f$ has finite dimension. Then there exists $\delta \in J$ such that $f * \delta = f$ (resp. $\delta * f = f$).*

PROOF. This is immediate from Theorem 19, from Corollary 1, and from the fact that hypothesis A implies hypothesis B.

§5. A lemma giving a lower bound for ω_ρ [26b]

We recall from section 1 that if ρ is a finite-dimensional irreducible representation of G and if $\mathfrak{C} \in \mathfrak{Z}$ is the Casimir operator, then $\rho(\mathfrak{C}) = \omega_\rho I_\rho$, where I_ρ is the identity of the representation space for ρ. First we deal with the case when G is semi-simple and ρ is absolutely irreducible. We need the following known fact [26b : p. 301] (cf. also [16b : pp. 275–279]) :

LEMMA 20. *Let k be a field of characteristic zero and let $\mathfrak{g}$ be the three-dimensional Lie algebra over k with basis elements X, Y, and H*

satisfying $[H, X]=X$, $[H, Y]=-Y$, $[X, Y]=H$. *Let* σ *be any finite-dimensional representation of* $\mathfrak{g}$. *Then* $\sigma(X)\sigma(Y)$ *and* $\sigma(Y)\sigma(X)$ *are semi-simple matrices, with all eigenvalues being of the form* $q/2$, *where* q *is a non-negative (rational) integer.*

PROOF. By direct calculation, one verifies the Killing form to be non-degenerate; hence, $\mathfrak{g}$ is semi-simple. Therefore, any finite-dimensional representation of $\mathfrak{g}$ is fully reducible, and so we may assume σ to be irreducible.

First, assume k is algebraically closed and let V be the representation space of σ. If μ is any eigenvalue of $\sigma(H)$, let l be the largest integer $\geqq 0$ such that $\mu+l$ is an eigenvalue of $\sigma(H)$, and let $\psi_0 \neq 0$ be such that $\sigma(H)\psi_0=(\mu+l)\psi_0$, $\psi_0 \in V$. If $r \geqq 1$, define $\psi_r=\sigma(Y)^r\psi_0$. Then one easily proves by induction, using the fact that $\sigma([H, Y]) =-\sigma(Y)$, that $\sigma(H)\psi_r=(\mu+l-r)\psi_r$, $r \geqq 0$. Since $\dim V$ is finite, we have $\psi_r=0$ for some r; let J be the least integer (of course $\geqq 0$) for which $\psi_{J+1}=0$, and put $\psi_{-1}=0$. Then it follows by induction on r that

$$\sigma(X)\psi_r=\left(\mu+l-\frac{1}{2}(r-1)\right)r\psi_{r-1}, \quad r \geqq 0, \tag{28}$$

since $\sigma(X)\psi_0=0$ as a consequence of the fact that $\mu+l+1$ is not an eigenvalue of $\sigma(H)$. Applying (28) with $r=J+1$, and using the fact that $(J+1)\psi_J \neq 0$, one obtains $\mu+l=\frac{1}{2}J$, and therefore

$$\sigma(H)\psi_r=\left(\frac{1}{2}J-r\right)\psi_r,$$

$$\sigma(X)\psi_r=\frac{1}{2}(J-r+1)r\psi_{r-1}.$$

Thus, $\psi_0, \cdots, \psi_J$ are a basis for a space $\neq \{0\}$ invariant under $\sigma(\mathfrak{g})$, which can be no other than V. Since

$$\sigma(X)\sigma(Y)\psi_r=\frac{1}{2}(J-r)(r+1)\psi_r, \quad \text{and}$$

$$\sigma(Y)\sigma(X)\psi_r=\frac{1}{2}(J-r+1)r\psi_r$$

for $0 \leqq r \leqq J$, it follows that $\sigma(X)\sigma(Y)$ and $\sigma(Y)\sigma(X)$ are in diagonal form with respect to this basis and have non-negative, half-integral eigenvalues.

If k is not algebraically closed, denote by $\bar{k}$ its algebraic closure. Then $\mathfrak{g}$ may be replaced by $\mathfrak{g}_{\bar{k}}=\mathfrak{g}\otimes_k\bar{k}$ and σ extended to a representation $\bar{\sigma}$ of it. One sees from the above that $\bar{\sigma}(X)\bar{\sigma}(Y)$ and $\bar{\sigma}(Y)\bar{\sigma}(X)$ are semi-simple and have non-negative, half-integral eigenvalues, and since $\bar{\sigma}(X)=\sigma(X)$ and $\bar{\sigma}(Y)=\sigma(Y)$, the entire proof is complete.

Now let $\mathfrak{g}$ be a semi-simple Lie algebra over a field k, let B be the Killing form, $\{X_i\}_{1\leq i\leq n}$ a basis of $\mathfrak{g}$, $g_{ij}=B(X_i, X_j)$, $(g^{ij})=$ matrix inverse to (g_{ij}). Then $X^i=\sum_j g^{ij}X_j$ is a basis dual to $\{X_i\}$, and $\mathfrak{C}=\sum_i X_iX^i$ is the Casimir element of the center $\mathcal{Z}$ of $U(\mathfrak{g})$. If ρ is an irreducible representation of $\mathfrak{g}$, then $\rho(\mathfrak{C})=\omega I$, where $\omega=\omega_\rho$ and $I=I_{d_\rho}$.

The main results we need about ω_ρ are that it is positive and that it is bounded below by the square of the length of the highest weight of ρ.

First assume k to be algebraically closed.

Choose a Cartan subalgebra $\mathfrak{h}$ of $\mathfrak{g}$, dim $\mathfrak{h}=l$, let $\{\alpha_1, \cdots, \alpha_{n-l}\}=\Sigma$ be the set of all roots, let Σ_+ (resp. Σ_-) be the set of positive (resp. negative) roots with respect to some ordering, and let $\Delta=\{\alpha_1, \cdots, \alpha_l\}$ be the set of simple, positive roots. For each $\alpha\in\Sigma$, choose $X_\alpha\in\mathfrak{g}_\alpha$, H_α, etc., as in Chapter 7, section 5, and let $H_i=H_{\alpha_i}$. Then [27 : Chap. III, § 4] $\alpha(H_\alpha)>0$ and $B(H_\alpha, H)=\alpha(H)$ for all $\alpha\in\Sigma$, $H\in\mathfrak{h}$. By Chapter 8, section 7, $B(X_\alpha, X_\beta)=0$ if $\alpha+\beta\neq 0$ and $B(X_\alpha, X_{-\alpha})=1$, while $B(\mathfrak{h}, X_\alpha)=0$ if $\alpha\neq 0$. So if we choose $X_i=H_i$, $1\leq i\leq l$, $X_{i+l}=X_{\alpha_i}$, $i=1, \cdots, n-l$, as a basis of $\mathfrak{g}$, then the dual basis is given by

$$X^i=H^i=\sum_{j=1}^{l} g^{ij}H_j, \quad i=1, \cdots, l$$

$$X^\alpha=X_{-\alpha}, \quad \alpha\in\Sigma.$$

Hence

$$\rho(\mathfrak{C})=\sum_{i,j=1}^{l} g^{ij}\rho(H_i)\rho(H_j)+\sum_{\alpha\in\Sigma}\rho(X_\alpha)\rho(X_{-\alpha}).$$

Since $\alpha(H_\alpha)>0$, there exists $a>0$ such that $a^2=\alpha(H_\alpha)$. Let $X=a^{-1}X_\alpha$, $Y=a^{-1}X_{-\alpha}$, $a^{-2}H_\alpha=H$. Then X, Y, and H satisfy the hypotheses of the preceding lemma, so that $\rho(X)\rho(Y)$ and $\rho(Y)\rho(X)$ have half-integral, non-negative eigenvalues, hence $\rho(X_\alpha)\rho(X_{-\alpha})$ and $\rho(X_{-\alpha})\rho(X_\alpha)$ have non-negative eigenvalues, since $a^2=\alpha(H_\alpha)>0$ (actually, by [27 :

p. 145, last line] $\alpha(H_\alpha)$ is rational, and in consequence all the eigenvalues are rational——a fact we don't really need here).

If Λ is any weight of ρ, let V_Λ be the corresponding weight space in V; let Λ be a fixed weight $\neq 0$ and define

$$P=\{\alpha\in\Sigma_+ \mid \Lambda(H_\alpha)\geqq 0\},$$

$$N=\{\alpha\in\Sigma_+ \mid \Lambda(H_\alpha)<0\}.$$

Since $[X_\alpha, X_{-\alpha}]=H_\alpha$, we obtain easily

$$\sum_{\alpha\in\Sigma}\rho(X_\alpha)\rho(X_{-\alpha})=\sum_{\alpha\in P}\{2\rho(X_{-\alpha})\rho(X_\alpha)+\rho(H_\alpha)\}$$
$$+\sum_{\alpha\in N}\{2\rho(X_\alpha)\rho(X_{-\alpha})-\rho(H_\alpha)\}.$$

Let

$$\Gamma=\rho(\mathfrak{C})-\sum_{i,j=1}^{l}g^{ij}\rho(H_i)\rho(H_j)-\sum_{\alpha\in P}\rho(H_\alpha)+\sum_{\alpha\in N}\rho(H_\alpha).$$

Then

$$\Gamma=2\sum_{\alpha\in P}\rho(X_{-\alpha})\rho(X_\alpha)+2\sum_{\alpha\in N}\rho(X_\alpha)\rho(X_{-\alpha}).$$

Since $\rho(\mathfrak{C})$ is a scalar multiple of the identity, and $\rho(\mathfrak{h})V_\Lambda\subset V_\Lambda$, we have $\Gamma V_\Lambda\subset V_\Lambda$, and $\rho(H)|_{V_\Lambda}=\Lambda(H)\cdot Id_{V_\Lambda}$. Hence, if $\Gamma_\Lambda=\Gamma|_{V_\Lambda}$, we have

$$\Gamma_\Lambda=\{\omega-\sum_{i,j=1}^{l}g^{ij}\Lambda(H_i)\Lambda(H_j)-\sum_{\alpha\in P}\Lambda(H_\alpha)+\sum_{\alpha\in N}\Lambda(H_\alpha)\}Id_{V_\Lambda}$$
$$=\{\omega-|\Lambda|^2-\sum_{\alpha>0}|\Lambda(H_\alpha)|\}Id_{V_\Lambda},$$

and

$$\operatorname{tr}(\Gamma_\Lambda)=\sum_{\alpha\in P}2\operatorname{tr}\{\rho(X_\alpha)\rho(X_{-\alpha})\}|_{V_\Lambda}+2\sum_{\alpha\in N}\operatorname{tr}\{\rho(X_\alpha)\rho(X_{-\alpha})\}|_{V_\Lambda}.$$

Now since the eigenvalues of $\rho(X_\alpha)\rho(X_{-\alpha})$ and of $\rho(X_{-\alpha})\rho(X_\alpha)$ are non-negative (and, by the way, rational), and each of them stabilizes V_Λ, their restrictions to V_Λ have the same properties, thus $\operatorname{tr}\Gamma_\Lambda\geqq 0$, and so

$$\omega\geqq|\Lambda|^2+\sum_{\alpha>0}|\Lambda(H_\alpha)|. \tag{29}$$

Let ϕ be a non-zero weight vector for the highest weight Λ_0. (We know that $m_{\Lambda_0}=\dim V_{\Lambda_0}=1$.) Then for each $\alpha\in\Sigma$

$$\rho(H)\rho(X_\alpha)\phi=\{\Lambda_0(H)+\alpha(H)\}\rho(X_\alpha)\phi$$

and $\rho(X_\alpha)\phi=0$ for all $\alpha\in\Sigma_+$. Then

$$\rho(\mathfrak{C}) = \sum_{i,j=1}^{l} g^{ij}\rho(H_i)\rho(H_j) + \sum_{\alpha>0} \{\rho(X_{-\alpha})\rho(X_\alpha) + \rho(H_\alpha)\},$$

$$\rho(\mathfrak{C})\phi = \{|\Lambda_0|^2 + \sum_{\alpha>0} \Lambda_0(H_\alpha)\}\phi,$$

and hence $\omega = \omega_\rho = |\Lambda_0|^2 + \sum_{\alpha>0} \Lambda_0(H_\alpha)$, thus, because of (29), we actually have $\Lambda_0(H_\alpha) \geqq 0$ for all $\alpha > 0$ and $\omega_\rho = |\Lambda_0|^2 + \sum_{\alpha>0} |\Lambda_0(H_\alpha)|$. Thus we have

LEMMA 21. *Let Λ_0 be the highest weight of ρ. Then $\omega_\rho > 0$ if $\rho \neq 0$ and $\omega_\rho \geqq |\Lambda_0|^2$.*

If k is not algebraically closed, let $\bar{k}$ be its algebraic closure, assume that ρ is irreducible over k, let $\bar{\mathfrak{g}} = \mathfrak{g} \otimes_k \bar{k}$, and extend ρ to a representation $\bar{\rho}$ of $\bar{\mathfrak{g}}$. Then, since $\bar{\mathfrak{g}}$ is semi-simple, $\bar{\rho}$ is fully reducible, $\bar{\rho} = \bigoplus \rho_i$. But we still know that $\rho(\mathfrak{C}) = \omega_\rho I_\rho$; and since for any base $\{X_j\}$ of $\mathfrak{g}$, this is also a base for $\bar{\mathfrak{g}}$ and $\bar{\rho}(X_j) = \rho(X_j)$ for all j, it follows that $\omega_\rho = \omega_{\rho_i}$ for all i, and hence we have

COROLLARY 1. *If ρ is irreducible over k but not absolutely irreducible, then $\omega_\rho = \omega_{\rho_i} > 0$, where ρ_i is any of the irreducible components of the extension of ρ to the algebraic closure of k.*

In particular, we have

COROLLARY 2. *Let K be a compact, semi-simple Lie group and let ρ be an irreducible representation of K by real matrices. Then over $\boldsymbol{C}$, ρ is either irreducible or is the sum of an irreducible representation and of its complex conjugate, and in either case, $\omega_\rho > 0$ and ω_ρ is bounded below by an absolute constant (independent of ρ) times the squared length of the highest weight of an(y) absolutely irreducible summand of ρ.*

§ 6. Convergence of Fourier series [26d: §§ 3, 4]

Let K be a compact Lie group, let $\mathcal{E}$ be the set of equivalence classes of finite-dimensional, irreducible representations of K, and let $\mathfrak{k}$ be the Lie algebra of K. We may write $\mathfrak{k} = \mathfrak{k}' + \mathfrak{c}$, where $\mathfrak{c}$ is the center of $\mathfrak{k}$ and $\mathfrak{k}'$ is the derived algebra which is semi-simple. If $\mathfrak{h}$ is a Cartan subalgebra of $\mathfrak{k}$, then $\mathfrak{h} = \mathfrak{h}' + \mathfrak{c}$, where $\mathfrak{h}' = \mathfrak{k}' \cap \mathfrak{h}$ is a Cartan subalgebra of $\mathfrak{k}'$.

Let ρ be a finite-dimensional, irreducible, unitary representation of K. By Schur's lemma, every element of $\mathfrak{c}$ is represented under $d\rho$ by scalars that must be purely imaginary since K is compact. The highest weight of ρ is a linear functional on $\mathfrak{h}$ and if Λ_0 is the highest weight, then $\dim V_{\Lambda_0}=1$. The weights of all representations of K form a lattice.

Let Q be a positive-definite quadratic form on $\mathfrak{k}$ invariant under $Ad\,K$. We may choose Q to be the sum of the negative of the Killing form on $\mathfrak{k}'$ and of any positive-definite quadratic form on $\mathfrak{c}$. Let $X_1, \cdots, X_r$ be an orthonormal basis of $\mathfrak{k}$ with respect to Q such that $X_1, \cdots, X_s$, $s \leqq r$, is a basis of $\mathfrak{k}'$, and $X_{s+1}, \cdots, X_r$ is a basis of $\mathfrak{c}$. Denote by Ω the element of the universal enveloping algebra $\mathfrak{X}$ of $\mathfrak{k}$ given by

$$\Omega = 1 - X_1^2 - X_2^2 - \cdots - X_r^2.$$

If $\mathfrak{C}$ is the Casimir operator of $\mathfrak{k}'$, we have

$$\Omega = 1 + \mathfrak{C} - \sum_{j=s+1}^{r} X_j^2.$$

Clearly Ω is in the center of $\mathfrak{X}$. Therefore, if ρ is an irreducible representation of K and if we denote again by ρ the associated representations of $\mathfrak{k}$ and of $\mathfrak{X}$, we have

$$\rho(\Omega) = c(\rho)\rho(e), \tag{30}$$

where of course $\rho(e)$ is the identity transformation of the representation space. By our earlier remarks, $\rho(X_j)$ is a pure imaginary scalar, $s+1 \leqq j \leqq r$, so that $\rho(X_j^2)$ is a negative real number times the identity. Using this together with Corollary 2 of Lemma 21 of the preceding section, we see that $c(\rho) > 0$ and that $c(\rho)$ is bounded below by the square of the length of the highest weight of ρ (which must be a pure-imaginary-valued real-linear functional on $\mathfrak{c}$). To calculate $c(\rho)$, we denote by χ the character of ρ, $\chi(k) = \mathrm{tr}(\rho(k))$. If $X \in \mathfrak{k}$, a direct calculation based on the definitions and on the fact that ρ is a representation gives

$$X \cdot \chi(k) = \mathrm{tr}(\rho(k)\rho(X)),$$

hence

$$\begin{aligned} \Omega\chi(k) &= \mathrm{tr}(\rho(k)\rho(\Omega)) = c(\rho)\mathrm{tr}(\rho(k)) \\ &= c(\rho)\chi(k); \end{aligned} \tag{31}$$

in other words, χ is an eigenfunction of the differential operator Ω, having eigenvalue equal to $c(\rho)$. Since $c(\rho)$ is real, one has

$$\Omega\bar{\chi}=c(\rho)\bar{\chi}, \quad \Omega\bar{\alpha}=c(\rho)\bar{\alpha}.$$

Let K_0 be the connected component of K containing e. Because K is compact, the index $N=[K: K_0]$ is finite. Let $\mathcal{E}_0$ be the set of equivalence classes of irreducible, finite-dimensional representations of K_0. If $\mathfrak{d}\in\mathcal{E}$ and $\mathfrak{d}_0\in\mathcal{E}_0$, let $[\mathfrak{d}:\mathfrak{d}_0]$ be the number of times $\mathfrak{d}_0$ occurs in the restriction of $\mathfrak{d}$ to K_0. If $\mathfrak{d}_0\in\mathcal{E}_0$, let $\mathcal{E}(\mathfrak{d}_0)=\{\mathfrak{d}\in\mathcal{E}\mid[\mathfrak{d}:\mathfrak{d}_0]\geqq 1\}$. Let $\rho_0\in\mathfrak{d}_0$ and let (ρ, V) be the induced representation of K coming from ρ_0 (V is the representation space of ρ). Then $\dim V=N\dim(\mathfrak{d}_0)$. Let $d(\mathfrak{d})$ (resp. $d(\mathfrak{d}_0)$) denote $\dim(\mathfrak{d})$ (resp. $\dim(\mathfrak{d}_0)$). By the Frobenius reciprocity theorem (Chapter 8, section 5), a class $\mathfrak{d}\in\mathcal{E}$ occurs in ρ precisely $[\mathfrak{d}:\mathfrak{d}_0]$ times. Thus we have

$$\sum_{\mathfrak{d}\in\mathcal{E}(\mathfrak{d}_0)}[\mathfrak{d}:\mathfrak{d}_0]d(\mathfrak{d})=Nd(\mathfrak{d}_0). \tag{32}$$

On the other hand, since the Lie algebra of K_0 is also $\mathfrak{k}$, it is clear from the definitions that $c(\rho_0)=c(\rho)$ whenever $\rho\in\mathfrak{d}$, $\rho_0\in\mathfrak{d}_0$, and $[\mathfrak{d}:\mathfrak{d}_0]\geqq 1$.

Now we let $c(\rho)=c(\mathfrak{d})$, $c(\rho_0)=c(\mathfrak{d}_0)$, when $\rho\in\mathfrak{d}$, $\rho_0\in\mathfrak{d}_0$; then $c(\mathfrak{d})=c(\mathfrak{d}_0)$. Our first objective is to prove

LEMMA 22. *The series*

$$\sum_{\mathfrak{d}\in\mathcal{E}} c(\mathfrak{d})^{-m}d(\mathfrak{d})^2$$

converges when m is large enough.

PROOF. From (32) we have

$$\begin{aligned}\sum_{\mathfrak{d}\in\mathcal{E}} c(\mathfrak{d})^{-m}d(\mathfrak{d})^2 &\leqq \sum_{\mathfrak{d}_0\in\mathcal{E}_0}\{c(\mathfrak{d}_0)^{-m}\sum_{\mathfrak{d}\in\mathcal{E}(\mathfrak{d}_0)} d(\mathfrak{d})^2\}\\ &\leqq N^2\sum_{\mathfrak{d}_0\in\mathcal{E}_0} c(\mathfrak{d}_0)^{-m}d(\mathfrak{d}_0)^2,\end{aligned} \tag{33}$$

so it is enough to treat the case when $K=K_0$, i.e., when K is connected.

But $d(\mathfrak{d})$ is a polynomial function of fixed degree in the coordinates of the highest weight in the root lattice (Chapter 8, section 8), and $c(\mathfrak{d})$ is positive and $\geqq|\Lambda_0|^2$, where Λ_0 is the highest weight. From this the truth of the lemma's assertion is obvious.

Consider a series:

$$\sum_{j \in J} v_j, \tag{34}$$

where $v_j \in V$, J is a countable indexing set, and V is (as usual) a complete, complex, locally convex topological linear space with a Hausdorff topology. The series (34) is said to converge if its finite partial sums converge in the usual sense to a limit $\mathfrak{s}$, which is then called the sum of the series. The series is said to converge absolutely if the series of non-negative terms

$$\sum_{j \in J} \nu(v_j) \tag{35}$$

converges for every continuous semi-norm ν on V. It is easy to see that absolute convergence implies convergence.

Let ρ be a representation of the compact group Lie K on V and for $\mathfrak{d} \in \mathcal{E}$, define

$$E_{\mathfrak{d}} v = \rho(\bar{\alpha}_{\mathfrak{d}}) v.$$

THEOREM 20. [26d: § 3]. *Let $v \in V^\infty$. Then the Fourier series*

$$\sum_{\mathfrak{d} \in \mathcal{E}} E_{\mathfrak{d}} v \tag{36}$$

converges absolutely to the limit v.

PROOF. Since $\Omega \bar{\alpha}_{\mathfrak{d}} = c(\mathfrak{d}) \bar{\alpha}_{\mathfrak{d}}$, it follows from Corollary 2 of Lemma 15 that for $v \in V^\infty$ and for any $\mathfrak{d} \in \mathcal{E}$ we have

$$E_{\mathfrak{d}} \rho(\Omega) v = c(\mathfrak{d}) E_{\mathfrak{d}} v. \tag{37}$$

LEMMA 23. *Let $\nu \in \mathcal{S}$, the set of all continuous semi-norms on V. Then there exists $\nu_0 \in \mathcal{S}$ such that*

$$\nu(E_{\mathfrak{d}} v) \leqq c(\mathfrak{d})^{-m} d(\mathfrak{d})^2 \nu_0(\rho(\Omega^m) v)$$

for every $\mathfrak{d} \in \mathcal{E}$, any non-negative integer m, and any $v \in V^\infty$.

PROOF. Since K is compact and the representation ρ is continuous, there exists, by Chapter 8, section 2, Lemma 2, $\nu_0 \in \mathcal{S}$ for which

$$\nu(\rho(k) u) \leqq \nu_0(u)$$

holds for all $k \in K$ and $u \in V$. Consequently,

(38) $$\nu(E_{\mathfrak{d}}u)=\nu(\rho(\bar{\alpha}_{\mathfrak{d}})u)\leqq d(\mathfrak{d})^2\nu_0(u),$$

since for every $k\in K$ we have $|\alpha_{\mathfrak{d}}(k)|=|d(\mathfrak{d})\chi_{\mathfrak{d}}(k)|\leqq d(\mathfrak{d})^2$. Now iteration of (37) gives

(39) $$E_{\mathfrak{d}}v=c(\mathfrak{d})^{-m}E_{\mathfrak{d}}\rho(\Omega^m)v, \quad m\geqq 0,$$

for $v\in V^\infty$, from which the lemma follows.

It is now obvious from Lemmas 22 and 23 that the series (36) converges absolutely. Let v_0 be its sum. We must prove that $v=v_0$. Put $u=v-v_0$. If $\mathfrak{d}_0\in\mathcal{E}$, it follows from the continuity of $E_{\mathfrak{d}_0}$ and from the fact that $E_{\mathfrak{d}_0}E_{\mathfrak{d}}=0$ for $\mathfrak{d}\neq\mathfrak{d}_0$, by the orthogonality relations (19) of Chapter 8, that $E_{\mathfrak{d}_0}v=E_{\mathfrak{d}_0}v_0$ and hence $E_{\mathfrak{d}_0}u=0$. Thus, $E_{\mathfrak{d}}u=0$ for all $\mathfrak{d}\in\mathcal{E}$.

In order to prove $u=0$, it is sufficient to show that $\nu(u)=0$ for each $\nu\in\mathcal{S}$, because V is Hausdorff [64 : Thm. 1, p. 26]. Let $\nu\in\mathcal{S}$ and let ν_0 be determined as in Lemma 23. Let $\varepsilon>0$ be given; then there exists a neighborhood $\mathcal{N}$ of the identity $e\in K$ such that $\nu(\rho(k)u-u)\leqq\varepsilon$ for $k\in\mathcal{N}$. Let f be a continuous function on K with $\mathcal{S}(f)\subset\mathcal{N}$ such that $f\geqq 0$ and $\int_K f(k)dk=1$. Then $\nu(\rho(f)u-u)\leqq\varepsilon$, as follows from the condition on $\mathcal{N}$. By the Peter-Weyl theorem, we may choose β a continuous function on K of the form

(40) $$\beta=\sum_{\mathfrak{d}\in F}\bar{\alpha}_{\mathfrak{d}}*\beta=\sum_{\mathfrak{d}\in F}E_{\mathfrak{d}}\beta,$$

where F is a finite subset of $\mathcal{E}$, such that $|\beta-f|\leqq\varepsilon$ on K. We have

$$\nu(\rho(f)u-\rho(\beta)u)\leqq\varepsilon\sup_{k\in K}\nu(\rho(k)u)\leqq\varepsilon\nu_0(u),$$

and so

$$\nu(\rho(\beta)u-u)\leqq\varepsilon(\nu_0(u)+1);$$

and from the fact that $\bar{\alpha}_{\mathfrak{d}}*\beta=\beta*\bar{\alpha}_{\mathfrak{d}}$, because $\bar{\alpha}_{\mathfrak{d}}$ is a central function, we have

$$E_{\mathfrak{d}}\rho(\beta)u=\rho(\beta)E_{\mathfrak{d}}u=0,$$

so that from (40) we obtain $\rho(\beta)u=0$; in other words, $\nu(u)=\nu(-u)\leqq\varepsilon(\nu_0(u)+1)$, and letting $\varepsilon\to 0$, we obtain $\nu(u)=0$, completing the proof.

It is instructive to note how, in the case when K is the circle $\boldsymbol{R}/\boldsymbol{Z}$ (i.e., is one-dimensional and commutative), this is related to the usual proof that the Fourier series of a twice continuously differentiable periodic function converges to that function. Namely, in this case, $\Omega=1-\partial^2/\partial x^2$, each class $\mathfrak{d}$ is one-dimensional, and the classes are in one-to-one correspondence with the integers $\boldsymbol{Z}$: If $n\in\boldsymbol{Z}$, then the character χ_n of the corresponding class is $e^{2\pi inx}$, $x\in\boldsymbol{R}$, and $\Omega\chi_n=(1+4\pi^2n^2)\chi_n$, or $c(n)=1+4\pi^2n^2$ and the equation (37) becomes for $v=f\in C^2(K)$:

$$\int_0^1 e^{-2\pi inx}(f(x)-f''(x))dx=(1+4\pi^2n^2)\int_0^1 e^{-2\pi inx}f(x)dx, \tag{41}$$

where f'' is the second derivative of f, which is equivalent to the formula

$$a_n=\int_0^1 e^{-2\pi inx}f(x)dx=-(4\pi^2n^2)^{-1}\int_0^1 e^{-2\pi inx}f''(x)dx, \tag{42}$$

obtainable through integration by parts, and which gives an estimate for the coefficients a_n good enough to ensure absolute convergence of the Fourier series

$$\sum_{n=-\infty}^{+\infty} a_n e^{2\pi inx}.$$

The same kind of thing, of course, holds for periodic functions of any finite number of real variables having continuous partial derivatives up to sufficiently high order. This will be used in Chapter 13 and has been used in the proof of Theorem 17 (Chap. 8).

§7. Hua's determination of an orthonormal basis of $\mathcal{O}^2(D)$

According to Chapter 4, section 3, the space $H=\mathcal{O}^2(D)$ of square-integrable holomorphic functions on a domain $D\subset\boldsymbol{C}^n$ is a (separable) Hilbert space and thus must have an orthonormal basis. When D is a symmetric bounded domain, Theorem 20 suggests a method for finding such a basis, and this has been exploited by Hua [28b] to find such a basis when D is a bounded symmetric domain of classical type.

We now explain part of the ideas involved. Return to the definitions and notation of Chapter 7, section 7; then $D=\zeta(G_R^0)$ is realized as a bounded domain in $\mathfrak{p}^+$, K is the stability group of the origin $0\in D$, and K acts by linear transformations on $\mathfrak{p}^+$. More precisely, $\mathfrak{p}^+$ is the Lie algebra of P^+ and is normalized by K, and if $k \in K$, $z \in D$, then $z \cdot k = Ad(k^{-1})(z)$. This representation of K may be extended to an action of K on the holomorphic functions on any open set D' in $\mathfrak{p}^+$ stable under K, and in particular to an action on H and to an action on the space of polynomial functions, which is contained in H. Denote by ρ this action of K on H. The main idea suggested by Theorem 20 and utilized by Hua is to decompose H into the direct sum of the spaces $H_{\mathfrak{d}}$, where $\mathfrak{d}$ runs over the classes of finite-dimensional irreducible representations of K. It is clear that ρ is unitary and that the spaces $H_{\mathfrak{d}}$ are mutually orthogonal for distinct $\mathfrak{d}$.

One obtains a preliminary decomposition of H by considering the action of the center of K. There is no significant loss in assuming D to be irreducible. Then each $z \in Z$, the center of K, is represented on $\mathfrak{p}^+$ by a scalar multiple of the identity, and the non-zero eigenspaces of $\rho(Z)$ in H are the spaces $\{\boldsymbol{P}_m\}_{m=0,1,2,\cdots}$, where $\boldsymbol{P}_m$ denotes the space of homogeneous polynomials of degree m. Each space $\boldsymbol{P}_m$, in turn, may be decomposed under $\rho(K)$, and if we denote by ρ_m the restriction of ρ to $\boldsymbol{P}_m$, then ρ_m is the m-th symmetric tensor power of ρ_1, and ρ_1 is the adjoint representation of K on $\mathfrak{p}^+$. The papers of Hua [loc. cit.] are largely devoted to computing the decomposition of the representations ρ_m into irreducible ones by finding explicit polynomial bases for the irreducible subspaces of $\boldsymbol{P}_m$, and to normalizing the orthogonal polynomials so obtained to have unit length. Finally, one is able in this way effectively to compute the kernel function. Although Hua's calculations are only for classical type domains, there seems to be no practical obstacle to doing the same thing for the two exceptional irreducible domains. It should be remarked, by the way, that the so-called Fourier expansion of an element of H turns out to be essentially nothing else than its power series development at 0, which converges throughout D, since the latter is "centered" around the origin; the remarkable point is that the series, which may cease to be convergent at points of ∂D, nevertheless converges in $L^2(D)$. Moreover, it is clear that one likewise has a representation of K on $\mathcal{O}^p(D)$, $1 \leqq p \leqq \infty$, and it follows from Theorem 20 that if $v \in \mathcal{O}^p(D)$, then the Fourier series for v converges to v in the sense of L^p, hence, by Chapter 4, section 2, converges to v normally on D. In particular, we see that $\mathcal{O}^p(D) \cap \mathcal{O}^q(D)$ is dense in $\mathcal{O}^p(D)$ and in $\mathcal{O}^q(D)$ for $1 \leqq p \leqq q \leqq \infty$.

Another use of Theorem 20 is to be found in the proof of Theorem 3, pp. 15–16 of [26e], where one also uses a further result, to be dealt with in section 2 of the next chapter.

CHAPTER 10
FUNCTIONAL ANALYSIS FOR AUTOMORPHIC FORMS

§1. Lemmas on operator algebras [22b]

We recall the definitions of algebraic irreducibility, complete irreducibility, and topological irreducibility from Chapter 8, section 2. The definitions there were for representations of an associative algebra $\mathfrak{A}$ over the complex numbers. If ρ is a representation of a topological group G, then ρ will be said to have a certain property of irreducibility if the naturally associated representation of the algebra $\mathcal{K}(G)$ (with convolution product) has that property.

Let $x \mapsto \rho(x)$ be a completely irreducible representation of a complex associative algebra $\mathfrak{A}$ by an algebra of bounded operators on a Banach space $\mathcal{H}$. It is easy to see that ρ is topologically irreducible. In fact, if such were not the case, then there is a closed subspace V of $\mathcal{H}$ such that $\rho(x)V \subset V$ for every $x \in \mathfrak{A}$, and there is a point $x_0 \in \mathcal{H} - V$. By the Hahn-Banach theorem, there exists a bounded linear functional f on $\mathcal{H}$ which is not identically zero on V. Define $T(v) = f(v) \cdot x_0$, $v \in \mathcal{H}$. Then T is bounded and maps V onto the line spanned by x_0. Clearly such an operator could not be a strong limit of operators $\rho(x)$, $x \in \mathfrak{A}$. When $\mathcal{H}$ is a Hilbert space, there is a converse [22a, 45].

LEMMA 1 (*von Neumann*). *Let G be a topological group and let ρ be a unitary representation of G on a Hilbert space $\mathcal{H}$. Assume that ρ is topologically irreducible. Then ρ is completely irreducible.*

PROOF. Let $\mathfrak{A}$ be the algebra $\mathcal{K}(G)$ (with convolution product) and denote again by ρ the natural extension to $\mathfrak{A}$ of the representation ρ of G. Let

$$\mathfrak{A}_0 = \{\rho(x) \mid x \in \mathfrak{A}\}.$$

Clearly $\rho(e)$ is the identity transformation of $\mathcal{H}$; since ρ is unitary, it is clear that if $a \in \mathfrak{A}_0$, then the adjoint a^* of a with respect to the given inner product on $\mathcal{H}$ is also in $\mathfrak{A}_0$. Therefore, if V is a subspace of $\mathcal{H}$ such that $a \cdot V \subset V$ for all $a \in \mathfrak{A}_0$, then the orthogonal complement $V^{\perp}$ of V is also stable under $\mathfrak{A}_0$.

Let $\mathfrak{A}'$ be the set of all bounded operators a' on $\mathcal{H}$ such that $a'a = aa'$ for all $a \in \mathfrak{A}_0$. Let T be a bounded operator on $\mathcal{H}$ such that $Ta' = a'T$ for all $a' \in \mathfrak{A}'$. We claim that if V is closed and stable under $\mathfrak{A}_0$, then $TV \subset V$. In fact, it is clear that the operator S defined on $\mathcal{H} = V \oplus V^{\perp}$ by $S|V = Id_V$, $S|V^{\perp} = 0$ belongs to $\mathfrak{A}'$, since both V and $V^{\perp}$ are stable under $\mathfrak{A}_0$, while if $v \in V$, we may write $Tv = v' + v''$ with $v' \in V$, $v'' \in V^{\perp}$, and then $v' + v'' = Tv = TSv = STv = S(v' + v'') = v'$, so that $v'' = 0$ and $Tv = v' \in V$. We let

$$\mathfrak{A}'' = \{T \mid T \text{ is a bounded operator, } Ta' = a'T \text{ for all } a' \in \mathfrak{A}'\}.$$

To complete the proof of Lemma 1, we need two auxiliary lemmas [22a].

LEMMA 2. *If T is a bounded operator on $\mathcal{H}$, then $T \in \mathfrak{A}''$ if and only if for each $x \in \mathcal{H}$ and any $\varepsilon > 0$, there exists $a \in \mathfrak{A}_0$ such that*

$$\|Tx - ax\| < \varepsilon. \tag{1}$$

PROOF. Suppose $T \in \mathfrak{A}''$. Fix $x \in \mathcal{H}$ and $\varepsilon > 0$. The closure of the set of all ax, $a \in \mathfrak{A}_0$, is invariant under $\mathfrak{A}_0$, hence is stable under T. In particular, there must exist $a \in \mathfrak{A}_0$ such that $\|Tx - ax\| < \varepsilon$.

Conversely, suppose (1) is fulfilled for each $x \in \mathcal{H}$, $\varepsilon > 0$, for a suitable choice of $a \in \mathfrak{A}_0$. It follows that if P is a projection belonging to $\mathfrak{A}'$, and if x, ε, and a satisfy (1), then we must have

$$\|PTx - aPx\| < \varepsilon. \tag{2}$$

If $Px = 0$, this implies $PTx = 0$; if $Px = x$, then $Id. - P = P'$ is also a projection in $\mathfrak{A}'$ ($1 \in \mathfrak{A}'$) and we obtain $P'Tx = 0$, or $PTx = Tx = TPx$. Combining these two cases, we see that T commutes with all projections in $\mathfrak{A}'$; and then it follows from the approximation of any element of $\mathfrak{A}'$ by a linear combination of projections in $\mathfrak{A}'$ [41 : p. 95] (noting that $\mathfrak{A}'$ is closed in the topology of the operator norm) that T commutes with every element of $\mathfrak{A}'$, i.e., $T \in \mathfrak{A}''$.

LEMMA 3. *With notation as above, let $T \in \mathfrak{A}''$, let $\varepsilon > 0$ be given, and let $\{x_n\}$ be a countable set of points of $\mathcal{H}$ such that*

(3) $$\sum_n \|x_n\|^2 < +\infty.$$

Then there exists $a \in \mathfrak{A}_0$ *such that*

(4) $$\sum_n \| Tx_n - ax_n \|^2 < \varepsilon.$$

PROOF. Let $\mathcal{G}$ be the Hilbert space of sequences $\{x_n\}$ satisfying (3), and if A is a continuous bounded operator on $\mathcal{H}$, let U_A be the operator on $\mathcal{G}$ defined by $U_A\{x_n\} = \{Ax_n\}$. Then U gives a representation of $\mathfrak{A}_0$ by an algebra $\mathfrak{B}$ of bounded operators on $\mathcal{G}$ that contains the identity and is closed under taking adjoints. By the preceding lemma, it is clearly sufficient to show that if $T \in \mathfrak{A}''$, then $U_T \in (\mathfrak{B}')' = \mathfrak{B}''$, the double commutor algebra of $\mathfrak{B}$.

We may put any bounded operator S on $\mathcal{G}$ in the form of an infinite matrix $\{S_{pq}\}$, where each S_{pq} is a continuous bounded operator in $\mathcal{H}$ such that if $S\{x_n\} = \{y_n\}$, then $y_p = \sum_q S_{pq}x_q$; it is easy to see that $S \in \mathfrak{B}'$ (the commutor of $\mathfrak{B}$) if and only if $S_{pq} \in \mathfrak{A}'$ for all p, q. Therefore $T \in \mathfrak{A}''$ clearly implies $U_T \in \mathfrak{B}''$, which proves the lemma.

Now we complete the proof of Lemma 1. We have assumed ρ is a topologically irreducible unitary representation. Therefore, if V is any non-zero subspace of $\mathcal{H}$ stable under $\mathfrak{A}_0 = \rho(\mathfrak{A})$, its closure must be all of $\mathcal{H}$. If $x_0 \in \mathcal{H}$, $x_0 \neq 0$, then $\rho(e)x_0 = x_0$, so that $\mathfrak{A}_0 x_0 \neq \{0\}$ and $Cl(\mathfrak{A}_0 x_0) = \mathcal{H}$. Therefore, if T is a given bounded operator on $\mathcal{H}$, and if $\varepsilon > 0$ is given, there exists $a \in \mathfrak{A}_0$ such that

$$\| Tx_0 - ax_0 \| < \varepsilon.$$

Hence, $T \in \mathfrak{A}''$ by Lemma 2. Then it follows from Lemma 3 that T is a strong limit of operators $a \in \mathfrak{A}_0$, completing the proof.

LEMMA 4 (*Kaplansky* [34]). *Let M_n be the algebra of $n \times n$ matrices over a field k. Then for a suitable choice of positive integer r, there exists a non-commutative polynomial P in r variables with the following properties:*

(P1) $P(A_1, \cdots, A_r) = 0$ *for all* $A_1, \cdots, A_r \in M_n$;
(P2) *there exist* $A_1, \cdots, A_r \in M_{n+1}$ *such that* $P(A_1, \cdots, A_r) \neq 0$; *and*
(P3) $P(X_1, \cdots, X_r)$ *depends linearly on each of the variables* X_i.

SKETCH OF PROOF (*based on* [34]): If $x_1, \cdots, x_r$ are arbitrary elements in an associative algebra $\mathfrak{A}$, let

$$[x_1, \cdots, x_r] = \sum_{\pi} \operatorname{sgn}(\pi) x_{\pi(1)} \cdots x_{\pi(r)},$$

where π runs over all permutations of $\{1, \cdots, r\}$ and $\operatorname{sgn}(\pi)$ is $+1$ or -1 according as π is even or odd. If $\mathfrak{A}$ has dimension $k-1$ over its center, it is clear that $[x_1, \cdots, x_k]=0$, for all $x_1, \cdots, x_k \in \mathfrak{A}$. Now if $\mathfrak{A}=M_n$, let $r(n)$ be the smallest integer such that $[x_1, \cdots, x_r]=0$ for all r-tuples of elements $(x_1, \cdots, x_r)$ of $\mathfrak{A}$. Clearly $r(n) \leqq n^2+1$ and $r(1)=2$. It is proven in [34] that $r(n) \geqq r(n-1)+2$. From this it follows that P defined by $P(X_1, \cdots, X_r)=[X_1, \cdots, X_r]$ with $r=r(n)$ satisfies the requirements of the lemma.

LEMMA 5 (*Kaplansky-Godement* [22b]). *Let $\mathfrak{A}$ be an associative algebra over $\boldsymbol{C}$ and let n be a positive integer such that for any $\alpha \in \mathfrak{A}$ there exists a representation ρ of $\mathfrak{A}$ with* $\dim(\rho) \leqq n$ *such that* $\rho(\alpha) \neq 0$. *Then each completely irreducible representation of $\mathfrak{A}$ is of dimension $\leqq n$.*

PROOF. Let P be a polynomial as in Lemma 4, and let $x_1, \cdots, x_r \in \mathfrak{A}$. Now $P(x_1, \cdots, x_r) \in \mathfrak{A}$ and if $P(x_1, \cdots, x_r) \neq 0$, then there exists a representation ρ of $\mathfrak{A}$ by $n \times n$ matrices such that

$$0 \neq \rho(P(x_1, \cdots, x_r)) = P(\rho(x_1), \cdots, \rho(x_r)),$$

an evident contradiction. Hence we must have $P(x_1, \cdots, x_r)=0$ for all $x_1, \cdots, x_r \in \mathfrak{A}$.

Now let $x \mapsto \rho(x)$ be a completely irreducible representation of $\mathfrak{A}$ on some Banach space $\mathscr{H}$. Then

$$P(\rho(x_1), \cdots, \rho(x_r)) = \rho(P(x_1, \cdots, x_r)) = \rho(0) = 0.$$

Let $A_1, \cdots, A_r$ be any bounded operators on $\mathscr{H}$; if we approximate A_1 by operators $\rho(x_1)$, we obtain by (P3) and by continuity of P that $P(A_1, \rho(x_2), \cdots, \rho(x_r))=0$; proceeding similarly with each of the other variables in turn, we obtain $P(A_1, \cdots, A_r)=0$; assuming $\mathscr{H}$ contains an $(n+1)$-dimensional subspace $\mathfrak{E}$ and denoting by $A_1, \cdots, A_r$ arbitrary operators on $\mathfrak{E}$, each of which is the restriction to $\mathfrak{E}$ of some bounded operator on $\mathscr{H}$ (Hahn-Banach theorem), we obtain $P(A_1, \cdots, A_r) \equiv 0$, contradicting (P2). Thus we have $\dim \mathscr{H} \leqq n$, proving the lemma.

LEMMA 6 (*Dieudonné*). *Every algebraically irreducible representation of an associative complex algebra $\mathfrak{A}$ by bounded operators on a complex Banach space is completely irreducible.*

PROOF. Let ρ be the algebraically irreducible representation of $\mathfrak{A}$ on the complex Banach space $\mathscr{H}$ and let M be the set of all linear endomorphisms (bounded or not) which commute with every $\rho(x)$, $x\in\mathfrak{A}$. Clearly, M contains the subalgebra M_0 of scalar multiplications by complex numbers. We first show that $M=M_0$.

Let $a\in\mathscr{H}$, $a\neq 0$, and for $U\in M$, define

$$||U||=\inf_{x,\ \rho(x)a=Ua}||\rho(x)||$$

(since $\bigcup_{x\in\mathfrak{A}}\rho(x)a=\mathscr{H}$, because ρ is *algebraically* irreducible, the set of $x\in\mathfrak{A}$ such that $\rho(x)a=Ua$ is not empty). If $y\in\mathfrak{A}$ is such that $\rho(y)a=Va$, $z=x+y$, then clearly

$$(U+V)a=\rho(z)a=\rho(x)a+\rho(y)a,$$

from which $||U+V||\leqq||U||+||V||$, while

$$UVa=\rho(x)\rho(y)a=\rho(xy)a,$$

so that $||UV||\leqq||U||\cdot||V||$; similarly, $||\lambda U||=|\lambda|\cdot||U||$ for $\lambda\in\boldsymbol{C}$. Thus M is made into a normed algebra. If $\alpha\in M$, $\alpha\neq 0$, then $\alpha:\mathscr{H}\to\mathscr{H}$ is a $\boldsymbol{C}$-isomorphism of the simple $\mathfrak{A}$-module $\mathscr{H}$ onto itself (Schur's lemma). Therefore, M is a division algebra over $\boldsymbol{C}$ (possibly non-commutative and possibly of infinite dimension over $\boldsymbol{C}$). But (Gelfand-Mazur [64 : pp. 128–130]) any normed division algebra over $\boldsymbol{C}$ is isometrically isomorphic to $\boldsymbol{C}$, hence, M_0 is all of M. Thus $M=M_0=\boldsymbol{C}$.

Now by [32a : Thm. 6, p. 232] with $\mathfrak{A}=\mathscr{H}$ and $\mathfrak{D}=M$, given any bounded operator T on $\mathscr{H}$ and any finite set $\{a_1,\cdots,a_n\}\subset\mathscr{H}$, there exists some $x\in\mathfrak{A}$ such that $\rho(x)a_i=Ta_i$, $1\leqq i\leqq n$ (in the algebraic context, we can forget about the $\varepsilon>0$!). This proves Lemma 6.

Let G be a locally compact, unimodular group and let K be a compact subgroup. As before, let $\mathcal{E}$ denote the family of equivalence classes of finite-dimensional, irreducible representations of K, let $\mathfrak{d}\in\mathcal{E}$, let $d(\mathfrak{d})=\dim(\mathfrak{d})$, $\chi=\chi_{\mathfrak{d}}$, the character of $\mathfrak{d}$. In the rest of this part, product of functions will mean convolution product, i.e., product in the measure-theoretic sense (cf. Chapter 8, section 1).

The orthogonality relations imply $\chi\chi = d(\mathfrak{d})^{-1}\chi$, so if we put $\alpha_{\mathfrak{d}} = d(\mathfrak{d})\chi$, we have $\alpha_{\mathfrak{d}}\alpha_{\mathfrak{d}} = \alpha_{\mathfrak{d}}$. Now a measure on K is considered as one on G and a continuous function on K is made, as explained before, into a measure on K by multiplying it by the (normalized) Haar measure dk on K. Then put

$$L(\mathfrak{d}) = \{f \in \mathcal{K}(G) \mid \bar{\alpha}_{\mathfrak{d}} f = f\bar{\alpha}_{\mathfrak{d}} = f\}. \tag{5}$$

Clearly $L(\mathfrak{d})$ is a subalgebra of $\mathcal{K}(G)$ under the convolution product, and $f \mapsto \bar{\alpha}_{\mathfrak{d}} f \bar{\alpha}_{\mathfrak{d}}$ is a projection of $\mathcal{K}(G)$ onto $L(\mathfrak{d})$. In fact, one sees easily that for any $f \in \mathcal{K}(G)$ we have $\bar{\alpha}_{\mathfrak{d}} f \bar{\alpha}_{\mathfrak{d}} \in L(\mathfrak{d})$, and if $f \in L(\mathfrak{d})$, then $\bar{\alpha}_{\mathfrak{d}} f \bar{\alpha}_{\mathfrak{d}} = f$.

Let ρ be a representation (Chapter 8, section 2) of G on a Banach space $\mathcal{H}$. As before, let $E_{\mathfrak{d}} = \rho(\bar{\alpha}_{\mathfrak{d}})$, the projection of $\mathcal{H}$ onto its $\mathfrak{d}$-isotypic subspace $\mathcal{H}_{\mathfrak{d}}$. Evidently $\mathcal{H}_{\mathfrak{d}}$ is closed in $\mathcal{H}$, and $\mathcal{H}_{\mathfrak{d}}$ is quasi-semi-simple (Chapter 9, section 2) as a K-module under the representation ρ. If $\dim \mathcal{H}_{\mathfrak{d}}$ is finite, then $\dim \mathcal{H}_{\mathfrak{d}} = p \cdot d(\mathfrak{d})$ for some integer $p \geqq 0$, and in this case we say $\mathfrak{d}$ is contained p times in ρ. As before, ρ gives rise to a representation, again denoted by ρ, of $\mathcal{K}(G)$ as an algebra under the convolution product, operating on $\mathcal{H}$ by bounded transformations, and we have

$$\rho(\bar{\alpha}_{\mathfrak{d}} f \bar{\alpha}_{\mathfrak{d}}) = E_{\mathfrak{d}} \rho(f) E_{\mathfrak{d}}; \tag{6}$$

therefore, $\rho(f) \cdot \mathcal{H}_{\mathfrak{d}} \subset \mathcal{H}_{\mathfrak{d}}$ for $f \in L(\mathfrak{d})$ and so, fixing $\mathfrak{d}$, and denoting the restriction of $\rho(f)$ to $\mathcal{H}_{\mathfrak{d}}$ by $\tilde{\rho}(f)$, we obtain a representation $\tilde{\rho}$ of $L(\mathfrak{d})$ on $H_{\mathfrak{d}}$. If $f \in L(\mathfrak{d})$, then $\rho(f) = 0$ obviously implies $\tilde{\rho}(f) = 0$; if, conversely, $f \in L(\mathfrak{d})$ and $\tilde{\rho}(f) = 0$, then (6) implies that $\rho(f) = 0$.

More generally, ρ determines a representation of the algebra $M(G)$ of measures on G with compact support, and if $\mu \in M(G)$, then $\rho(\mu)$ is a bounded operator on $\mathcal{H}$. Complete irreducibility has been defined for the representation ρ of G to mean the complete irreducibility of the subalgebra $\rho(\mathcal{K}(G))$ of $\rho(M(G))$, which of course implies the density of the former in the latter with respect to the topology of pointwise convergence for operators, and implies trivially the complete irreducibility of $\rho(M(G))$.

LEMMA 7. *Assume ρ is completely irreducible; then the representation $\tilde{\rho}$ of $L(\mathfrak{d})$ on $\mathcal{H}_{\mathfrak{d}}$ is completely irreducible.*

PROOF. Let T be a bounded operator on $\mathcal{H}_{\mathfrak{d}}$ and extend it to a bounded operator on $\mathcal{H}$ by defining

$$T\cdot v = TE_{\mathfrak{d}}\cdot v, \quad v\in\mathcal{H}.$$

Let $\mathcal{A}$ be the set of pairs (F, ε), where F is a finite subset of $\mathcal{H}$ and $\varepsilon>0$. We write $(F', \varepsilon')\succ(F, \varepsilon)$ if $F\subset F'$ and $\varepsilon'<\varepsilon$. Now there exists a generalized sequence $\{f_a\}_{a\in\mathcal{A}}\subset\mathcal{K}(G)$ such that $\rho(f_a)$ converges strongly on $\mathcal{H}$ to T in the sense that if $a=(F, \varepsilon)\in\mathcal{A}$, then

$$\|\rho(f_a)\cdot x - T\cdot x\|<\varepsilon, \quad x\in F.$$

Putting $g_a=\bar{\alpha}_{\mathfrak{d}}f_a\bar{\alpha}_{\mathfrak{d}}$, we have $g_a\in L(\mathfrak{d})$ and we see that for $x\in\mathcal{H}_{\mathfrak{d}}$ we have

$$\tilde{\rho}(g_a)\cdot x = E_{\mathfrak{d}}\rho(f_a)E_{\mathfrak{d}}\cdot x\to E_{\mathfrak{d}}TE_{\mathfrak{d}}\cdot x = T(E_{\mathfrak{d}}\cdot x) = T\cdot x,$$

where $\to$ indicates convergence in the Banach space norm. Thus $\tilde{\rho}(g_a)$ converges strongly to T on $\mathcal{H}_{\mathfrak{d}}$, as desired.

DEFINITION. *A set Ω of representations of G is said to be complete if for every non-zero $f\in\mathcal{K}(G)$, there exists $\rho\in\Omega$ such that $\rho(f)\neq 0$.*

LEMMA 8. *Let Ω be a complete set of representations of G, and, other notation being as above, let p be a non-negative integer. Assume $\mathfrak{d}$ is contained at most p times in each $\rho\in\Omega$. Then $\mathfrak{d}$ is contained at most p times in any completely irreducible representation of G.*

PROOF. Let ρ be completely irreducible. If $f\in L(\mathfrak{d})$, then $\rho(f)=0$ if and only if $\tilde{\rho}(f)=0$, so that

$$\tilde{\Omega}=\{\tilde{\rho}\mid\rho\in\Omega\}$$

is a complete set of representations of $L(\mathfrak{d})$, and if $\tilde{\rho}\in\tilde{\Omega}$, then $\dim\tilde{\rho}\leqq p\dim(\mathfrak{d})$; hence, by Lemma 5, every completely irreducible representation of $L(\mathfrak{d})$ has dimension $\leqq p\dim(\mathfrak{d})$. But if ρ_1 is any completely irreducible representation of G on a Banach space $\mathcal{H}_1$, then $\tilde{\rho}_1$ is a completely irreducible representation of $L(\mathfrak{d})$ (by Lemma 7), hence $\dim\mathcal{H}_{1\mathfrak{d}}\leqq p\dim(\mathfrak{d})$, which completes the proof.

LEMMA 9. *Let G be a connected, semi-simple Lie group which has a faithful, finite-dimensional representation. Then the set of finite-dimensional irreducible representations of G is complete.*

PROOF. Let $\mathfrak{A}$ be the vector space spanned by all the coefficient functions of finite-dimensional representations of G. $\mathfrak{A}$ is an algebra because the product of two coefficients is a coefficient (tensor products) and is closed under complex conjugation (conjugate representations); if $g, g' \in G$, $g \neq g'$, then there exists $f \in \mathfrak{A}$ such that $f(g) \neq f(g')$. Hence (Stone-Weierstrass) $\mathfrak{A}$ is dense in $\mathcal{K}(G)$ in the compact-uniform topology. On the other hand, since G is semi-simple, any finite-dimensional representation is a direct sum of irreducible representations, hence, $\mathfrak{A}$ is generated as an algebra by the coefficients of irreducible, finite-dimensional representations of G (comp. Chapter 8, section 6). From these facts, given any $h \in \mathcal{K}(G)$, there exists $f \in \mathfrak{A}$ such that $(f, h) = \int_G f\bar{h} \neq 0$; hence, there exists a finite-dimensional irreducible representation ρ of G such that $\rho(h) \neq 0$; therefore, the set of finite-dimensional, irreducible representations of G is complete, and this completes the proof.

This and the preceding lemma make it clear that if G is a connected, semi-simple Lie group with a faithful, finite-dimensional representation, then the number of times any $\mathfrak{d}$ occurs in any infinite-dimensional, completely irreducible representation of G is bounded by the number of times it occurs in finite-dimensional, irreducible representations. Now if G is connected and semi-simple, has a faithful, finite-dimensional representation, and K is a maximal compact subgroup of G, then $G = N \cdot K$, where N is a connected, solvable, closed subgroup of G (Iwasawa decomposition—Chapter 7, section 5). Fix this notation and if $g \in G$, write $g = n \cdot k$, where $n = n(g) \in N$ and $k = k(g) \in K$ are uniquely determined.

DEFINITION. *Let N be as defined above and let $n \mapsto \alpha(n)$ be a character (i.e., a one-dimensional representation) of N; denote by $\mathcal{H}^\alpha$ the vector space of continuous complex functions θ on G such that $\theta(ng) = \alpha(n)\theta(g)$, $n \in N$, $g \in G$; and define a representation $g \mapsto \rho^\alpha(g)$ of G on $\mathcal{H}^\alpha$ by letting*

$$\rho^\alpha(g)\theta(g') = \theta(g'g), \quad \theta \in \mathcal{H}^\alpha.$$

Then ρ^α is called the representation of G induced by α, and the set of representations induced in this way by all the characters of N is denoted by Ω_N.

Note that if $G = N \cdot K$, as above, $\mathcal{H}^\alpha$ can be naturally identified with $\mathcal{K}(K)$ by assigning to each $\theta \in \mathcal{H}^\alpha$ its restriction to K; since $K \cap N = \{e\}$, the values on K may be arbitrarily assigned, by Tietze's extension theorem, subject only to the requirement of continuity. Clearly ρ^α is a continuous representation of G on $\mathcal{H}^\alpha$, the latter being assigned the compact-uniform topology.

LEMMA 10. *Let N and G be as above. Then each finite-dimensional, irreducible representation of G is a direct summand of some member of Ω_N.*

PROOF. Let σ be a finite-dimensional, irreducible representation of G on $\mathcal{H}$ and let σ' be the representation of G on the algebraic dual space $\mathcal{H}'$ of $\mathcal{H}$, contragredient to σ; let $\rho = \sigma' \mid N$. Then ρ contains a one-dimensional representation α^{-1} (possibly trivial) of N and there exists $a' \in \mathcal{H}'$, $a' \neq 0$, such that $\rho(n)a' = \alpha(n)^{-1}a'$ for all $n \in N$ (Lie-Kolchin theorem (cf. [6a])). If $a \in \mathcal{H}$, define the function θ_a on G by $\theta_a(g) = (\sigma(g)\cdot a, a')$; θ_a is a linear function of a and for $n \in N$ we have

$$\theta_a(ng) = (\sigma(ng)a, a') = (\sigma(n)\sigma(g)a, a') = (\sigma(g)a, \sigma'(n)^{-1}a')$$

$$= (\sigma(g)a, \rho(n)^{-1}a') = \alpha(n)(\rho(g)a, a') = \alpha(n)\theta_a(g),$$

thus $\theta_a \in \mathcal{H}^\alpha$, and $a \mapsto \theta_a$ is a linear mapping of $\mathcal{H}$ into $\mathcal{H}^\alpha$. Since $\theta_{\sigma(g)a}(g') = \theta_a(g'g)$, this linear mapping is clearly a homomorphism of G-modules which is not identically zero. Since $\mathcal{H}$ is a simple G-module, it follows that $a \mapsto \theta_a$ is an injection of $\mathcal{H}$, as a G-module, into $\mathcal{H}^\alpha$, thus $\sigma \subset \rho^\alpha$, as desired, since G is semi-simple.

Now if $f \in \mathcal{K}(G)$, $f \neq 0$, let σ be a finite-dimensional representation of G such that $\sigma(f) \neq 0$, and let α be as above such that $\sigma \subset \rho^\alpha$. Then $\rho^\alpha(f)$ stabilizes the image $\theta(\mathcal{H})$ under $a \mapsto \theta_a$ of the representation space $\mathcal{H}$ of σ, hence, the restriction of $\rho^\alpha(f)$ to $\theta(H)$ is $\sigma(f) \neq 0$; therefore, $\rho^\alpha(f) \neq 0$. In other words, Ω_N is a complete set of representations of G. If $G = N \cdot K$ (Iwasawa decomposition), we may identify any $\mathcal{H}^\alpha$ with $\mathcal{K}(K)$ as indicated above, and then $\rho^\alpha \mid K$ is the same as the regular representation of K on $\mathcal{K}(K)$ by virtue of the relation

$$\rho^\alpha(k)f(k') = f(k'k),$$

and each $\mathfrak{d}$ is contained just $\dim(\mathfrak{d})$ times in the regular representation (Chapter 8, section 4). Hence we have

THEOREM 21. *Let G be a connected, semi-simple Lie group having a faithful, finite-dimensional representation, and let K be a maximal compact subgroup of G. Then any class $\mathfrak{d}$ of finite-dimensional irreducible representations of K is contained at most* $\dim(\mathfrak{d})$ *times in any completely irreducible representation ρ of G.*

This is an obvious consequence of what we have proved above.

A special case of the theorem is, of course, that which arises when ρ is an irreducible unitary representation of G.

§ 2. Some further results

As usual, a bounded operator T on a Banach space $\mathcal{H}$ is called compact or completely continuous if the image under it of the unit sphere of $\mathcal{H}$ has compact closure in $\mathcal{H}$. We have [21 : Chap. I, § 2, no. 3]

LEMMA 11. *Let G be a Lie group and let ρ be a unitary representation of G on a separable Hilbert space $\mathcal{H}$. Suppose that for each $\varphi \in C_c^\infty(G)$ the operator $\rho(\varphi)$ on $\mathcal{H}$ is compact. Then $\mathcal{H}$ is the Hilbert space direct sum of a countable family $\{\mathcal{H}_i\}$ of closed subspaces invariant under $\rho(G)$ such that the representation ρ_i of G on $\mathcal{H}_i$ so obtained is topologically irreducible and such that the number of irreducible ρ_i equivalent to a given one is finite.*

PROOF. [21 : loc. cit.]. We consider $\varphi \in C_c^\infty(G)$ satisfying $\varphi(g^{-1}) = \overline{\varphi(g)}$ for all $g \in G$. Then, for such φ, the operator $\rho(\varphi)$ is self-adjoint and is completely continuous by hypothesis. Therefore, $\rho(\varphi)$ has a countable spectrum without limit point other than 0, and all the points of that spectrum are real and of finite multiplicity. Thus

$$\mathcal{H} = \mathcal{H}_\varphi \oplus \sum_{k=1}^{\infty} \mathcal{H}_{\varphi, k}, \tag{7}$$

where $\mathcal{H}_\varphi = \{v \in \mathcal{H} \mid \rho(\varphi) \cdot v = 0\}$ and

$$\mathcal{H}_{\varphi, k} = \{v \in \mathcal{H} \mid \rho(\varphi) \cdot v = \lambda_k v\}$$

have finite dimension, where $\lambda_k = \lambda_k(\varphi)$ runs over the non-zero spectrum of $\rho(\varphi)$. Naturally, the equality in (7) means that the left side is the closure of the restricted, orthogonal, direct sum (briefly, Hilbert space direct sum) of the terms on the right.

Clearly, if $\mathcal{H}'$ is any subspace of $\mathcal{H}$ invariant under $\rho(G)$, we have an analogous decomposition

$$\mathcal{H}' = \mathcal{H}'_{\varphi} \oplus \sum_{k=1}^{\infty} \mathcal{H}'_{\varphi,k}, \tag{8}$$

where $\mathcal{H}'_{\varphi} = \mathcal{H}' \cap \mathcal{H}_{\varphi}$, $\mathcal{H}'_{\varphi,k} = \mathcal{H}' \cap \mathcal{H}_{\varphi,k}$.

Consider now all the subspaces $\mathcal{H}_{\varphi,k}$ as φ runs over all elements of $C_c^{\infty}(G)$ satisfying $\overline{\varphi(g)} = \varphi(g^{-1})$ and $\lambda_k = \lambda_k(\varphi)$ runs over the non-zero eigenvalues of each φ. We first show that $\mathcal{H}$ is the closure of the span of all these subspaces. In fact, let $\mathcal{H}_1$ be the closure of the sum of all these spaces, and if $\mathcal{H}_1 \neq \mathcal{H}$, let f be a non-zero element of $\mathcal{H}_1^{\perp}$. Then $f \in \bigcap_{\varphi} \mathcal{H}_{\varphi}$, so that $\rho(\varphi)f = 0$ for all $\varphi \in C_c^{\infty}(G)$ satisfying $\varphi(g^{-1}) = \overline{\varphi(g)}$.

As ρ is a continuous representation, it follows that given any $\varepsilon > 0$, there exists a neighborhood $\mathcal{N}$ of $e \in G$ such that

$$\|\rho(g)f - f\| < \varepsilon \|f\|, \quad g \in \mathcal{N}. \tag{9}$$

We may now choose φ of the given type such that $\mathcal{S}(\varphi) \subset \mathcal{N}$, such that $\varphi \geqq 0$ everywhere, and such that

$$\int_G \varphi(g)dg = 1. \tag{10}$$

Then

$$\|\rho(\varphi)f - f\| = \|\int_{\mathcal{N}} \varphi(g)(\rho(g)f - f)dg\| < \varepsilon \|f\|,$$

because of (8) and (9), and since $\rho(\varphi)f = 0$, we obtain $\|f\| < \varepsilon \|f\|$, for each $\varepsilon > 0$. This is an evident contradiction, and so $\mathcal{H}_1 = \mathcal{H}$.

It follows that any closed invariant subspace $\mathcal{H}'$ of $\mathcal{H}$ has a non-zero intersection with at least one of the spaces $\mathcal{H}_{\varphi,k}$.

Fix one of the spaces $\mathcal{H}_{\varphi,k}$ and consider all the intersections of it with all *closed* invariant subspaces of $\mathcal{H}$. From among these, choose one $\neq \{0\}$ of minimum dimension and call it $\mathcal{H}'_{\varphi,k}$. Let $\mathcal{H}'$ run through all closed invariant subspaces of $\mathcal{H}$ such that $\mathcal{H}' \cap \mathcal{H}_{\varphi,k} = \mathcal{H}'_{\varphi,k}$ and let $\mathcal{H}_1$ be the intersection of all these; clearly, $\mathcal{H}_1$ may be characterized as the *minimal, closed,* invariant subspace of $\mathcal{H}$ containing $\mathcal{H}'_{\varphi,k}$.

Now $\mathcal{H}_1$ is irreducible as a closed G-module; for if it had a proper, closed, invariant subspace $\mathcal{H}_{11}$, then it would be the direct sum of $\mathcal{H}_{11}$ and $\mathcal{H}_{11}^{\perp}$ and as each of these is invariant under G, each must be invariant under $\rho(\varphi)$, so that $\mathcal{H}'_{\varphi,k}$ would be the direct sum of its intersections with each; one way or the other, this leads to a contradiction of the minimality of $\mathcal{H}'_{\varphi,k}$ or the minimality of $\mathcal{H}_1$.

Thus, $\mathcal{H}_1$ is irreducible in the sense that it contains no proper *closed* G-submodule, and we may write

$$\mathcal{H} = \mathcal{H}_1 \oplus \mathcal{H}_1^{\perp}.$$

Now consider all families $\{\mathcal{H}_\alpha\}$ consisting of closed, mutually orthogonal, irreducible $\rho(G)$-invariant subspaces of $\mathcal{H}$. If $\{\mathcal{H}_\alpha\}$ and $\{\mathcal{H}'_{\alpha'}\}$ are two such families, we write $\{\mathcal{H}_\alpha\} \prec \{\mathcal{H}'_{\alpha'}\}$ if every $\mathcal{H}_\alpha$ is equal to some $\mathcal{H}'_{\alpha'}$. The number of subspaces occurring in each family is at most countable because we have assumed $\mathcal{H}$ to be separable. It is now a simple matter to apply Zorn's lemma to obtain a maximal such family (though the actual role of Zorn's lemma is more apparent than real); if $\mathcal{H}$ were not the closure $\mathcal{H}_2$ of the restricted direct sum of the members of the maximal family, we could write

$$\mathcal{H} = \mathcal{H}_2 \oplus \mathcal{H}_2^{\perp},$$

and applying the preceding arguments to $\mathcal{H}_2^{\perp}$ we could enlarge the supposedly maximal family $\mathcal{H}_2$, obtaining a contradiction. Thus we have shown that $\mathcal{H}$ is the closure of the restricted direct sum of closed, irreducible, invariant subspaces.

Now, if $\mathcal{H}_l$ is a closed, irreducible, invariant subspace of $\mathcal{H}$, its intersection with $\mathcal{H}_{\varphi,k}$ is different from $\{0\}$ for some φ and some k; and if $\mathcal{H}_m$ is a $\rho(G)$-module equivalent to $\mathcal{H}_l$, then $\mathcal{H}_{\varphi,k} \cap \mathcal{H}_m \neq \{0\}$, too. Thus, the number of closed, irreducible, invariant subspaces of $\mathcal{H}$ equivalent to $\mathcal{H}_l$ as $\rho(G)$-modules is finite since that number is $\leqq \dim(\mathcal{H}_{\varphi,k})$, which we know to be finite. This completes the proof of Lemma 11.

At this point, we have verified most of the facts that are prerequisite to a reading of the first chapter of [26e], in which it is proved that the spaces of automorphic forms of certain types are of finite dimension; these types include the holomorphic automorphic forms of a fixed positive weight with respect to an arithmetic group acting on a bounded symmetric domain. Omitted here are the proofs of certain special additional facts about universal enveloping algebras. However, we insert here the proof of a lemma that we shall need later; its proof may be found in [52] or in [26e].

LEMMA 12. *Let (X, μ) be a locally compact measure space, where μ is a positive measure such that $\mu(X) < +\infty$. Let $\mathfrak{F}$ be a closed subspace of $L^2(X, \mu)$ which is contained in $L^\infty(X, \mu)$ (the space of μ-essentially bounded functions on X). Then*

$$\dim \mathfrak{F} < +\infty.$$

PROOF. (This proof is credited to Hörmander in [26e].) It is no loss of generality to assume $\mu(X)=1$. If $f \in L^\infty(X, \mu)$, it follows from a trivial calculation that $\|f\|_2 \leqq \|f\|_\infty$. Let $\mathfrak{F}^2$ (resp. $\mathfrak{F}^\infty$) denote the space $\mathfrak{F}$ supplied with the topology it inherits as a subspace of $L^2(X, \mu)$ (resp. of $L^\infty(X, \mu)$). Then the identity mapping of $\mathfrak{F}$ onto itself is a continuous and hence bounded mapping of $\mathfrak{F}^\infty$ onto $\mathfrak{F}^2$; it follows from [41: Theorem, p. 18] that the same mapping is a bounded mapping of $\mathfrak{F}^2$ onto $\mathfrak{F}^\infty$, hence there is a positive constant c such that $\|f\|_\infty \leqq c\|f\|_2$ for all $f \in \mathfrak{F}$. Therefore, if $f_1, \cdots, f_n \in \mathfrak{F}$ are mutually orthogonal elements of length 1 of $\mathfrak{F}^2$, we have

$$\left|\sum_{j=1}^n a_j f_j(x)\right| \leqq \left\|\sum_{j=1}^n a_j f_j\right\|_\infty \tag{11}$$

for all n-tuples of complex numbers $(a_1, \cdots, a_n)$ and for x in the complement of a set of measure zero. By Hölder's inequality and from the choice of the constant c we have

$$\left\|\sum_{j=1}^n a_j f_j\right\|_\infty \leqq c\left\|\sum_{j=1}^n a_j f_j\right\|_2 \leqq c\left(\sum_{j=1}^n |a_j|^2\right)^{1/2}. \tag{12}$$

Combining (11) and (12) and taking $a_j = \overline{f_j(x)}$ we obtain

$$\sum_{j=1}^n |f_j(x)|^2 \leqq c^2 \tag{13}$$

almost everywhere, and integrating over X we obtain $n \leqq c^2$, hence $\dim \mathfrak{F} \leqq c^2$.

This lemma is used in proving the finiteness of the dimensions of spaces of automorphic forms. The main difficulty in its application is to see that a given subspace of $L^\infty(X, \mu)$ is actually a closed subspace of $L^2(X, \mu)$. This is true in the case of spaces of cusp forms which are characterized by certain "closed" conditions expressed by the vanishing of certain Fourier coefficients (q.v. infra, Chapter 11, section 4). This lemma may also be used to verify the finiteness of the dimension of the space of theta functions of a given type, but determining in this way the dimension (which is easy to find by other means) is more difficult.

§ 3. L^p-spaces on G [52: Exps. 9, 10]

Let D be a bounded symmetric domain in $\boldsymbol{C}^n$ and let $G = \mathrm{Hol}(D)$. Denote by K a maximal compact subgroup of G and fix a representation ρ of K in a finite-dimensional vector space V. Though there is no great difficulty in treating the general case, we shall for simplicity henceforth assume that dim $V=1$. Then ρ is a character (of absolute value one) of the center Z of K. Now K is the isotropy group in G of a unique point $z_0 \in D$. We are going to further restrict ρ by assuming

$$\rho(k) = j(z_0, k)^l, \tag{14}$$

for some positive integer l.

If f is a complex-valued, measurable function on D, we let $'f$ be the complex-valued, measurable function on G defined by

$$'f(g) = f(z_0 \cdot g) j(z_0, g)^l, \quad g \in G. \tag{15}$$

It may at once be verified that $'f$ satisfies

$$'f(kg) = \rho(k) \cdot {'f}(g), \quad k \in K. \tag{16}$$

Thus, the assignment $f \mapsto {'f}$ establishes a one-to-one correspondence between complex-valued, measurable functions on D and complex-valued, measurable functions on G satisfying (16); denote the complex linear space composed of the latter by $\mathcal{M}(\rho)$. If, in particular, Γ is a discrete subgroup of G and f is a (not necessarily holomorphic) automorphic form of weight l with respect to Γ, satisfying (by definition)

$$f(z \cdot \gamma) j(z, \gamma)^l = f(z), \quad \gamma \in \Gamma, \; z \in D, \tag{17}$$

then $'f$ satisfies

$$'f(g\gamma) = {'f}(g), \quad \gamma \in \Gamma, \; g \in G. \tag{18}$$

Let dg be a suitably normalized Haar measure on G. We define a norm $\| \;\|_p$, $1 \leqq p < \infty$, on the measurable complex-valued functions on G as usual by

$$\|f\|_p = \left(\int_G |f(g)|^p dg \right)^{1/p}, \tag{19}$$

and as in the last section define $||f||_\infty$ to be the essential supremum of $|f|$ on G. Let L^p, $1 \leqq p \leqq \infty$, be the space of measurable f for which $||f||_p < +\infty$, and put

$$L^p(\rho) = \mathcal{M}(\rho) \cap L^p.$$

If f is a complex, measurable function on D, then in order for $'f \in L^p(\rho)$ it is necessary and sufficient that

$$\int_D |f(z) \boldsymbol{K}_D(z)^{-l/2}|^p d\beta_z < +\infty, \tag{20}$$

where $\boldsymbol{K}_D(z)$ is the Bergmann kernel function (Chapter 4, section 3) on D and $d\beta_z$ is the G-invariant Bergmann volume element on D; of course, the integral in (19) is a constant multiple of that in (20), and now we take the Haar measure on G such as to make that constant unity.

As explained in Chapter 4, $L^p(\rho)$ is a Banach space and $L^2(\rho)$, a Hilbert space. We denote by $\mathcal{H}^p(\rho)$ the subspace of $L^p(\rho)$ coming from holomorphic functions on D. It is a closed subspace as we know from earlier results. We note here some facts about the kernel function $\boldsymbol{K}_\rho$ of $\mathcal{H}^2(\rho)$. If $\{f_\nu\}$ is an orthonormal basis of $\mathcal{H}^2(\rho)$, then from general facts about kernel functions [48] (cf. Chapter 4), one sees that

$$\boldsymbol{K}_\rho(g_1, g_2) = \sum_\nu f_\nu(g_1)\overline{f_\nu(g_2)}, \quad g_1, g_2 \in G, \tag{21}$$

and

$$\boldsymbol{K}_\rho(g_1 g, g_2 g) = \boldsymbol{K}_\rho(g_1, g_2), \tag{22}$$

since G is unimodular and right translation is measure-preserving. Therefore, $\boldsymbol{K}_\rho$ may be viewed as a function of one variable in putting

$$\boldsymbol{K}_\rho(g_1, g_2) = \boldsymbol{K}_\rho'(g_1 g_2^{-1}). \tag{23}$$

On the other hand, from (16) and (21) we see that

$$\boldsymbol{K}_\rho(k_1 g_1, k_2 g_2) = \rho(k_1)\overline{\rho(k_2)}\boldsymbol{K}_\rho(g_1, g_2), \tag{24}$$

for $k_1, k_2 \in K$, and therefore

$$\boldsymbol{K}_\rho'(k_1 g k_2^{-1}) = \rho(k_1)\overline{\rho(k_2)}\boldsymbol{K}_\rho'(g), \quad k_1, k_2 \in K. \tag{25}$$

Therefore, in order to determine the absolute value of $\boldsymbol{K}_\rho$, it is sufficient to calculate $\boldsymbol{K}_\rho'(a)$, where $a \in A$, and $G = KAK$ is the Cartan decomposition of G.

Let $\mathcal{H}_\rho$ be the Hilbert space of holomorphic functions f on D

satisfying (20), i.e., such that $'f \in \mathcal{H}^2(\rho)$, and let $\boldsymbol{K}_\rho{}^*$ be its kernel function. Following the ideas of Chapter 4, section 3, we see that $\boldsymbol{K}_\rho{}^*(z, \zeta)$ satisfies

$$\boldsymbol{K}_\rho{}^*(zg, \zeta g) j(z, g)^l \overline{j(\zeta, g)^l} = \boldsymbol{K}_\rho{}^*(z, \zeta), \quad g \in \mathrm{Hol}(D),$$

so that $\boldsymbol{K}_\rho{}^*(z, z) = c \cdot \boldsymbol{K}_D(z, z)^l$, where c is a constant. Put $f(z, \zeta) = \boldsymbol{K}_\rho{}^*(z, \zeta) - c \cdot \boldsymbol{K}_D(z, \zeta)^l$. Then f is a complex analytic function of $(z, \zeta) \in D \times D^-$ and hence is a complex analytic function of two new variables, $z+\zeta$ and $i(z-\zeta)$, and is zero when $\bar{z} = \zeta$, which is precisely when these new variables are real; thus, $f \equiv 0$, or $\boldsymbol{K}_\rho{}^*(z, \zeta) \equiv c \cdot \boldsymbol{K}_D(z, \zeta)^l$. Now using (15), where $'f$ is allowed to run over an orthonormal basis of $\mathcal{H}^2(\rho)$, one obtains

$$\boldsymbol{K}_\rho{}'(g) = c \cdot j(z_0, g)^l \boldsymbol{K}_D(z_0 \cdot g, z_0)^l, \tag{26}$$

and if we take D to be the Harish-Chandra bounded realization of $K \backslash G$, $z_0 = 0$, and K to be the stability group of 0, then $\boldsymbol{K}_D(0 \cdot g, 0)$ is a constant and for $a \in A$ the calculation of $j(0, a)^l$ may be reduced to the corresponding calculation for a product of unit discs, in number equal to $\dim A$, if we note that the exponent of the result should be multiplied by a certain positive constant to allow for the multiplicities of the weight spaces in the adjoint representation of A. The result of the calculation is [3] that $j(0, a)^l$ is a product of negative powers of hyperbolic cosines, with (negative) exponents that depend linearly on l. It follows that we have

LEMMA 13. *There exists a positive integer l_0 such that if $l \geqq l_0$, then*

$$\boldsymbol{K}_\rho{}' \in L^p(\rho), \quad 1 \leqq p \leqq \infty.$$

(Comp. Lemma 3 of Chapter 11, section 1.) In fact, since $\boldsymbol{K}_D(z, z')$ is a holomorphic function of z, it follows that we actually have $\boldsymbol{K}_\rho{}' \in \mathcal{H}^p(\rho)$. Then by standard inequalities [60a : § 13] we obtain

COROLLARY. *$\mathcal{H}^p(\rho) \subset \mathcal{H}^\infty(\rho)$, $1 \leqq p < +\infty$, if $l > l_0$.*

PROOF. Let $f \in \mathcal{H}^p(\rho)$ and let $q > 0$ be determined by $1/p + 1/q = 1$. The integral which represents $\boldsymbol{K}_\rho{}' * g$, for $g \in \mathcal{H}^p(\rho)$, converges absolutely, and the mapping $g \mapsto \boldsymbol{K}_\rho{}' * g$ is a continuous mapping of $L^p(\rho)$ into itself because $\boldsymbol{K}_\rho{}' \in L^1(\rho)$. On the other hand, $\mathcal{H}^2(\rho) \cap \mathcal{H}^p(\rho)$ is dense in $\mathcal{H}^p(\rho)$, because it contains the image of the space of polynomials, and $\boldsymbol{K}_\rho' * g = g$ for $g \in \mathscr{H}^2(\rho)$. Thus $\boldsymbol{K}_\rho' * f = f$, and so, finally,

$$\|f\|_\infty \leqq \|\boldsymbol{K}_\rho'\|_q \|f\|_p.$$

Now let Γ be a discrete subgroup of G and if $f \in \mathscr{M}(\rho)$ and $f(g\gamma) = f(g)$ for all $\gamma \in \Gamma$, $g \in G$, define

$$\|f\|_{\Gamma, p} = \left(\int_{G/\Gamma} |f(\dot{g})|^p d\dot{g}\right)^{1/p}, \quad 1 \leqq p < \infty,$$

where $d\dot{g}$ is the G-invariant measure on G/Γ coming from dg. Define $\|f\|_{\Gamma,\infty}$ as the essential supremum of $|f|$ on G. Finally, denote by $\mathscr{H}^p_\Gamma(\rho)$ the space of $f \in \mathscr{M}(\rho)$ coming from holomorphic functions on D and satisfying

$$f(g\gamma) = f(g), \quad g \in G, \ \gamma \in \Gamma, \quad \text{and} \quad \|f\|_{\Gamma, p} < +\infty. \tag{27}$$

If G/Γ has finite volume, then it is clear that we have

$$\mathscr{H}^\infty_\Gamma(\rho) \subset \mathscr{H}^p_\Gamma(\rho) \subset \mathscr{H}^1_\Gamma(\rho), \quad 1 \leqq p \leqq \infty. \tag{28}$$

Thus, in particular, $\mathscr{H}^\infty_\Gamma(\rho)$ is a subspace of $\mathscr{H}^2_\Gamma(\rho)$, and if we could prove that it is a *closed* subspace, then it would follow from Lemma 12 that $\dim \mathscr{H}^\infty_\Gamma(\rho) < +\infty$. In fact, when l is large enough, all the spaces $\mathscr{H}^p_\Gamma(\rho)$ coincide (Satake [52 : Exp. 9 (Supp.)]). We shall indicate a proof of this in Chapter 11, but for now recall a special case.

THEOREM 22. *If G/Γ is compact, all the spaces $\mathscr{H}^p_\Gamma(\rho)$ coincide, $1 \leqq p \leqq \infty$, and $\dim(\mathscr{H}^\infty_\Gamma(\rho)) < +\infty$.*

PROOF. In fact, in this case each of the spaces $\mathscr{H}^p_\Gamma(\rho)$ consists of continuous functions which are right-invariant under Γ, hence bounded. Therefore each of the spaces consists of the same set of functions, namely, those which are obtained from holomorphic automorphic forms of weight l on D with respect to Γ.

Now, convergence in any L^p-norm implies, for holomorphic functions, uniform convergence on any compact subset of D. Therefore $\mathscr{H}^p_\Gamma(\rho)$ is a closed subspace of $L^2(G/\Gamma)$, and one may apply Lemma 12 to the measure space $(G/\Gamma, d\dot{g})$.

COROLLARY. *If l is any positive integer, the automorphic forms of weight l on D with respect to Γ form a space of finite dimension if D/Γ is compact.*

CHAPTER 11
CONSTRUCTION OF AUTOMORPHIC FORMS

§1. Poincaré series [3 : §5]

Let G be a reductive Lie group with finitely many components and having a finite-dimensional, faithful, linear representation. Let K be a maximal compact subgroup of G, and let Γ be a discrete subgroup of G. Let $\mathfrak{g}$ be the Lie algebra of G, let $\boldsymbol{U}=\boldsymbol{U}(\mathfrak{g})$ be its universal enveloping algebra, let $\mathfrak{k}$ be the Lie algebra of K, and let $\mathfrak{X}$ be the universal enveloping algebra of $\mathfrak{k}$.

Let V be a finite-dimensional complex vector space and let $\mathcal{F}=\mathcal{F}(G, V)$ denote the vector space of measurable mappings from G to V. We assume V to be supplied with a positive-definite norm $\| \ \|$, and if $f\in F$, let $\|f\|(g)=\|f(g)\|$ for $g\in G$. For other notation, we refer to Chapter 9, especially section 4.

THEOREM 23. *Let $f\in\mathcal{F}$ be of class C^∞ such that $\|f\|$ is (absolutely) integrable on G. We assume that f is $\mathcal{Z}(\mathfrak{g})$-finite (i.e., that f is annihilated by an ideal of finite codimension in $\mathcal{Z}(\mathfrak{g})$ under the representation l) and is finite on the right with respect to K. Then the series*

$$\mathcal{P}_f(g)=\sum_{\gamma\in\Gamma} f(g\cdot\gamma) \tag{1}$$

and

$$\mathcal{P}_{\|f\|}(g)=\sum_{\gamma\in\Gamma} \|f(g\cdot\gamma)\| \tag{2}$$

converge normally on G and the functions $\mathcal{P}_f$ and $\mathcal{P}_{\|f\|}$ are bounded on G.

PROOF. It is sufficient to prove this for $\mathcal{P}_{\|f\|}$. As Γ is discrete, there exists a symmetric ($\mathcal{N}=\mathcal{N}^{-1}$), compact neighborhood $\mathcal{N}$ of e such that $\Gamma\cap\mathcal{N}^2=\{e\}$. By Corollary 2 of Theorem 19, there exists $\alpha\in C_c^\infty(G)$ such that $\mathcal{S}(\alpha)\subset\mathcal{N}$, such that $\alpha(kgk^{-1})=\alpha(g)$ for all $k\in K$, and such that $f*\alpha=f$, thus

$$f(g\gamma)=f*\alpha(g\gamma)=\int_G f(g\gamma h^{-1})\alpha(h)dh \tag{3}$$
$$=\int_G f(gh^{-1})\alpha(h\gamma)dh, \quad g\in G,\ \gamma\in\Gamma.$$

Let $M=\sup_{g\in G}|\alpha(g)|$. Since $\mathcal{S}(\alpha)\subset\mathfrak{N}=\mathfrak{N}^{-1}$, we have

$$\|f(g\gamma)\|\leqq M\cdot\int_{\mathfrak{N}\cdot\gamma^{-1}}\|f(gh^{-1})\|dh. \tag{4}$$

Since $\gamma\neq\gamma'$, γ, $\gamma'\in\Gamma$, implies that $\mathfrak{N}\gamma\cap\mathfrak{N}\gamma'$ is empty, we have

$$\sum_{\gamma\in\Gamma}\|f(g\gamma)\|\leqq M\cdot\int_G\|f(gh^{-1})\|dh\leqq M\cdot\|f\|_1.$$

Hence the series converge and $\mathcal{P}_{\|f\|}$ is bounded on G. Let C be a compact subset of G, and let $\varepsilon>0$ be given. Then there exists a compact subset D of G such that

$$\int_{G-D}\|f(h)\|dh\leqq\varepsilon. \tag{5}$$

It follows from (4) by a change of variable of integration, since G is unimodular, that

$$\|f(g\gamma)\|\leqq M\cdot\int_{g\gamma\mathfrak{N}}\|f(h)\|dh; \tag{6}$$

and if A is the finite subset of $\gamma\in\Gamma$ such that $C\gamma\mathfrak{N}\cap D$ is non-empty, then for any fixed $g\in C$, the sets $g\gamma\mathfrak{N}$, $\gamma\in\Gamma-A$, are disjoint subsets of $G-D$ because $\Gamma\cap\mathfrak{N}^2=\{e\}$; therefore,

$$\sum_{\gamma\in\Gamma-A}\|f(g\gamma)\|\leqq M\cdot\int_{G-D}\|f(h)\|dh\leqq M\cdot\varepsilon, \quad g\in C.$$

It follows that the series in (2) converges uniformly on C.

The series $\mathcal{P}_f$ is called a Poincaré series.

Let D be a bounded symmetric domain in $\boldsymbol{C}^n$. It is known that $\mathrm{Hol}(D)^0=G$ is always isomorphic as a Lie group to the identity component of the group of real points of a connected, semi-simple, linear algebraic group defined over $\boldsymbol{R}$. Thus we may employ the notation of Chapter 7, section 7, except that what is there called G_1 shall here be denoted by G, and we no longer assume D to be irreducible, thus, $\mathfrak{g}$ need not be simple. Then $\mathfrak{g}$ is the direct sum of simple components, its root system is the union of the root systems of the simple components, and the same is true for the compact and non-compact roots, so that $\mathfrak{p}^+$ and $\mathfrak{p}^-$ are the sums of the corresponding subspaces of the simple components of $\mathfrak{g}_C$, etc. Now $\zeta(G)$ is a bounded domain in the complex vector space $\mathfrak{p}^+$ and K is the stability group in G of the origin $0 \in \mathfrak{p}^+$ and since $\zeta(G)$ is holomorphically isomorphic to D, we simply take $D = \zeta(G)$.

The operation of G on D is as follows: If $\bar{p} \in D \subset \mathfrak{p}^+$, let $p = \exp(\bar{p})$; so there exists $g_1 \in G$, $g_1 = p^- k_0 p$, $p^- \in \mathfrak{p}^-$, $k_0 \in K_C$, and if $g \in G$, then G contains $g_1 g = p_2^- k_{02} p_2$, $p_2 = \exp(\bar{p}_2)$, $\bar{p}_2 \in \mathfrak{p}^+$, $k_{02} \in K_C$, $p_2^- \in P^-$, and we write (cf. p. 99)

$$\bar{p}_2 = \bar{p} \cdot g, \quad k_{02} = \mu(\bar{p}, g). \tag{7}$$

Viewing g as a transformation of D and identifying the tangent space to D at $\bar{p}$ with $\mathfrak{p}^+$, one may by direct calculation based on the definitions [3 : p. 454] determine the differential of g at $\bar{p}$; the result of the calculation is that it is (cf. Chapter 7, section 7)

$$\operatorname{Jac}(\bar{p}, g) = Ad_{\mathfrak{p}^+}(k_{02}^{-1}); \tag{8}$$

therefore, the Jacobian determinant is $j(\bar{p}, g) = \det(Ad_{\mathfrak{p}^+}(k_{02}^{-1}))$. For other details, we refer the reader to [3 : § 1; 27 : Chap. VIII].

Let ρ be a unitary representation of K on a finite-dimensional, complex vector space V and assume that ρ has been extended to a rational representation of K_C (if ρ has such an extension, it is unique, since K is Zariski-dense in K_C). We write $\mu_\rho(z, g) = \rho(\mu(z, g))$, $z \in D$, $g \in G$. Since $\mu(z, k) = k$, $k \in K$, we have in particular $\mu_\rho(z, k) = \rho(k)$. For the moment we generalize the procedure of Chapter 10, section 3; namely, if F is a function on D with values in V, we associate to it the function $f \in \mathscr{F}(G, V)$ defined by $f(g) = \mu_\rho(0, g) F(\zeta(g))$, $g \in G$; it is easy to see that $f(kg) = \rho(k) f(g)$, which as in Chapter 9, section 4, we express by saying that f is of type ρ on the left, and in particular f is K-finite on the left. If F is of class C^∞, so is f.

LEMMA 1. [52; 3]. a) *F is a holomorphic mapping if and only if $Y \cdot f = 0$ for all $Y \in \mathfrak{p}^-$ (viewed as a set of complex right-invariant differential operators on G).*

b) *Let $f: G \to V$ be of type ρ on the left and of class C^∞ such that $Y \cdot f = 0$ for all $Y \in \mathfrak{p}^-$. Then f is $\mathscr{Z}(\mathfrak{g})$-finite.*

PROOF. a) With the normalization of E_μ taken in Chapter 7, section 7, the elements

(9) $$Y_\mu = i(E_\mu - E_{-\mu}) \quad \text{and} \quad X_\mu = E_\mu + E_{-\mu},$$

where μ runs through the positive non-compact roots, are a basis of $\mathfrak{p}$, while $E_{-\mu} = \frac{1}{2}(X_\mu + iY_\mu)$ runs through a basis of $\mathfrak{p}^-$ and $E_\mu = \frac{1}{2}(X_\mu - iY_\mu)$, through a basis of $\mathfrak{p}^+$, so that the elements of $\mathfrak{p}^-$ are the linear combinations with constant coefficients of the partial differential operators $\partial/\partial \bar{z}_j$, where z_j are complex coordinates in $\mathfrak{p}^+$; thus, for $F \in C^\infty(D)$, $YF=0$ for all $Y \in \mathfrak{p}^-$ is just the system of Cauchy-Riemann equations (p. 11) for F to be complex analytic. So to prove a), it is sufficient to prove

(10) $$Yf(g) = \mu_\rho(0, g) \cdot YF(\zeta(g)), \quad Y \in \mathfrak{p}^-, \ g \in G.$$

Let $Y \in \mathfrak{g}$. Then $Y = Y_- + Y_0 + Y_+$, $Y_\pm \in \mathfrak{p}^\pm$, $Y_0 \in \mathfrak{k}$. By expanding $\exp(tY)$ according to powers of t, one sees that if we write

$$e^{tY} = \exp(tY) = p_t^- k_0(t) p_t^+,$$

in the usual notation, then $Y_0 = \frac{d}{dt} k_0(t)\,|_{t=0}$. Then one has

(11) $$\begin{aligned} Yf(e) &= \frac{d}{dt}\{\mu_\rho(0, e^{tY}) \cdot F(0 \cdot e^{tY})\}\,|_{t=0} \\ &= \frac{d}{dt}(\mu_\rho(0, e^{tY}))\,|_{t=0} \cdot F(0) + \mu_\rho(0, e)\frac{d}{dt}F(0 \cdot e^{tY})\,|_{t=0}. \end{aligned}$$

Now, $\mu(0, e^{tY}) = k_0(t)$, $\mu_\rho(0, e^{tY}) = \rho(k_0(t))$, so that

$$\frac{d}{dt}\mu_\rho(0, e^{tY})\,|_{t=0} = d\rho\left(\frac{d}{dt}k_0(t)\,|_{t=0}\right) = d\rho(Y_0),$$

hence (11) becomes

(12) $$Yf(e) = d\rho(Y_0) \cdot F(0) + Y \cdot F(0),$$

and by linearity this then holds for all $Y \in \mathfrak{g}_{\mathbb{C}}$. If $Y \in \mathfrak{p}^-$, then $Y_0 = 0$ and we obtain (10) for $g = e$. Replacing F by the function F' defined by $F'(x) = \mu_\rho(x, g)F(x \cdot g)$, we see that F' is the function on D associated to the function $f' = r_g f \in \mathcal{F}$. Since $\mu_\rho(x, g)$ is holomorphic in $x = \zeta(h)$, where h varies in G and g is fixed, and since Y.(function of $\zeta(h)) = 0$, $Y \in \mathfrak{p}^-$, is just the system of Cauchy-Riemann equations, we obtain

(13) $$YF'(0) = \mu_\rho(0, g)YF(0 \cdot g),$$

while

$$(14) \qquad Y(r_g f)(e) = \frac{d}{dt}(f(\exp(tY)\cdot g))\,|_{t=0} = Yf(g).$$

Combining (14), (12) (with $Y_0 = 0$), and (13), we obtain for $Y \in \mathfrak{p}^-$:

$$Yf(g) = Yf'(e) = YF'(0) = \mu_\rho(0, g)\, YF(0\cdot g),$$

which proves (10) and hence a).

b) Let h_0 be the element of the center of $\mathfrak{k}$ such that $[h_0, X^\pm] = i^{\pm 1}\cdot X^\pm$, $X^\pm \in \mathfrak{p}^\pm$. (That such an h_0 exists is easily seen from the fact that in each simple component of $\mathfrak{g}_{\mathbb{C}}$, the adjoint operation of the elements of the center of $\mathfrak{k}$ is given by scalar multiplications in the corresponding simple components of $\mathfrak{p}^+$ and $\mathfrak{p}^-$ [27 : Chap. VIII]). As we shall see, to prove b), it will suffice to prove

LEMMA 2. *Let $(\rho, V) = \oplus(\rho_\nu, V_\nu)$, ρ_ν being irreducible. If F and f correspond as before, if F is holomorphic, and if $Z \in \boldsymbol{U}$ commutes with every element of $\mathfrak{X}$, then (viewing Z as a right-invariant differential operator) we have*

$$(15) \qquad Z\cdot f = \sum_\nu \lambda_\nu(Z)\cdot f,$$

where $f = \sum f_\nu$, f_ν denoting the composition of f with the projection of V onto its direct summand V_ν, and where λ_ν is a complex-valued linear function of Z, independent of f.

PROOF. [52 : Exp. 10]. It is enough to treat the case when ρ is irreducible and to show that then f is an eigenfunction of Z such that the eigenvalue is a linear function of Z and does not depend on f.

Let $\{X_i\}$ be a basis of $\mathfrak{p}^+$, $\{Y_i\}$, one of $\mathfrak{p}^-$. From facts previously proved about the universal enveloping algebra (Chapter 9, section 1), and since $\mathfrak{k}$ normalizes $\mathfrak{p}^\pm$, we may write

$$(16) \qquad Z = \sum_{m,n} X^m K_{m,n} Y^n,$$

the sum being finite and where $X^m = X_1^{m_1}\cdots X_r^{m_r}$, $Y^n = Y_1^{n_1}\cdots Y_r^{n_r}$, $K_{m,n} \in \mathfrak{X}$, and in this form, the expression in (16) is *unique*. If $m = (m_1, \cdots, m_r)$, define $|m| = \sum m_i$. Then we have

$$0 = [h_0, Z] = \sum_{m,n} (|m| - |n|)\, i X^m K_{m,n} Y^n,$$

so that $K_{m,n}=0$ unless $|m|=|n|$. Therefore, $Z=K_{0,0}+XP$, with $X\in \boldsymbol{U}$ and $P\in\mathfrak{p}^-$, and since F is holomorphic, part a) shows that

$$Z\cdot f=K_{0,0}f. \tag{17}$$

Now, Z commutes with every element of $\mathfrak{X}$, hence for $A\in\mathfrak{k}$,

$$0=[A, Z]=[A, K_{0,0}]+[A, X]P+X[A, P],$$

and since A normalizes $\mathfrak{p}^-$, the sum of the last two terms is in $\boldsymbol{U}\cdot\mathfrak{p}^-$, while the first term is in $\mathfrak{X}$, so that $[A, K_{0,0}]=0$; that is, $K_{0,0}$ is in the center of $\mathfrak{X}$, and $Z\cdot f=K_{0,0}f$. If $A\in\mathfrak{k}$, we have

$$Af(g)=\frac{d}{dt}f(e^{tA}g)\,|_{t=0}=\frac{d}{dt}\rho(e^{tA})f(g)\,|_{t=0}=d\rho(A)\cdot f(g), \tag{18}$$

where $d\rho$ is the representation of $\mathfrak{k}$ in End(V) associated to ρ, which may uniquely be extended to $\mathfrak{X}$, so that

$$K_{0,0}f=d\rho(K_{0,0})f. \tag{19}$$

Now, $d\rho(K_{0,0})$ commutes with $d\rho(\mathfrak{X})$, and since ρ is irreducible, $d\rho(K_{0,0})$ is a scalar, which we denote by $\lambda(Z)$. Clearly λ is a linear function of Z and is independent of f, and this completes the proof of Lemma 2.

Now to prove Lemma 1, one need only note from Lemma 2 that if $\mathcal{Z}$ is the center of $\boldsymbol{U}$, then $\mathcal{Z}\cdot f$ is finite-dimensional, its dimension being equal to or less than the number of summands V_ν of V.

LEMMA 3. *Let $j(\ ,g)$ be the functional determinant on D for $g\in G$. Then the function $j(0, g)^l$ is in $L^1(G)$ for every integer $l\geqq 2$.*

PROOF. If dx is the usual Euclidean measure on D and if $d\mu$ is the invariant Bergmann measure, then

$$d\mu(x)=|j(0, g)|^{-2}dx, \quad 0\cdot g=x. \tag{20}$$

Since K is the stability group of 0, and $|j(0, k)|=1$ for $k\in K$, the absolute value of $j(0, g)$, for g such that $0\cdot g=x$, depends only on x, and we put $f(x)=|j(0, g)|$. It follows that

$$\int_G |j(0, g)|^l dg=c\cdot\int_D f(x)^l d\mu(x)=c\cdot\int_D f(x)^{l-2}dx, \tag{21}$$

where c is a finite positive constant.

On the other hand, we see from Chapter 4, section 4, that $|j(0, g)|$

is bounded on G, hence f is bounded on D, so that the integral in (21) is finite, proving the lemma.

Now let F be a polynomial mapping from D into V, let $l \geqq 2$ be an integer, and let Γ be a discrete subgroup of G. Put

$$\mathcal{P}_F(x) = \mathcal{P}(x) = \sum_{\gamma \in \Gamma} j(x, \gamma)^l F(x \cdot \gamma), \quad x \in D.$$

We see already from Chapter 5, section 2, that this series converges normally on D. We may now obtain this again and more besides by applying Theorem 23. Namely, define $f: G \to V$ by

$$f(g) = j(0, g)^l F(\zeta(g));$$

then

$$\mathcal{P}_f(g) = \sum_{\gamma \in \Gamma} f(g \cdot \gamma) = j(0, g)^l \cdot P(\zeta(g)), \tag{22}$$

and with

$$\mathcal{P}_{\|f\|}(g) = \sum_{\gamma \in \Gamma} \|f(g \cdot \gamma)\|,$$

we have

THEOREM 24. *The series in* (22) *is a Poincaré series in the previous sense, and* $\mathcal{P}_{\|f\|}$ *is bounded on* G.

PROOF. Since $f(kg) = j(0, k)^l f(g)$, f is of finite type on the left with respect to K. Hence, by Lemma 1, b), f is $\mathcal{Z}$-finite. Since F is given by polynomials, it is bounded, hence by Lemma 3, $\|f\|$ is absolutely integrable on G.

We have

$$j(0, gk) = j(0, g) j(0 \cdot g, k) = j(0, g) j(0, k), \quad k \in K, \ g \in G,$$

since k acts by a linear transformation on D. For the same reason, the set of translates $F(zk) = r_k F(z)$ are polynomials of the same degree as F, hence f is K-finite on the right. Therefore, all the assumptions of Theorem 23 are satisfied, so $\mathcal{P}_{\|f\|}$ is indeed bounded on G, and this completes the proof.

Thus we have shown that the series $\mathcal{P}_f$ in (22), for $l \geqq 2$, is in $\mathcal{H}^\infty_\Gamma(\rho)$ (at least in our narrower context of the last chapter, where we restricted our considerations to the case $\dim V = 1$, $\rho(k) = j(0, k)^l$). It is our eventual purpose to indicate a proof in the case $\dim V = 1$ that, when l is large enough, the spaces $\mathcal{H}^p_\Gamma(\rho)$, $1 \leqq p \leqq \infty$, all coincide with each other, with the space of "cusp forms", and with the space of (scalar) Poincaré series ($\dim V = 1$).

§2. Godement's criterion for the convergence of Eisenstein series [6e; 3; 22c]

Let G be a connected, semi-simple, linear algebraic group defined over a subfield k of $\boldsymbol{R}$; denote by P a k-parabolic subgroup of G containing a maximal k-split torus S of G, and by P_0 a minimal k-parabolic subgroup of G such that $S \subset P_0 \subset P$. Then P_0 defines an ordering on the k-roots of G with respect to S: $\alpha \in {}_k\Sigma$ is positive if and only if the subgroup of G generated by the α-eigenspace of the adjoint representation of S is contained in P_0. Let ${}_k\Delta$ denote the simple positive k-roots in ${}_k\Sigma$. There is a subset θ of ${}_k\Delta$ such that if we define

$$S_\theta = (\bigcap_{\alpha \in \theta} \ker \alpha)^0,$$

then P is generated by $Z(S_\theta)$ and by the unipotent radical U_0 of P_0. We write $P = {}_kP_\theta$. In particular, $P_0 = {}_kP_\emptyset$. (For this and what follows, cf. Chapter 7, sections 2 and 3.) Generally, when $k = \boldsymbol{Q}$, we omit it as a subscript on Δ, P_θ, etc.

We now suppose ${}_k\Sigma$ is a subset of a Euclidean space E of dimension equal to the k-rank of G and supplied with an inner product $(\ ,\ )$ invariant under the Weyl group ${}_kW = N(S)/Z(S)$; then ${}_kW$ is generated by the reflections r_α in the planes perpendicular to the elements $\alpha \in {}_k\Delta$. Let $\boldsymbol{X}^*(S)$ be the character group of S; since S is k-split, we have $\boldsymbol{X}^*(S) = \boldsymbol{X}^*(S)_k$. We view $\boldsymbol{X}^*(S)$ as a subset of $\boldsymbol{X}^*(S) \otimes \boldsymbol{Q}$ and the latter as being contained in E. For each $\alpha \in {}_k\Delta$, we have defined (Chapter 7, section 3) an element $\boldsymbol{m}_\alpha$ of $\boldsymbol{X}^*(S) \otimes \boldsymbol{Q}$; it satisfies

$$(23) \qquad (\boldsymbol{m}_\alpha, \alpha) = c_\alpha(\alpha, \alpha), \quad (\boldsymbol{m}_\alpha, \beta) = 0, \quad \alpha, \beta \in {}_k\Delta,\ \alpha \neq \beta,$$

where c_α is a positive rational number [9: p. 141]; we denote by Λ_α the smallest integral multiple $d_\alpha \boldsymbol{m}_\alpha$ of it which is the highest k-weight of an irreducible, strongly k-rational representation ρ_α of G. Then ${}_kP_{({}_k\Delta - \{\alpha\})}$ is the stability group of the highest weight line of ρ_α. The elements $\{\Lambda_\alpha\}$ of $\boldsymbol{X}^*(S)$ are called the fundamental dominant k-weights of G with respect to S and they form a basis of E.

Let $\chi_0=\det(Ad_{\mathfrak{U}_0})$, where $\mathfrak{U}_0$ is the Lie algebra of U_0. As an element of $X^*(S)$, $\chi_0=\sum_\alpha e_\alpha\Lambda_\alpha$, and one may verify from properties of ${}_kW$ that each e_α is strictly positive. In fact, χ_0 is a positive scalar multiple of the sum of all the positive C-roots restricted to S, and so it is clear from Chapter 8, section 8, Lemma 4, that $(\chi_0, \alpha)>0$ for every $\alpha\in{}_k\Delta$.

More generally, let $P={}_kP_\theta$, let U_θ be the unipotent radical of P, and let $\theta'={}_k\Delta-\theta$. Let $\chi_\theta=\det(Ad_{\mathfrak{U}_\theta})|S_\theta$, where $\mathfrak{U}_\theta$ is the Lie algebra of U_θ. Then it is easy to see that χ_θ is the restriction to S_θ of an element $\sum_{\alpha\in\theta'} e_\alpha\Lambda_\alpha$ of $X^*(S)$, where e_α are certain rational numbers. One may show all $e_\alpha>0$, but this is more difficult. For us it will be sufficient to determine e_α in particular cases of interest. Let $A=S_{\theta R}^0$, $Z_G(S_\theta)=S_\theta\cdot M_\theta$, where $S_\theta\cap M_\theta$ is finite, so that

$$G_R=K\cdot P_R=K\cdot M_R\cdot A\cdot U_R, \tag{24}$$

with $M=M_\theta$, $U=U_\theta$, and K maximal compact in G_R; and if $g\in G_R$ we may write $g=kmau$, corresponding to the factorization in (24). Then km, a, and u are uniquely determined 1) because $K\cap P_R\subset M_R$, 2) because A and U_R are connected and torsion-free, and 3) because of known facts about the Iwasawa decomposition. Hence, we may write $a=a(g)$. Clearly $a(kg)=a(g)$, $k\in K$, $g\in G_R$.

Now let Γ be a discrete subgroup of G_R (which in our applications will be arithmetic) and let Γ_∞ be a subgroup of $\Gamma\cap MU$. Let $\{s_\alpha\}_{\alpha\in\theta'}$ be a set of complex numbers, and if $p\in P$, let $p^{\Lambda_s}=\prod_{\alpha\in\theta'}|p^{\Lambda_\alpha}|^{s_\alpha}$. Then $a(p)^{\Lambda_s}=p^{\Lambda_s}$. The following result is due to Godement [cf. 6e; 22c]. The same applies to succeeding results, e.g., Theorem 25.

LEMMA 4. *Suppose that*

1) $a(\gamma)^{\Lambda_\alpha}\geqq\varepsilon>0$, $\gamma\in\Gamma'$, $\alpha\in\theta'$;
2) $(MU)_R/\Gamma_\infty$ *has finite measure; and*
3) $\operatorname{Re} s_\alpha>e_\alpha$, $\alpha\in\theta'$.

Then the series

$$E(g, s)=\sum_{\gamma\in\Gamma/\Gamma_\infty} a(g\gamma)^{-\Lambda_s} \tag{25}$$

converges normally on G_R.

PROOF. M is contained in the kernel of every k-character of $Z(S)$

and a unipotent group has no non-trivial rational characters. Hence $a(g\gamma)^{\Lambda_\alpha}=a(g)^{\Lambda_\alpha}$ for all $\gamma\in\Gamma_\infty$, so that the series is a well-defined sum. Moreover, MU is unimodular since $\det(Ad_{\mathfrak{u}})$ is a k-character of M. In what follows, all considerations are for elements of $G_{\boldsymbol{R}}$, and so we drop the subscript $\boldsymbol{R}$ from $M_{\boldsymbol{R}}$ and from $U_{\boldsymbol{R}}$.

To prove convergence, we may assume all $s_\alpha\in\boldsymbol{R}$. Let $\mathfrak{N}$ be a compact subset of $G_{\boldsymbol{R}}$. Since P is parabolic, $G_{\boldsymbol{R}}/P_{\boldsymbol{R}}$ is compact, and since $f(c,g)=a(cg)/a(g)$ is continuous, everywhere strictly positive on $\mathfrak{N}\times G_{\boldsymbol{R}}$ ($c\in\mathfrak{N}$, $g\in G_{\boldsymbol{R}}$), and right-invariant under $P_{\boldsymbol{R}}$, we see that there exist d, $d'>0$ such that

$$da(g)^{-\Lambda_s}\leqq a(cg)^{-\Lambda_s}\leqq d'a(g)^{-\Lambda_s},\quad g\in G_{\boldsymbol{R}},\ c\in\mathfrak{N}. \tag{26}$$

Hence, normal convergence on $G_{\boldsymbol{R}}$ is equivalent to convergence at any one point, say at the identity e. And by the same inequalities, convergence at e is equivalent to convergence of the series

$$\sum_{\gamma\in\Gamma/\Gamma_\infty}\int_{\mathfrak{N}}a(g\gamma)^{-\Lambda_s}dg, \tag{27}$$

where $\mathfrak{N}$ is a compact neighborhood of e and dg is a Haar measure on $G_{\boldsymbol{R}}$. Choose $\mathfrak{N}$ so small that $(\mathfrak{N}\cdot\mathfrak{N}^{-1})\cap\Gamma=\{e\}$; then the series (27) is majorized by the integral

$$I_0=\int_{\mathfrak{N}\Gamma/\Gamma_\infty}a(g)^{-\Lambda_s}dg.$$

Define $A(t)=\{a\in A\mid a^{\Lambda_\beta}\geqq t \text{ for all } \beta\in\theta'\}$. By hypothesis, $\Gamma\subset KMA(t')U$ for some $t'>0$ and since $\mathfrak{N}$ is compact, $\mathfrak{N}K\subset KMA(t'')U$ for some $t''>0$, hence, there exists $t>0$ such that $\mathfrak{N}\Gamma\subset KMA(t)U=KA(t)MU$. On the other hand, by hypothesis 2), there exists a subset ω of MU having finite measure such that $MU=\omega\Gamma_\infty$. Therefore,

$$I_0\leqq\varepsilon'\cdot\int_{KA(t)\omega}a(g)^{-\Lambda_s}dg,$$

for some $\varepsilon'>0$. Let dk, da, and dv be the Haar measures on K, A, and MU respectively; then $dg=dk\,da\,dv$, and $da\,dv=\chi(a)dv\,da$, where $\chi(a)=\det(Ad_{\mathfrak{u}}(a))$, and $\det(Ad_{\mathfrak{u}})=\sum e_\alpha\Lambda_\alpha$, as explained above. Hence,

$$\int_{KA(t)\omega}a(g)^{-\Lambda_s}dg=\text{const.}\int_K dk\int_\omega dv\int_{A(t)}a^{-\Lambda_s+\chi}da,$$

so that to prove convergence, it will suffice to prove that

$$\int_{A(t)} a^{-\Lambda_s+\chi} da$$

converges. However, A is the direct product of one-dimensional subgroups A_α such that Λ_β is trivial on A_α for $\beta \in \theta' - \{\alpha\}$. Hence, the above integral is the product of integrals

$$\int_{A_\beta(t)} a_\beta^{(e_\beta - s_\beta)} da_\beta = c_\beta \cdot \int_t^{+\infty} t_\beta^{(e_\beta - s_\beta)-1} dt_\beta,$$

where c_β is a constant, and each integral on the right here converges since $e_\beta - s_\beta < 0$. (Here we have used the fact that a Haar measure on A_β is dt/t.)

THEOREM 25. *Let $f: G_R \to \boldsymbol{C}$ be of class C^∞ and suppose, with notation as above, that*

i) *hypotheses* 1)–3) *of Lemma 4 are satisfied,*
ii) $f(g\gamma) = f(g)$ *for all* $\gamma \in \Gamma_\infty$,
iii) $|f(gp)| p^{\Lambda_s}$ *is bounded for g in a fixed compact set,* $p \in P_R$.

Then $E_f(g) = \sum_{\gamma \in \Gamma/\Gamma_\infty} f(g\gamma)$ *converges normally on* G_R.

PROOF. First of all, $G_R = K \cdot P_R$. Then for $k \in K$, $p \in P_R$, and $g = kp$, it follows from previous remarks that

$$f(g) a(g)^{\Lambda_s} = f(kp) p^{\Lambda_s}$$

and therefore $|f(g) a(g)^{\Lambda_s}|$ is bounded for $g \in G_R$ because of iii). Therefore, the series for E_f is majorized by a constant times

$$\sum_{\gamma \in \Gamma/\Gamma_\infty} a(g\gamma)^{-\Lambda_s},$$

which converges normally by Lemma 4. This proves the theorem.

THEOREM 26. *Let G be a linear, semi-simple, algebraic group defined over $\boldsymbol{Q}$ and let Γ be an arithmetic subgroup of G_R. Let P be a $\boldsymbol{Q}$-parabolic subgroup of G, define M, U, θ, and θ' as before, and let Γ_∞ be a subgroup of finite index in $(MU)_R \cap \Gamma$. Then the hypotheses 1) and 2) of Lemma 4 are satisfied.*

PROOF. Verification of 2): Since Γ is an arithmetic subgroup of G_R and MU is defined over $\boldsymbol{Q}$, it follows that $\Gamma \cap MU$ is an arithmetic subgroup of $(MU)_{\boldsymbol{R}}$. Moreover, as remarked in the proof of the lemma itself, MU has no non-trivial rational character defined over $\boldsymbol{Q}$. Therefore [8 : p. 522], $(MU)_{\boldsymbol{R}}/(\Gamma \cap MU)$ has finite invariant measure, and so the same is true of $(MU)_{\boldsymbol{R}}/\Gamma_{\infty}$.

Verification of 1): Let $\beta \in \theta'$ and let ρ_β be an (absolutely) irreducible, strongly $\boldsymbol{Q}$-rational representation of G with highest weight Λ_β. Since $\beta \in \theta'$, we may assume $P_{\Delta-\{\beta\}} \supset P$. Let F_β be the representation space of ρ_β, and let e_β be a generator of the $\boldsymbol{Q}$-rational highest weight line stabilized by $P_{\Delta-\{\beta\}}$ [9 : § 12]. Then $\rho_\beta(p)\cdot e_\beta = p^{\Lambda_\beta}\cdot e_\beta$, $p \in P$. It is known (and not hard to see) that there exists a Γ-invariant lattice L_β in $F_{\beta\boldsymbol{Q}}$. We may assume $e_\beta \in L_\beta - \{0\}$, hence $\rho_\beta(\Gamma)e_\beta \subset L_\beta - \{0\}$; therefore, there exists $\varepsilon > 0$ such that $\|\rho_\beta(\gamma)e_\beta\| > \varepsilon$ for all $\gamma \in \Gamma$, where $\|\ \|$ is a norm on F_β with respect to which $\rho_\beta(K)$ is unitary. Put $\gamma = kma(\gamma)u$, where $k \in K$, etc. Since M and U are contained in P and have no non-trivial $\boldsymbol{Q}$-character, we have

$$\varepsilon < \|\rho_\beta(\gamma)e_\beta\| = \|\rho_\beta(a(\gamma))e_\beta\| = a(\gamma)^{\Lambda_\beta},$$

which proves 1) is satisfied.

If $f: G_{\boldsymbol{R}} \to C$ satisfies $f(gp) = f(g)p^{-\Lambda_s}$, then f satisfies the hypotheses ii) and iii) of Theorem 25: iii) is obviously satisfied and ii) follows from the fact that Γ_∞ is contained in MU, and the latter, having no non-trivial $\boldsymbol{Q}$-character and being connected, lies in the kernel of all $\boldsymbol{Q}$-characters of P.

AN EXAMPLE (cf. Chapter 6, section 5). Let $G = Sp(n, \boldsymbol{C})$ (n is the rank) and let

$$P = \left\{ \begin{pmatrix} A & 0 \\ C & D \end{pmatrix} \in G \text{ (where } A, C, \text{ and } D \text{ are } n \times n \text{ matrices)} \right\}.$$

Then P is a maximal $\boldsymbol{Q}$-parabolic subgroup of G. Let

$$f(g) = f\left(\begin{pmatrix} A & B \\ C & D \end{pmatrix}\right) = \det(Bi + D)^{-m}.$$

So $f(g\gamma) = \det(ZB_\gamma + D_\gamma)^{-m} f(g)$, where $\gamma = \begin{pmatrix} A_\gamma & B_\gamma \\ C_\gamma & D_\gamma \end{pmatrix}$ and $Z = (iE)\cdot g = (Bi + D)^{-1}(Ai + C)$. Then, with respect to a suitable basis of the Cartan subalgebra of diagonal matrices in g, θ consists of all the compact simple roots $c_i - c_{i+1}$, $i = 1, \cdots, n-1$, and θ' consists of a single element, the non-compact simple root $2c_n$, and we have

$$S_\theta = \left\{ \begin{pmatrix} sE & 0 \\ 0 & s^{-1}E \end{pmatrix} \right\},$$

$$\Lambda_{2c_n} = 2(c_1 + \cdots + c_n) = 2nc_n + \sum_{i=1}^{n-1} a_i(c_i - c_{i+1}).$$

If we put

$$a_s = \begin{pmatrix} sE & 0 \\ 0 & s^{-1}E \end{pmatrix},$$

then $\det(Ad_{\mathfrak{u}} a_s) = s^{n(n+1)}$, while $\det(s^{-1}E)^{-m} = s^{mn}$, hence, the convergence condition is $m > n+1$. Now $\Gamma_\infty = \Gamma \cap P$ will play the role of Γ_∞ in the preceding discussion and

$$(28) \qquad E_f(g) = \sum_{\gamma \in \Gamma/\Gamma_\infty} f(g\gamma) = f(g) \sum_{\{B,\, D\}} \det(ZB + D)^{-m},$$

where $\{B, D\}$ runs over a maximal set of non-right-associated, integral, primitive, symmetric pairs of matrices, and this gives the usual result on convergence of Eisenstein series for the Siegel modular group.

§ 3. Poincaré-Eisenstein series [3; 6e]

Returning to the general situation of section 2 and the notation there, let H be a subgroup of finite index in G_R, let $\Delta(p, s) = p^{-\Lambda_s}$, $p \in P_R$, and let f' be a continuous, complex-valued function on H. We have [3 : p. 497]:

THEOREM 27. *Let G be defined over $\mathbf{Q}$ and let Γ be an arithmetic subgroup of H. Assume that $f'(g\gamma) = f'(g)$, $g \in H$, $\gamma \in \Gamma_\infty$, and that*

$$m(g) = \sup_{p \in P \cap H} |f'(g \cdot p) \Delta(p, s)^{-1}|$$

is finite for every $g \in H$ and is bounded on every compact set in H. If $\operatorname{Re} s_\alpha > e_\alpha$ for each $\alpha \in \theta'$, then

$$E_{f'}(x) = \sum_{\gamma \in \Gamma/\Gamma_\infty} f'(x \cdot \gamma), \quad x \in H,$$

converges normally on H.

PROOF. But for the inessential replacement of G_R by its subgroup H, this is merely a trivial paraphrasing of the results of Theorems 25 and 26, in view of the fact that the condition

$$\operatorname{Re} s_\alpha > e_\alpha, \quad \alpha \in \theta',$$

is just condition 3) of Lemma 4. Q.E.D.

The normal subgroup $S_\theta \cdot U$ of P is called the split radical of P; we let B be a connected, normal $\boldsymbol{Q}$-subgroup of P containing it and let $C=P/B$. Then C is a reductive, connected $\boldsymbol{Q}$-group which has no non-trivial rational character defined over $\boldsymbol{Q}$. The natural mapping of P onto C maps P_R onto a subgroup of finite index of C_R. This follows from the facts that C_R has finitely many components and that for a non-singular, algebraic $\boldsymbol{R}$-variety X, $\dim_R X_R = \dim X$ (cf. [61]), hence that

$$\dim_R(P_R/B_R) = \dim_R P_R - \dim_R B_R = \dim_R C_R.$$

Therefore, the natural mapping maps $H \cap P$ onto a subgroup of C_R, likewise of finite index because $H \cap P \subset G_R \cap P = P_R$, and

$$[P_R : H \cap P] \leqq [G_R : H] < +\infty.$$

Furthermore, if f is a complex-valued function on H such that

$$|f(h \cdot c)| = |f(h) \cdot \Delta(c, s)|, \quad h \in H, \ c \in B_R \cap H,$$

then the restriction to $H \cap P$ of $|f(h) \cdot \Delta(h, s)^{-1}|$ is right-invariant under $B \cap H$ and may be viewed as a function on the open subgroup C_1 of C_R defined by

$$C_1 = (H \cap P)/(H \cap B).$$

Let $\mathfrak{c}_1$ be the Lie algebra of C_1 and let $\boldsymbol{U}^{(1)}$ be its universal enveloping algebra. Let Γ and Γ_∞ be as previously defined and put $\Gamma_0 = \Gamma_\infty \cap B$. Assume that $f(h \cdot \gamma) = f(h)$ for $h \in H$, $\gamma \in \Gamma_0$. One defines the "Poincaré-Eisenstein" series E_f by

$$E_f(h) = \sum_{\gamma \in \Gamma/\Gamma_0} f(h \cdot \gamma), \quad h \in H. \tag{29}$$

THEOREM 28. *Assume that C_R^0 has compact center and let $f : H \to \boldsymbol{C}$ and $f' : C_1 \to \boldsymbol{C}$ be continuous functions as above such that*

(i) $|f(k \cdot h \cdot b)| = |f(h)| \cdot \dot{\Delta}(b, s)$, *for* $k \in K$, $h \in H$, $b \in B \cap H$, *where* $s = (s_\alpha)_{\alpha \in \theta'}$ *is real, and*
(ii) $f' \in L^1(C_1)$ *is* $\mathcal{Z}(\mathfrak{c}_1)$*-finite (where* $\mathcal{Z}(\mathfrak{c}_1)$ *is the center of* $\boldsymbol{U}^{(1)}$*), and is of finite type on the right with respect to some maximal compact subgroup of* C_1, *and* $|f'(\pi(p))|$ *is equal to* $|f(p) \cdot \Delta(p, s)^{-1}|$ *for*

$p\in P\cap H$, where $\pi: P\to C$ is the natural mapping.

Then if $s_\alpha > e_\alpha$, $\alpha\in\theta'$, the series (29) converges normally on H.

PROOF. We have, for $h\in H$, $b\in B\cap H$, $p\in P\cap H$:

$$|f(hbp)|=|f(hpb')|=|f(hp)|\Delta(b', s),$$

where $b'=p^{-1}bp$, hence $\Delta(b', s)=\Delta(p, s)^{-1}\Delta(b, s)\Delta(p, s)=\Delta(b, s)$; thus, $|f(hbp)|=|f(hp)|\Delta(b, s)$. Since Δ takes the constant value one on Γ_0 and Γ_0 is normal in Γ_∞, we obtain

$$|f(h\cdot\tau\cdot\sigma)|=|f(h\cdot\sigma)|,\quad h\in H,\ \tau\in\Gamma_0,\ \sigma\in\Gamma_\infty.$$

Define $\mathcal{P}_{|f|}(g)=\sum_{\eta\in\Gamma_\infty/\Gamma_0}|f(g\cdot\eta)|$, $g\in H$, and

$$\mathcal{P}_{|f'|}(\pi(p))=\sum_{\eta\in\Gamma_\infty/\Gamma_0}|f'(\pi(p\cdot\eta))|,\quad p\in P\cap H. \tag{30}$$

Because of (ii) and Theorem 23, the series in (30) converges normally and represents a bounded function on $P\cap H$. On the other hand, if $g\in H$, we may write $g=k_g\cdot p_g$, $k_g\in K$, $p_g\in P\cap H$, and the factorization of g in this way is unique up to an element of $K\cap P$. By (i) and (ii), one has

$$\mathcal{P}_{|f|}(g\cdot p)\Delta(p, s)^{-1}=\Delta(p_g, s)\mathcal{P}_{|f'|}(\pi(p_g\cdot p)). \tag{31}$$

Therefore, the left side of (31) is bounded for $p\in P\cap H$ and p_g in a compact set. Moreover, it is clear that $\mathcal{P}_f$ is right-invariant under Γ_∞. Thus we have

$$E_f(h)=\sum_{\gamma\in\Gamma/\Gamma_\infty}\mathcal{P}_f(h\gamma),$$

and if

$$E_{|f|}(h)=\sum_{\gamma\in\Gamma/\Gamma_\infty}\mathcal{P}_{|f|}(h\gamma),$$

then Theorem 27 implies the latter series converges normally on H, whence the theorem.

§4. Boundary components and partial Cayley transforms [3; 39a, b; 46]

Let D be a bounded symmetric domain in $\boldsymbol{C}^n$ and let $\partial D=\bar{D}-D$ be the (proper) boundary of D. Denote by G_R^0 the identity component of Hol(D) and by r the (real) rank of G_R^0. Then ∂D is the union of r orbits $\omega_0, \cdots, \omega_{r-1}$ of G_R^0. Each orbit ω_j is the disjoint union of subsets

F of ω_j with the following properties:

1) If $p \in F$, then there exists a neighborhood $\mathfrak{N}$ of p in $\boldsymbol{C}^n$ such that $F \cap \mathfrak{N}$ is a complex submanifold of $\mathfrak{N}$; and
2) if σ is the unit disc in $\boldsymbol{C}^n$ and if f is a holomorphic mapping of σ into $\boldsymbol{C}^n$ such that $f(\sigma) \subset \bar{D}$ and $f(\sigma) \cap F$ is non-empty, then $f(\sigma) \subset F$.

Such a set F is called a boundary component of D. If $g \in G_R^0$, then either $F \cdot g = F$ or $F \cdot g \cap F$ is empty. As suggested by our choice of notation, we assume the group G_R^0 to be the identity component of the group of real points of a real, semi-simple, linear algebraic group G. Then, if F is a boundary component of D, the group

$$N(F) = \{g \in G_R^0 \mid F \cdot g = F\}$$

is an $\boldsymbol{R}$-parabolic subgroup of G_R^0 and contains

$$Z(F) = \{g \in N(F) \mid x \cdot g = x \quad \text{for all} \quad x \in F\}$$

as a normal subgroup. Moreover, $G(F) = N(F)/Z(F)$ is a subgroup of finite index of $\mathrm{Hol}(F)$. The assignment $F \mapsto N(F)$ is an injection of the set of boundary components of D into the set of $\boldsymbol{R}$-parabolic subgroups of G_R^0. (Cf. [3 : § 1].)

Now assume that G is defined over $\boldsymbol{Q}$. A boundary component F is called "rational" if $N(F)$ is defined over $\boldsymbol{Q}$ (more correctly, if the Zariski-closure $N(F)_C$ of $N(F)$ in G is defined over $\boldsymbol{Q}$). In that case, there exists a connected, normal $\boldsymbol{Q}$-subgroup B of $N(F)_C$, contained in $Z(F)_C{}^0$ and containing the unipotent radical $U(F)_C$ of $N(F)_C$ such that $Z(F)/(B_R \cap G_R^0)$ is compact. This follows [3 : §§ 2, 3] from examination of the cases of root systems arising for Hermitian symmetric spaces considered as a subclass of the class of Riemannian symmetric spaces. It may be remarked that in general, for the construction of the Satake compactification of the quotient X/Γ of a symmetric space X by an arithmetic group Γ, one would impose the existence of the group B as an additional condition; but when $X = D$, a bounded symmetric domain, *or* when $Z(F)_C$ is itself defined over $\boldsymbol{Q}$, that condition is implied by the requirement that $N(F)_C$ be defined over $\boldsymbol{Q}$.

In this and succeeding sections, we endeavor to present a descriptive sketch, stripped of many technical details for which we refer the reader to [3; 46; 39b; 2e], of how one may give a useful description of the spaces $\mathcal{H}_{\Gamma}^{p}(\rho)$ and use it to prove some of the assertions made about these spaces in Chapter 10, section 3. For this, we introduce again in this section the partial Cayley transformations discussed in Chapter 7, section 7. The discussion of automorphic forms will be continued in succeeding sections.

We first discuss the case when D is irreducible. If $b \in \{1, \cdots, r\}$, there is a naturally associated [3 : § 1] boundary component $F = F_b$ of D, and we denote by c_b or c_F the element

$$\prod_{i=1}^{b} \exp\left(\frac{\pi}{4}(E_{-i} - E_i)\right) \tag{32}$$

(defined in Chapter 7) of $\exp(\mathfrak{p}_C) \subset G_C$. It is unnecessary for our purposes here to specify F as a certain point set of ∂D; the facts about F that we do need will be explained as they are required. The element c_F is called the "partial Cayley transform with respect to F". As in Chapter 6, sections 6 and 7, and Chapter 7, section 7, one has $G_R^0 \cdot c_b \subset P^- K_C P^+$, $1 \leqq b \leqq r$. Using ζ as before to denote the projection of $P^- K_C P^+$ onto $\mathfrak{p}^+ = \boldsymbol{C}^n$, we define

$$D_F = \zeta(G_R^0 \cdot c_F). \tag{33}$$

The real rank of $\mathrm{Hol}(F)$ is $r - b$. If we view D as imbedded in its compact dual D^c [62 : p. 260] on which $\boldsymbol{C}^n$ is a Zariski-open set, then operation on the right by c_F (G_C acts on D^c) transforms F into the variety "at infinity", $D^c - \boldsymbol{C}^n$, and D onto the unbounded domain D_F in $\boldsymbol{C}^n$. Each element of $Z(F)$ operates on D_F by linear affine transformations. The boundary components F_b, $1 \leqq b \leqq r$, are mutually incident in the sense that if $b < c$, then F_c is contained in the closure of F_b (and in fact is a boundary component of the latter, which is itself a bounded symmetric domain). It is useful to note here a certain hereditary property of the Cayley transforms. Let F and F' be two of the boundary components such that $F' \subset \partial F$. Then if $c_{FF'}$ is the Cayley transformation of F with respect to F', the element $c_{FF'}$, acting on F^c, may be viewed as the element of G_C defined by

$$c_F \cdot c_{FF'} = c_{F'}. \tag{34}$$

In terms of the domains, this reads

$$D_{F'} = D \cdot c_{F'} = D_F \cdot c_{FF'}. \tag{34'}$$

When D is not irreducible, then it is a product of irreducible domains, each boundary component F of D is a product of boundary components of the irreducible factors of D (whereby we count a domain as a boundary component of itself), and if we denote by c_F the product of the partial Cayley transformations for these, then $D \cdot c_F = D_F$ is an unbounded domain in the ambient space $\boldsymbol{C}^n$ and $Z(F)$ operates on D_F by linear affine transformations. The earlier remark about the hereditary nature of the Cayley transformations in the case when D is irreducible carries over *verbatim* to the present, general case. Moreover, there exists a complex analytic mapping $\tau = \tau_F$ of D_F onto F with the following properties (where we let any subgroup of G_R^0 operate on D_F through conjugation by c_F^{-1}, without making further mention of this):

1) Each $g \in N(F)$ permutes the fibers of τ.
2) Each $g \in Z(F)$ carries each fiber of τ onto itself, and $Z(F)$ is transitive on each fiber.
3) If π is the natural homomorphism of $N(F)$ onto $G(F)$, we have $\tau(z)\pi(q) = \tau(z \cdot q)$, $q \in N(F)$, $z \in D_F$.
4) If $U(F)$ is the unipotent radical of $N(F)$, then $U(F)$ operates on D_F by affine linear transformations with unipotent linear parts.

Let $j_{D_F}(w, g)$ be the Jacobian determinant of $g \in G_R^0$ at $w \in D_F$ and if $\bar{g} \in G(F)$, let $j_F(t, \bar{g})$ be the Jacobian determinant of $\bar{g}$ at $t \in F$. We have $N(F)/Z(F) = G(F)$ and one may write $N(F)$ as the almost semi-direct product of $Z(F)$ by a group $L(F)$ such that $G(F)$ is the quotient of $L(F)$ by its finite center. This follows from the facts that $U(F) \subset Z(F)$, that $N(F)$ is the semi-direct product of $U(F)$ and of a reductive group R, and that $G(F)$ is the quotient by its center of the product $L(F)$ of certain simple factors of R. Then $j_{D_F}(w, g)$, considered as a function of g, is a character of $Z(F)$ when restricted to the latter (it does not depend on w), and we let

$$j_{D_F}(*, g)^l = \Delta_F(g, l), \quad g \in Z(F). \tag{35}$$

Moreover, if $g' \in L(F)$, then $\pi(g') = p_F^- k_F p_F^+$, where $p_F^{\pm} \in P_F^{\pm}$, $k_F \in K_{FC}$, and $P_F^{\pm}$ resp. K_{FC} are related to $\mathrm{Hol}(F)^0$ in the same way as $P^{\pm}$ resp. K_C are to $\mathrm{Hol}(D)^0$; and (Chapter 7, section 7; (8), section 1, this chapter)

$j_F(0, \pi(g')) = \det(Ad_{\mathfrak{p}_F^+}(k_F^{-1}))$, where $\mathfrak{p}_F^{\pm}$ is the Lie algebra of $P_F^{\pm}$. As the latter is in reality a character on the center of K_{FC}, as the function $j_F(t, \bar{g})$ is a holomorphic function of $t \in F$, and as $K_{FC} \subset K_C$ and $P_F^{\pm} \subset P^{\pm}$, it may not be surprising, at least when D is irreducible, that an appropriate integral power of $j_F(\tau(w), \pi(g'))$ is equal to some integral power of $j_{D_F}(w, g')$. Now, in fact, by calculating the multiplicities of certain $\boldsymbol{R}$-roots in the adjoint representation, one may prove, even if D is not irreducible, but *with the assumption that* G *be* $\boldsymbol{Q}$*-simple* [3 : § 3], that there exist positive integers m_F and q_F, and a multi-index $\boldsymbol{n}_F$ of positive integers such that

$$(36)\qquad \begin{aligned} |j_{D_F}(w, g)|^{m_F} &= |j_F(\tau(w), \pi(g))|^{\boldsymbol{n}_F} |\Delta_F(b, m_F)| \\ &= |j_F(\tau(w), \pi(g))|^{\boldsymbol{n}_F} |\chi_F(b)|^{-q_F}, \quad w \in D_F,\ g \in N(F), \end{aligned}$$

where $\chi_F = \det(Ad_{\mathfrak{U}(F)})$ ($\mathfrak{U}(F)$ being the Lie algebra of $U(F)$), where $g = b \cdot g'$ with $b \in Z(F)$, $g' \in L(F)$, and where the expression $j_F(\cdots)^{\boldsymbol{n}_F}$ is a product of integral powers with positive exponents of the Jacobian determinants in the different irreducible factors of F. As a matter of fact, in many instances all those exponents (the integers in the multi-index $\boldsymbol{n}_F$) are equal; and as the general case differs from the case when the exponents are equal only in minor aspects, we may sometimes assume that case for simplicity of exposition.

The elements F of the set of mutually incident boundary components we consider will form a "flag" $\mathscr{F}$; the elements of $\mathscr{F}$ correspond to elements of a set of mutually incident $\boldsymbol{R}$-parabolic subgroups $P_F = N(F)_C$ of G whose intersection is an $\boldsymbol{R}$-parabolic subgroup of G, and these intersection conditions are sufficient to characterize what we mean by a "flag" of boundary components (two parabolic subgroups of G are "incident" if their intersection is again parabolic—— a concept essentially due to Tits——and, in fact, two boundary components are incident in the geometric sense that one is a boundary component of the other if and only if the associated $\boldsymbol{R}$-parabolic subgroups of G are incident). From now on we fix a maximal flag of boundary components, $\mathscr{F}_1$, so that the intersection of all the associated $\boldsymbol{R}$-parabolic subgroups $N(F)_C$, $F \in \mathscr{F}_1$, is a *minimal* $\boldsymbol{R}$-parabolic subgroup P_* of G. (Cf. Chapter 6 for some explanation of details in examples.) If $\mathscr{R}$ is a maximal, connected, $\boldsymbol{R}$-trigonalizable subgroup of $P_{*\boldsymbol{R}}$ with unipotent radical N and maximal $\boldsymbol{R}$-diagonalizable subgroup A, then $\mathscr{R}$ operates in a simply transitive manner on D, so that $G_R^0 = K\mathscr{R} = KAN = K \cdot P_{*R}$, where K is the stability group of 0 in G_R^0; one has $K \cap AN = \{e\}$, and one may even verify that $G_R^0 = KAN$ is an Iwasawa decomposition in the sense of Chapter 7, section 5, by using the relations among the root spaces in the Lie subalgebras $\mathfrak{p}^{\pm}$ and $\mathfrak{k}_C$ of $\mathfrak{g}_C$. We have $A = A_F' \cdot A_F''$ (direct product) and $N = N_F' \cdot N_F''$ (semi-direct product), where $A_F' = L(F) \cap A$, $A_F'' = Z(F) \cap A$, $N_F' = L(F) \cap N$, and $N_F'' = Z(F) \cap N$.

We now need certain facts about the realization of D_F as a Siegel domain of the third kind (cf. Chapter 7, section 7, and [46; 39b; 3]). Let W be the center of $U = U(F)_C$ and denote by E_1 the quotient group $W_R \backslash U_R$. The root system of U (cf. [3 : p. 449]) is such as to make it evident on inspection that U is two-stage nilpotent (if not already Abelian) and that W contains the derived group of U, and thus E_1 is Abelian (cf. section 5). We identify W and E_1 once for all with their Lie algebras through the exponential mapping. Then W may be identified with a complex vector subspace of $\mathfrak{p}^+$ and we write $W = W' + iW'$, where $W' = W_R$; we have $z + w' \in D_F$ when $z \in D_F$ and $w' \in W'$. On the other hand, $\dim_R E_1$ is even and one supplies E_1 with a variable complex structure depending on a parameter $t \in F$ by identifying E_1 with a complex subspace E of $\mathfrak{p}^+$ complementary to the direct sum of W and of $\mathfrak{p}_F^+$ through a real linear mapping that depends on t. For the details of these relationships, which are a bit complicated, we refer the reader to [46 : § 6] for examples and to [39b : §§ 6 and 7] for a treatment of the general case. Then there exists an open, convex cone $V \subset W'$ and for each $t \in F$, an $\boldsymbol{R}$-bilinear mapping L_t of $E \times E$ into W (that is defined using the fact that the derived group of U is contained in W to obtain a bilinear mapping of $E_1 \times E_1$ into W) such that the conditions 1) and 2) of Chapter 7, section 7 are satisfied, with the property that

$$D_F = \{(z, u, t) \in W \times E \times F \mid t \in F, \ \operatorname{Im} z - L_t(u, u) \in V\}. \tag{37}$$

The real vector space W' is supplied with a positive-definite inner product (,) with respect to which V is self-dual:

$$V = \{x \in W' \mid (x, y) > 0 \quad \text{for all} \quad y \in V\}.$$

For example, if W' is the space of $n \times n$ real, symmetric matrices and V is the cone of positive-definite ones, then we define $(X, Y) = \operatorname{tr}(XY)$;

if $W'=\boldsymbol{R}^n$ and $V=\{(x_1, \cdots, x_n)\,|\,x_1^2-x_2^2-\cdots-x_n^2>0,\ x_1>0\}$, we define $(x, y)=\sum_{i=1}^{n} x_i y_i$; and so on.

When it is necessary to distinguish between the objects connected with D_F and with $D_{F'}$, we supply them with the subscripts F and F' respectively.

A consequence of the so-called "hereditary property" of the Cayley transforms, which one may verify, is that if F' is a boundary component of F, then the subspaces $\mathfrak{p}_F{}^+$, $\mathfrak{p}_{F'}{}^+$, W_F, and $W_{F'}$ of $\mathfrak{p}^+$ are related by:

$$\mathfrak{p}_{F'}{}^+\subset\mathfrak{p}_F{}^+,\quad W_F'\subset W_{F'}',\quad F, F'\in\mathcal{F},\quad F'\subset\partial F. \tag{38}$$

A cylindrical set $\boldsymbol{S}(Q, r)$ is defined in D_F as follows: Let Q be a compact subset of F, let $r\in V$, and define

$$\boldsymbol{S}(Q, r)=\{(z, u, t)\in D_F\,|\,t\in Q,\ \operatorname{Im} z-\operatorname{Re} L_t(u, u)-r\in V\}.$$

According to [46 : pp. 28–29; 39b : § 7], the function $\operatorname{Im} z-\operatorname{Re} L_t(u, u)$ is invariant under U_R, hence $\boldsymbol{S}(Q, r)\cdot U_R=\boldsymbol{S}(Q, r)$. We supply W' with a partial ordering by: $x_1>x_2$ if $x_1-x_2\in V$.

The group $N(F)$ is a parabolic subgroup of G_R^0. If the latter is simple, then $N(F)$ is the semi-direct product of its unipotent radical U_R and of the centralizer $Z(S_F)$ of a one-dimensional $\boldsymbol{R}$-split torus S_F, provided F is a proper boundary component of D, i.e., $F\neq D$. There is a single, simple, positive $\boldsymbol{R}$-root α_F of G with respect to the chosen maximal $\boldsymbol{R}$-split torus S containing S_F such that $\alpha_F|S_F$ is non-trivial. Let A_F be the identity component of the group of real points of S_F. If $a\in A_F$, the action of a on D_F is given by

$$(z, u, t)\cdot a=(\lambda_F(a)\cdot z,\ \mu_F(a)\cdot u,\ t), \tag{39}$$

where λ_F and μ_F are strictly negative powers of α_F and $\lambda_F=\mu_F{}^2$. This follows by calculating the restrictions of the different positive, non-compact roots to A_F. (N.B. The compact and non-compact roots ordinarily are defined with respect to a Cartan subgroup H of K, which is also a Cartan subgroup of G_R^0. However, we combine the definition of section 5 (below, p. 213) and the notation of [3 : § 1] to introduce a notion of $\boldsymbol{R}$-non-compact root. The relation between the notions is that if $c=c_F$ is the Cayley transform, then $c^{-1}G_R^0c\cap H_c=c^{-1}H'c$, where

H' contains a maximal, connected, $\boldsymbol{R}$-diagonalizable subgroup A of $G_{\boldsymbol{R}}^0$; and the positive, non-compact ($\boldsymbol{R}$-)root spaces of $c^{-1}Ac$ are subspaces of $\mathfrak{p}^+$ and $\mathfrak{p}^+$ contains the unbounded realization $\zeta(G_{\boldsymbol{R}}^0 \cdot c) = D_F$ of D.)

If $G_{\boldsymbol{R}}^0$ is not simple, i.e., if D is not irreducible, then $D = \prod_j D_j$, $1 \leqq j \leqq l$, where each D_j is irreducible, and if F is a proper boundary component of D, then $F = \prod_j F_j$, where F_j is a boundary component of D_j, and F_j is proper for at least one j. Moreover, $N(F) = \prod_j N(F_j) \subset \prod_j G_{j\boldsymbol{R}}^0 = G_{\boldsymbol{R}}^0$, where $G_{j\boldsymbol{R}}^0$ is the identity component of $\mathrm{Hol}(D_j)$. If G is defined over $\boldsymbol{Q}$, and is $\boldsymbol{Q}$-*simple*, and if F is a proper *rational* boundary component of D, then $N(F)$ is a proper $\boldsymbol{Q}$-parabolic subgroup of $G_{\boldsymbol{R}}^0$. From this it follows that $N(F_j)$ is a proper $\boldsymbol{R}$-parabolic subgroup of $G_{j\boldsymbol{R}}^0$ and that F_j is a proper boundary component of D_j for each j (for if T_F is a maximal, central $\boldsymbol{Q}$-split torus in a reductive complement of the unipotent radical of $N(F)$, then the centralizer of T_F in $G_{\boldsymbol{R}}^0$ cannot contain any factor $G_{j\boldsymbol{R}}^0$ because the Zariski-closure of the product of all such factors would be a proper normal $\boldsymbol{Q}$-subgroup of G). *We henceforth assume G to be defined over $\boldsymbol{Q}$ and $\boldsymbol{Q}$-simple.*

We now pick a maximal flag $\mathfrak{F}$ of rational boundary components, so that the intersection of all the groups $N(F)_C$, $F \in \mathfrak{F}$, will be a *minimal* $\boldsymbol{Q}$-parabolic subgroup P_0 of G. Without loss of generality, we may assume that $P_* \subset P_0$ and $\mathfrak{F} \subset \mathfrak{F}_1$, the previously selected flag of "standard" boundary components. We take a maximal $\boldsymbol{Q}$-split torus $T = {}_{\boldsymbol{Q}}T$ of G contained in a maximal $\boldsymbol{R}$-split torus S of G such that P_* is generated by its unipotent radical and by the centralizer of S and such that P_0 is generated by its unipotent radical and by the centralizer of T; moreover, we take compatible orderings on the root systems ${}_{\boldsymbol{R}}\Sigma$ of S and ${}_{\boldsymbol{Q}}\Sigma$ of T. We know from the earlier discussion of this section and from Chapter 7, section 3, that this may be done. We have $S = \prod_j S_j$, where S_j is a maximal $\boldsymbol{R}$-split torus of G_j and G_j, $j = 1, \cdots, l$, are, as above, the simple factors of G. Let other notation also be as above. If $F \in \mathfrak{F}$ and is proper, then $P = P_F = N(F)_C = \prod_j N(F_j)_C$ is a maximal (proper) $\boldsymbol{Q}$-parabolic subgroup of G and is the semi-direct product of its unipotent radical $U(F)$ and of the centralizer $Z(T_F)$ of a one-dimensional $\boldsymbol{Q}$-split torus T_F contained in

T, which is also the centralizer of a unique l-dimensional $\boldsymbol{R}$-split subtorus S_F of S; and $T_F \subset S_F$. Let $A_F = S_{FR}^0$ and let $J = T_R^0$; then $A_F = \prod_j A_{F_j}$ (direct product), where $A_{F_j} = A_F \cap S_j$.

For each j, we indicate the corresponding objects relative to the simple factor G_j with a subscript j.

With D_{F_j} (an abuse of notation for $(D_j)_{F_j}$) described as in (37), we have a positive-definite inner product $(\, , \,)_j$ such that V_j is self-dual with respect to it. If we put

$$(x, y) = \sum_j (x_j, y_j)_j, \quad x = (x_j), \; y = (y_j) \in W', \tag{40}$$

then $V = V_1 \times \cdots \times V_l$ is self-dual with respect to this.

If $a = (a_1, \cdots, a_l) \in A_F = \prod_j A_{F_j}$ and if $(z, u, t) = \{(z_j, u_j, t_j)\}_{j=1,\cdots,l}$, then

$$(z, u, t) \cdot a = \{(\lambda_{F_j}(a_j) z_j, \; \mu_{F_j}(a_j) u_j, \; t_j)\}_{j=1,\cdots,l}. \tag{41}$$

It follows that $j_{D_F}((z, u, t), a)$ is a product of strictly negative powers of the quantities $\alpha_{F_i}(a_i)$:

$$j_{D_F}(w, a) = \prod_i \alpha_{F_i}(a_i)^{m_i}, \quad m_i < 0, \; i = 1, \cdots, l, \; w \in D_F. \tag{42}$$

As the centralizer of T_F meets each G_j in a *proper* closed subgroup of the latter and since $T_F \subset S_F$, we see that each $a \in T_{FR}^0$ different from the identity acts non-trivially on each factor D_{F_j} and we have

$$(z, u, t) \cdot a = \{(\lambda_{F_j}(a_j) z_j, \; \mu_{F_j}(a_j) u_j, \; t_j)\}_{j=1,\cdots,l}, \tag{43}$$

and

$$j_{D_F}(w, a) = \alpha_F(a)^m, \quad m < 0, \tag{44}$$

where α_F is the uniquely determined simple $\boldsymbol{Q}$-root with respect to T which does not vanish on T_F.

Moreover, if Γ is an arithmetic subgroup of G_K^0, then $P \cap \Gamma$ is an arithmetic subgroup of P_R and $U_R/(U \cap \Gamma)$ is compact. In particular, if $\gamma \in W' \cap \Gamma$, we have

$$(z, u, t) \cdot \gamma = (z + \lambda_\gamma, u, t), \tag{45}$$

where the elements λ_γ form a lattice Λ in W' such that W'/Λ is compact. Let Λ' be the lattice dual to Λ with respect to $(\, , \,)$:

$$\Lambda' = \{x \in V \mid (x, \Lambda) \subset \boldsymbol{Z}\}.$$

§5. **Fourier-Jacobi series** [46; 2e]

We continue with the notation and conventions in force at the end of the preceding section. In particular, Γ is an arithmetic subgroup of G_R^0.

Let f be a (holomorphic) automorphic form on D_F with respect to Γ. Then

$$f(z+\lambda, u, t)=f(z, u, t), \quad \lambda \in \Lambda,$$

so that f has a Fourier expansion (cf. Chapter 1 for notation):

$$f(z, u, t)=\sum_{\rho \in \Lambda'} \phi_\rho(u, t)e((\rho, z)), \tag{46}$$

where ϕ_ρ are holomorphic functions of u and t and (,) is the natural extension of the previous inner product (,) to $W'_C=W$. It follows from Koecher's principle [38; 46 : §15; 3 : §10.14; 2e] that in fact $\phi_\rho=0$ for $\rho \notin \bar{V}$, except when $\dim G=3$; we exclude the latter case, which has in principle been dealt with in Chapter 1, for the rest of this chapter. Thus if we put $\Lambda^+=\Lambda' \cap \bar{V}$, we have

$$f(z, u, t)=\sum_{\rho \in \Lambda^+} \phi_\rho(u, t)e((\rho, z)). \tag{47}$$

Since $f((z, u, t)\cdot\gamma)=f(z, u, t)$, $\gamma \in U \cap \Gamma$, it follows from the uniqueness of the Fourier coefficients that $\phi_0(u+\alpha, t)=\phi_0(u, t)$, where α runs through a lattice in E_{1R}; therefore, ϕ_0 does not depend on u by Liouville's theorem and we may write $\phi_0(u, t)=\phi_0(t)$. It follows from (47) that given any cylindrical set $\boldsymbol{S}(Q, r)$ and constant $K>0$, there exists [46 : p. 119] a convergent series $\sum_{\rho \in \Lambda^+} M_\rho$ of positive terms such that

$$|\phi_\rho(u, t)e((\rho, z))|<M_\rho, \quad (z, u, t) \in \boldsymbol{S}(Q, r), \ |u| \leqq K. \tag{48}$$

We let $\mathcal{A}=\{(u, t) \mid t \in Q, |u| \leqq K\}$; then $\mathcal{A}$ is compact. If $r \in V$, we let

$$C(r, \mathcal{A})=\{(z, u, t) \mid (z, u, t) \in \boldsymbol{S}(Q, r), \quad |u| \leqq K\};$$

$C(r, \mathcal{A})$ is called a restricted cylindrical set. Let $z_0=x_0+iy_0$, $y_0 \in V$, be such that

$$y_0-L_t(u, u)-r_0 \in V \quad \text{for all} \quad (u, t) \in \mathcal{A},$$

and let $r_1=y_0-r_0$. Put

$$N_\rho = \sup_{(u,\,t)\in\mathcal{A}} |\psi_\rho(u, t)| e^{-2\pi(\rho,\, y_0 - r_0)}.$$

Then $\sum_{\rho\in\Lambda^+} N_\rho < +\infty$. It follows that if $r - r_1 \in V$ and if $(z, u, t) \in \mathcal{C}(r, \mathcal{A})$, then we have

$$|f(z, u, t)| \leqq C_1 \cdot \max_{\rho\in\Lambda^+,\, \psi_\rho \neq 0} e^{-2\pi(\rho,\, r - r_1)}, \tag{49}$$

where C_1 is a positive constant depending only on r_1 and $\mathcal{A}$; the right side of (49) is finite because $(\rho, r - r_1) \geqq 0$ for all $\rho \in \Lambda^+$. On the other hand

$$f(z, u, t) = \psi_0(t) + \sum_{\rho\in\Lambda^+ - \{0\}} \psi_\rho(u, t) e((\rho, z)),$$

so that if r is sufficiently large, we obtain likewise, if ψ_0 is nowhere zero on Q,

$$|f(z, u, t)| \geqq C' \min_{t\in Q} |\psi_0(t)|, \quad (z, u, t) \in \mathcal{C}(r, \mathcal{A}), \tag{50}$$

where C' is a positive constant depending on Q, r, and K.

Returning to the maximal $\boldsymbol{Q}$-split torus T of G, we have $Z(T) = M \cdot T$, $M \cap T$ finite, and there is a unique, minimal $\boldsymbol{Q}$-parabolic subgroup P_0 of G containing $Z(T)$ and contained in P_F for each $F \in \mathfrak{F}$, as we have noted above. Let $d > 0$ and with $J = T^0_{\boldsymbol{R}}$, define

$$J_d = \{b \in J \mid \alpha(b) \leqq d \text{ for each simple } \boldsymbol{Q}\text{-root } \alpha\}.$$

If, moreover, ω is a compact subset of $M_{\boldsymbol{R}} \cdot R_u(P_0)_{\boldsymbol{R}}$, let

$$\mathfrak{S} = \mathfrak{S}_{d,\,\omega} = K J_d \omega,$$

and if $0 < d' < d$, let $\mathfrak{S}_{d',\,d,\,\omega} = \mathfrak{S}_{d,\,\omega} - \mathfrak{S}_{d',\,\omega}$. Then $\mathfrak{S}$ is called a ($\boldsymbol{Q}$-)Siegel set. It is known [6d] that if d and ω are large enough, then there are a finite number of $g \in G_{\boldsymbol{Q}}$, say, $g_1, \cdots, g_N$, such that

$$\mathfrak{S}^* = \bigcup_{\nu=1}^{N} \mathfrak{S}_{d,\,\omega} g_\nu \tag{51}$$

is a fundamental set for Γ in $G_{\boldsymbol{R}}$; i.e., $G_{\boldsymbol{R}} = \mathfrak{S}^* \cdot \Gamma$ and the set of $\gamma \in \Gamma$ such that $\mathfrak{S}^* \cap \mathfrak{S}^* \gamma$ is non-empty is finite.

If f is an automorphic form on D of weight l with respect to Γ, then f_F defined on D_F by

$$f_F(w) = f(w \cdot c_F^{-1}) j(w, c_F^{-1})^l \tag{52}$$

is an automorphic form on D_F of weight l with respect to Γ (operating on D_F by conjugation with c_F^{-1}). Thus, each f_F has a Fourier expansion

(53) $$f_F(z, u, t) = \sum_{\rho \in \Lambda^+} \phi_\rho{}^F(u, t)e((\rho, z)),$$

where the subscript F has been omitted on z, u, t, Λ^+, etc., but may be employed in the future to avoid confusion.

We recall the notation of the preceding section where $G_R^0 = KAN$ (Iwasawa decomposition), $A = A_F' A_F''$, etc. If F is a rational boundary component, we have

$$J = J_F' \cdot J_F'' \quad \text{(direct product)},$$

where $J_F' = J \cap A_F'$, $J_F'' = J \cap A_F''$.

PROPOSITION 1 (*Satake* [*52: Exp. 9 (Supp.)*]). *There exists a rational number $m > 0$ with the following property: Let p, $\infty \geqq p \geqq 1$, be given and let f be an automorphic form on D of weight l with respect to Γ. Define $\rho(k) = j_D(z, k)^l$, $k \in K$ (this is independent of z). Then if $pml > 1$, if ω has non-empty interior, and if $'f \in L^p(\mathfrak{S}_{d,\omega})$, we have $\phi_0{}^F \equiv 0$ for all $F \in \mathfrak{F}$.*

Conversely, if $\phi_0{}^F \equiv 0$ for all $F \in \mathfrak{F}$, then $'f \in L^\infty(\mathfrak{S}_{d,\omega})$ and hence, automatically, $'f \in L^p(\mathfrak{S}_{d,\omega})$ for all $p \geqq 1$.

(Here $L^p(\mathfrak{S})$ is defined with reference to the Haar measure restricted to $\mathfrak{S}$.)

PROOF. For each F, we define

(54) $$'f_F(g) = f_F(w_{0F} \cdot g) j(w_{0F}, g)^l, \quad g \in G_R^0,$$

where $w_{0F} = z_0 \cdot c_F$ with $z_0 = 0$. It follows from (15), Chapter 10, section 3, from (52), and from (54) above that $'f_F = 'f$ for all $F \in \mathfrak{F}$.

Assume F to be such that $\phi_0{}^F \not\equiv 0$. If $d' < d$ and $\omega' \subset \omega$, clearly $'f \in L^p(\mathfrak{S}_{d,\omega})$ implies that $'f \in L^p(\mathfrak{S}_{d',\omega'})$. Direct calculation shows that if d' and ω' are chosen small enough, then the image $w_{0F} \cdot \mathfrak{S}_{d',\omega'}$ of $\mathfrak{S}_{d',\omega'}$ in D_F is contained in a cylindrical set $\boldsymbol{S}(Q, r)$. According to (50), we may choose ω' appropriately and assume $|f_F| \geqq c > 0$ on $w_{0F} \cdot \mathfrak{S}_{d',\omega'}$, where c is a constant, because $|u|$ is bounded on the latter set. The Haar measure on $G_R^0 = KAN$ is of the form

$$dg = \sigma(a) dk\, da\, dn$$

(cf. [27 : Chap. X]). We have $A = J \cdot A_1$ and $J \cap A_1 = \{1\}$, where A_1 is the connected subgroup of A on which all the $\boldsymbol{Q}$-roots of S are trivial, and $\mathfrak{S}_{d',\omega'}$ contains a set of the form

$$K\cdot\beta\cdot\omega', \tag{55}$$

where

$$\beta=\beta_{d'',d'}=J_{d'}\cap\{a\in J\,|\,\alpha(a)\geqq d''>0 \text{ for every simple root } \alpha \text{ of } G \text{ coming from a root of } G(F)\},$$

$d'>d''>0$, and ω' is a relatively compact subset of A_1N with non-empty interior. Moreover, $\sigma(a)$ is a product of powers of $\alpha_i(a)$ with strictly positive exponents, where α_i runs through the simple $\boldsymbol{R}$-roots of G. Let $h=|'f|$. Then for $b\in\beta$, $x=a_1n\in\omega'$, we have

$$\begin{aligned} h(kbx)&\geqq c\,|j_{D_F}(w_{0F},kbx)|^l=c\,|j_{D_F}(w_{0F},bx)|^l \\ &=c\,|j_{D_F}(w_{0F},b)|^l|j_{D_F}(w_{0F}\cdot b,x)|^l. \end{aligned}\tag{56}$$

Now $b=b'b''$, $b'\in J_F'$, $b''\in J_F''$, $x=x'x''$ with $x'\in(A_1\cap A_F')N_F'$, $x''=a_1''n''\in(A_1\cap A_F'')N_F''\subset Z(F)$, and b', x', and x'' remain in relatively compact sets and $j_{D_F}(w,x'')$ is a continuous function of x'' and is independent of w, so that

$$|j_{D_F}(w_{0F},b)|^l\geqq(\text{const.})\cdot|\Delta(b'',l)|, \tag{57}$$

while

$$|j_{D_F}(w_{0F}\cdot b,x)|\geqq(\text{const.})\cdot|j_{D_F}(w_{0F}\cdot b,x'')|\geqq\text{const.} \tag{58}$$

It follows from (54), (56), (57), and (58) that

$$h(kbx)\geqq c_1\cdot|\Delta(b'',l)|,$$

where c_1 is a positive constant. We know [3 : §7] and can easily deduce from (41) that $|\Delta(b'',1)|$ is a product of powers of $\alpha_i(b'')$ with strictly negative exponents and $\Delta(b'',l)=\Delta(b'',1)^l$. The range of integration for b'' is $0<\alpha_i(b'')\leqq d'$ for the set of simple $\boldsymbol{Q}$-roots α_i appearing in Δ (with negative exponents) which are not roots of $G(F)$. Since $\int_0^\varepsilon x^{-\eta}dx$ diverges when $\eta\geqq 1$, it follows that there exists a positive rational number m such that the integral of h^p over $\mathfrak{S}_{d',\omega'}$ will diverge as soon as $pml>1$, and this, of course, implies the first part of the proposition for $p<+\infty$. But since a Siegel set has finite Haar measure [8], we have

(59) $$L^p(\mathfrak{S}) \supset L^\infty(\mathfrak{S}), \quad p \geqq 1,$$

and it follows at once that the conclusions of the first half of the proposition hold for $p=\infty$ and all positive l.

To sketch the proof of the second half of the proposition, we need some of the facts about Cayley transforms and about the Jordan algebra structure on W. Continuing our notation for the flags $\mathfrak{F}$ and $\mathfrak{F}_1$ and for the minimal parabolic groups P_* and P_0, we see that if F, $F' \in \mathfrak{F}$, and $F' \subset \partial F$, then the cone V_F will be the interior of $W_F' \cap \bar{V}_{F'}$, in view of (38).

We now describe the relationships between the root space structure of W', a certain Jordan algebra structure on W, and the cone V, where the subscript referring to a particular boundary component has been dropped. The relationships will be described without proofs; for some of the details we refer the reader to [3; 39a, b; 46]. First of all, for an irreducible component G_j of G, the relative $\boldsymbol{R}$-root system ${}_R\Sigma$ is either of type C or else is of type BC, the latter being obtained from a system of type C by adding the halves of the longest roots. Thus, in either case, we may isolate the subsystem of type C in ${}_R\Sigma$, and we denote it by ${}_R\Sigma_C$. In a root system $\Sigma=\{\pm c_i \pm c_j, \pm 2c_i\}$ of type C, the roots $\{c_i+c_j, 2c_i\}$ constitute the so-called positive, non-compact roots; we denote this set by Σ^*; if Σ is of type BC, the positive non-compact roots are the union of Σ^* and of the set $\{c_i\}$. Now if F_j is a boundary component of the irreducible factor D_j of D, let ${}_R\Sigma_C{}^*(F_j)$ denote the set of roots of $U(F_j)$ with respect to S_j lying in ${}_R\Sigma_C{}^*$. Then the subgroup of $U(F_j)$ generated by the root spaces corresponding to elements of ${}_R\Sigma_C{}^*(F_j)$ is just the center W_j of $U(F_j)$. As explained in [3: §2], the possibilities for the relative $\boldsymbol{Q}$-root system of G, when G is almost $\boldsymbol{Q}$-simple, are just the same as those for the $\boldsymbol{R}$-root system of G_j, and there (loc. cit.) the relationship between the cases for the $\boldsymbol{Q}$-roots and for the $\boldsymbol{R}$-roots is described. In particular, if F is a rational boundary component of D, and if ${}_Q\Sigma_C{}^*(F)$ is defined in a manner analogous to that for ${}_R\Sigma_C{}^*(F_j)$, then the corresponding subgroup of $U(F)$ is just the center $W=W_F$ of $U=U(F)$. Now, W_j and W have structures of complex Jordan algebras defined over $\boldsymbol{R}$ and over $\boldsymbol{Q}$ respectively, and W_{jR} and $W_R=W'$ are real Jordan subalgebras of them. A system of mutually orthogonal primitive idempotents in a Pierce decomposition of W_{jR} is given by a set of suitably normalized vectors spanning the (one-dimensional) $\boldsymbol{R}$-root spaces corresponding to the longest roots in ${}_K\Sigma_C{}^*(F_j)$, and the self-dual cone V_j in $W_{j,1} = W_j'$ is the interior of the set of squares (with respect to the Jordan algebra structure) in W_j'. If $\alpha \in {}_Q\Sigma_C{}^*(F)$, $F \in \mathfrak{F}$, is a longest root in ${}_Q\Sigma_C$, then the corresponding $\boldsymbol{Q}$-root space in W' is the sum of the $\boldsymbol{R}$-root spaces in $W' = \bigoplus_j W_j'$ corresponding to the $\boldsymbol{R}$-roots which restrict to the given longest $\boldsymbol{Q}$-root, in each simple factor; such a sum of $\boldsymbol{R}$-root spaces will be called the $\boldsymbol{Q}$-block corresponding to the given longest $\boldsymbol{Q}$-root. If $w \in W'$, then w may be written as a sum of components from the different root spaces, and we refer to those components from the root spaces of the longest $\boldsymbol{R}$-(or $\boldsymbol{Q}$-)roots as the "longest $\boldsymbol{R}$-(or $\boldsymbol{Q}$-)root components" of w; we choose a basis for W' composed of root vectors such that the longest root vectors lie in the closure $\bar{V}$ of V. Then if $w \in V$, it may be seen that its longest $\boldsymbol{R}$-root components are strictly positive multiples of the longest $\boldsymbol{R}$-root vectors.

Let $\mathfrak{S} = \mathfrak{S}_{d,\omega}$ be a $\boldsymbol{Q}$-Siegel set in $G_{\boldsymbol{R}}$, and denote by $\bar{\mathfrak{S}}$ its canonical image $w_{0F} \cdot \mathfrak{S}$ in D_F; the set of points $y \in V$ such that $y = \operatorname{Im} z$ for some point (z, u, t) of $\bar{\mathfrak{S}}$ will be denoted by $\mathfrak{R} = \mathfrak{R}_{d,\omega}$, and the mapping $(z, u, t) \mapsto \operatorname{Im} z$, which maps $\bar{\mathfrak{S}}$ onto $\mathfrak{R}$, will be denoted for our immediate needs by $\boldsymbol{\theta}_F$. If f_1 and f_2 are two positive, real-valued functions on a space X, we write $f_1 \prec f_2$ if there is a positive constant c such that $f_1(x) \leqq c f_2(x)$ for all $x \in X$, and we write $f_1 \asymp f_2$ if both $f_1 \prec f_2$ and $f_2 \prec f_1$ hold. We define two functions f_1 and f_2 on $\mathfrak{S} \times V$ as follows. If $v \in V$ and if $s = kb\eta \in \mathfrak{S}$, $k \in K$, $b \in J_d$, $\eta \in \omega$, define $f_1(s, v) = (\boldsymbol{\theta}_F(w_{0F} \cdot b), v)$ and $f_2(s, v) = (\boldsymbol{\theta}_F(w_{0F} \cdot s), v)$, where $(\ ,\ )$ is the inner product on $W' \times W'$. Of course, $w_{0F} \cdot s = w_{0F} \cdot b\eta$, and η runs in a compact set, while the simple $\boldsymbol{Q}$-roots are $\leqq d$ on b. The crucial fact we need from reduction theory is

$$f_1 \asymp f_2. \tag{60}$$

In order to verify this fact, one needs some explicit information about the expression of the inner product $(\ ,\ )$ in terms of root spaces in W and about the operations of J and of $M_{\boldsymbol{R}} \cdot R_u(P_0)_{\boldsymbol{R}}$ on D_F. As for the former, all necessary explicit information is contained in [39b: §7], from which it will be seen in particular that each of the $\boldsymbol{R}$-root spaces in W' is supplied with a positive inner product invariant under the compact part of $Z(S)_{\boldsymbol{R}}$, and that the Jordan algebra inner product $\operatorname{tr}(X \circ Y)$ is a linear combination of these inner products on the mutually orthogonal separate root spaces. The point w_{0F} is of the form $(ie_F, 0, 0)$, where e_F is a linear combination of the longest root vectors with positive coefficients. If $s \in J_d$, then $w_{0F} \cdot s = (ie_F(s), 0, 0)$, where $e_F(s)$ is a linear combination of the longest root vectors with strictly positive coefficients ranged in increasing order of magnitude (with increasing order position of the roots), such that the coefficient of the lowest, longest root vector of $U(F)$ is bounded below by a positive constant. Assuming this, if $\eta \in \omega$, if $w_{0F} \cdot s = (z, u, t)$, and if $y = \operatorname{Im} z$, then in y the coefficients of the longest root vectors in each simple factor are still arranged in increasing order of magnitude, and the coefficients of the shorter (positive, non-compact) root vectors in y are dominated by the former (in fashion analogous to the manner in which the diagonal elements of a Minkowski-reduced positive, symmetric matrix dominate the off-diagonal elements; in fact, the latter case, as related to the Siegel upper half-plane H_n is a prototype for what occurs in the general situation). Using these and related facts, one may then verify (60).

Let $\boldsymbol{N}$ denote the set of positive natural integers. A sequence $\{v_n\}$ of points in $V = V_F$ is said to tend to infinity interiorly in V if each of the longest root coefficients of v_n tends to $+\infty$, in such a way that if v' is any point of V and if v'' is the sum of the longest $\boldsymbol{R}$-root components of v', then $(v_n, v') \asymp (v_n, v'')$ on $\boldsymbol{N} \times V$. On the other hand, α_F appears with positive coefficient in each of the longest roots in ${}_Q\Sigma_C{}^*(F)$, and the restriction of α_{F_j} to T (viewed as diagonally imbedded in $S = \prod S_j$) contains α_F with a positive coefficient. Hence, if $\{b_n\}$ is a sequence in J_d such that $\alpha_F(b_n) \to 0$, then $\boldsymbol{\theta}_F(w_{0F} \cdot b_n)$ tends to infinity interiorly in V_F. It can then also be seen from (60) that for η varying arbitrarily in the compact set ω, we have

$$(61) \qquad (v, \boldsymbol{\theta}_F(w_{0F} \cdot b_n \cdot \eta)) \asymp (v, \boldsymbol{\theta}_F(w_{0F} \cdot b_n)), \quad (v, n, \eta) \in V_F \times \boldsymbol{N} \times \omega.$$

Since the constants which are implicit in the relation $\asymp$ in (61) depend only on d and ω, it follows by continuity that (61) holds for all $v \in \bar{V}_F$.

Now assume that

$$(62) \qquad \phi_0{}^F \equiv 0, \quad F \in \mathfrak{F}.$$

Let $F, F' \in \mathfrak{F}$ be such that $F' \subset \partial F$. By (38), W_F' is a subspace of $W_{F'}'$

and $\mathfrak{p}_{F'}{}^+$ is a subspace of $\mathfrak{p}_F{}^+$. We have the relation $c_{F'}=c_F c_{FF'}$, and as a holomorphic transformation of D_F onto $D_{F'}$, $c_{FF'}$ may be shown to have the following form (cf. the example of Chapter 6, section 5):

$$(z_F,\ u_F,\ t_F)\to(z_F',\ \cdots)$$

where $z_F'=z_F+\varphi(u_F,\ t_F)$, φ being a holomorphic vector-valued function. This implies that in the Fourier expansions of f_F and of $f_{F'}$ the coefficients of the coordinates of z_F that appear in the exponentials $e((\rho,\ z))$ are the same. We range the elements of the flag $\mathfrak{F}$ in order of decreasing dimension,

$$D=F_r\supset F_{r-1}\supset\cdots\supset F_0,\quad F_{i-1}\subset\partial F_i,$$

and write

$$\begin{aligned}f_{F_i}(z_{F_i},\ u_{F_i},\ t_{F_i})&=f_i(z_i,\ u_i,\ t_i)\\&=\psi_{0i}(t_i)+\sum_{\rho_i\in\bar{V}_i-\{0\}}\psi_{\rho_i}(u_i,\ t_i)e((\rho_i,\ z_i)).\end{aligned}$$

Since $\psi_{0i}\equiv 0$, $i<r$, and in particular $\psi_{0,\,r-1}\equiv 0$, it follows that if $F\in\mathfrak{F}$ and if $\rho\in\Lambda_F{}^+$ is such that $\psi_\rho{}^F\not\equiv 0$, then the highest, longest $\boldsymbol{R}$-root vectors (of which there may be more than one!) appear in ρ with strictly positive coefficients, of which the largest, call it $c(\rho)$, is bounded below because ρ varies in a lattice.

Assuming (62), suppose, contrary to the claim of Proposition 1 that $'f$ is not bounded on some Siegel set $\mathfrak{S}_{d,\,\omega}$. Then there exists a sequence $\{s_n\}$ in $\mathfrak{S}=\mathfrak{S}_{d,\,\omega}$ such that $|'f(s_n)|\to+\infty$. Write

$$s_n=k_n b_n\eta_n,\quad k_n\in K,\ b_n\in J_d,\ \eta_n\in\omega. \tag{63}$$

For each F, $\{\alpha_F(b_n)\}$ is a sequence of positive real numbers $\leqq d$, and, by taking a subsequence if necessary, we may assume that for each $F\in\mathfrak{F}$ either $\alpha_F(b_n)\to 0$ or $\alpha_F(b_n)$ converges to a positive, non-zero limit. We must have $\alpha_F(b_n)\to 0$ for some F, since otherwise s_n would have to remain in a compact set. Let $F\in\mathfrak{F}$ be of minimum dimension such that $\alpha_F(b_n)\to 0$ and fix this F. We have

$$'f(s_n)={}'f_F(s_n)=f_F(w_{0F}\cdot s_n)j_{D_F}(w_{0F},\ s_n)^l. \tag{64}$$

On the one hand calculations similar to those in the first half of the proof show that

$$|j_{D_F}(w_{0F},\ s_n)|\leqq c'\cdot\Delta(b_n,\ 1), \tag{65}$$

where c' is a positive constant. On the other hand, by our discussion above, we see that $\boldsymbol{\theta}_F(w_{0F}\cdot b_n)$ tends to infinity interiorly in V_F and thence from (61) we see that $\boldsymbol{\theta}_F(w_{0F}\cdot s_n)$ does, too. This means, in particular, that the coefficients $\{a(s_n)_j\}$ of the highest, longest $\boldsymbol{R}$-root vectors in $\boldsymbol{\theta}_F(w_{0F}\cdot s_n)$ are of the largest order of magnitude; and all the longest $\boldsymbol{R}$-root vectors lying in the same $\boldsymbol{Q}$-block are of the *same order of magnitude* because ω is a compact set; and $\{a(s_n)_j\}$ tends to $+\infty$ as n does. Therefore, by (61) and by the remarks following (62), we have

$$(\rho, \boldsymbol{\theta}_F(w_{0F}\cdot s_n)) \geqq c_2\cdot c(\rho)a(s_n) \geqq c_3 a(s_n), \tag{66}$$

where c_2 and c_3 are positive constants and where, say, $a(s_n)$ is the coefficient of a fixed, highest, longest $\boldsymbol{R}$-root vector (as remarked, they all satisfy the $\asymp$ relation with respect to each other). Combining (49) and (66) we see that for some further positive constants c_4 and c_5 we have

$$|f_F(w_{0F}\cdot s_n)| \leqq c_4 e^{-c_5 a(s_n)}, \tag{67}$$

while from (65) and from the fact that $a(s_n)$ is the coefficient of a highest, longest root vector we obtain, for still further constants c_6 and c_7,

$$|j_{D_F}(w_{0F}, s_n)|^l \leqq c_6 a(s_n)^{c_7} \tag{68}$$

because on J_d, the product of all the positive non-compact roots is majorized by a power of the highest one. Putting (64), (67), and (68) together, we obtain

$$|'f(s_n)| \to 0 \quad \text{as} \quad n \to \infty, \tag{69}$$

which is a contradiction.

This completes our *sketch* of the proof.

Let $G_Q{}^0 = G_R{}^0 \cap G_Q$. If $g_0 \in G_Q{}^0$, define f_{g_0} as the holomorphic function on D given by

$$f_{g_0}(z) = f(z\cdot g_0)j(z, g_0)^l. \tag{70}$$

It is at once verified that f_{g_0} is an automorphic form of weight l with respect to $g_0\Gamma g_0{}^{-1}$, and "raising" it to the group $G_R{}^0$, we get

$$'f_{g_0}(g) = 'f(gg_0). \tag{71}$$

Now f_{g_0} also has a Fourier expansion, and for each of the rational boundary components $F \in \mathfrak{F}$, we may define f_{g_0F}, etc. Then we have

$$f_{g_0F}(z, u, t) = \sum_{\lambda \in \Lambda_{g_0}^+} \psi_\lambda^{Fg_0^{-1}}(u, t) e((\lambda, z)),$$

where $\Lambda_{g_0}^+ = (W_F' \cap g_0 \Gamma g_0^{-1}) \cap V_F$, with certain obvious conventions regarding the identification of isomorphic objects.

Now $\mathfrak{S}^* = \bigcup_{\nu=1}^{N} \mathfrak{S}_{d,\omega} g_\nu$ is a fundamental set for Γ for suitably chosen d, ω, and $g_1, \cdots, g_N \in G_Q$. As $G_Q = Z({}_QT)_Q \cdot G_Q^0$ [57b], we can in fact take each $g_\nu \in G_Q^0$. Since $'f$ is right-invariant under Γ, it follows from the proposition that:

A) There exists a positive constant m such that if $1 \leqq p \leqq +\infty$ and if $pml > 1$, then all the 0th Fourier coefficients $\psi_0^{Fg_\nu^{-1}}$ vanish, $F \in \mathfrak{F}$, $g_\nu \in G_Q^0$.
B) If $\psi_0^{Fg_\nu^{-1}} \equiv 0$ for all $F \in \mathfrak{F}$, $\nu = 1, \cdots, N$, then $f \in L_\Gamma^p(\rho)$ for all p, $1 \leqq p \leqq \infty$.

The conditions $\psi_0^{Fg_\nu^{-1}} \equiv 0$ are closed in the uniform topology on $L_\Gamma^\infty(\rho)$ as well as in the L^2-topology on $\mathcal{H}_\Gamma^2(\rho)$ which is a closed subspace of $L_\Gamma^2(\rho)$. Therefore, $\mathcal{H}_\Gamma^\infty(\rho)$ is a closed subspace of $\mathcal{H}_\Gamma^2(\rho)$ and $\mathcal{H}_\Gamma^\infty(\rho) = \mathcal{H}_\Gamma^p(\rho)$ for all $p \geqq 1$, if l is in both cases large enough. This completes the proofs of all the related statements of Chapter 10 (but for the omission of certain details in the proof of Proposition 1). In particular we have

THEOREM 29. $\dim(\mathcal{H}_\Gamma^2(\rho)) < +\infty$ *for all* $l > 0$.

PROOF. This holds for sufficiently large l by Proposition 1 and by section 3 of Chapter 10. But since the product of an automorphic form of weight l by one of weight l' is an automorphic form of weight $l + l'$, the truth of the assertion for general l follows at once.

It should be noted that for the greater part our proof of Theorem 29 follows the same lines as the proof of a more general result, Theorem 1 of [26e]. The main difference is that we have exploited the relationship of automorphic forms on the group to automorphic forms on the domains D_F, and the complex analytic Fourier expansion of the latter, in contrast to the algebraic ideas of [26e].

§ 6. Poincaré-Eisenstein series (cont'd) [3]

With the standing assumption that G is defined over $\boldsymbol{Q}$ and almost $\boldsymbol{Q}$-simple, let F be a rational boundary component of D and define D_F, etc., as in the preceding sections. In particular, π will denote the canonical mapping of $N(F)$ onto $G(F)=Z(F)\backslash N(F)$ and $\tau: D_F \to F$ will be the holomorphic mapping discussed previously. According to section 4, there exists a connected, normal $\boldsymbol{Q}$-subgroup B of $P=N(F)_C$ contained in $Z(F)_C{}^0$ and containing the unipotent radical $U=U(F)_C$ of P such that $Z(F)/(B\cap G_R{}^0)$ is compact. Then B is the semi-direct product of the split radical $S_F\cdot U$ of P and of a reductive $\boldsymbol{Q}$-group H, which in turn may be written as the almost-direct product of its connected center H' and of its derived group $\mathfrak{D}H$. We hasten to point out that S_F is defined over $\boldsymbol{Q}$, but not $\boldsymbol{Q}$-split, and although H' is a $\boldsymbol{Q}$-torus, it has no non-trivial $\boldsymbol{Q}$-characters. In fact, H'_R is compact: Let $A_F=S_{FR}{}^0$ as before. Then, by the choice of B, $Z(F)/U_R$ is the almost-direct product of B_R/U_R and of a compact group, while $Z(F)/A_FU_R$ modulo its derived group is compact (from facts about the $\boldsymbol{R}$-structure and boundary components [3: §§ 1-3]), hence B_R/U_R modulo its derived group is compact, hence H'_R is compact. Now if $\chi=\det(Ad_{\mathfrak{u}(F)_C})$, then χ and $j_{D_F}(*,\ )$ are both characters on $Z(F)$, so they are both one on $\mathfrak{D}H$ and on U, and by (36) of section 4, a negative rational power of one equals the other on A_F so that

$$\chi(g)^{-q_F}=j_{D_F}(*,g)^{m_F}, \quad g\in \mathfrak{D}H\cdot A_F\cdot U_R.$$

Let Γ be an arithmetic subgroup of $G_R^0=\mathrm{Hol}(D)^0$, and define Γ_∞ and $\Gamma_0=\Gamma_\infty\cap B$ as in sections 2 and 3. Then Γ_0 is [6d: § 7.13] an arithmetic subgroup of B_R commensurable with

$$(\Gamma_0\cap H')\cdot(\Gamma_0\cap\mathfrak{D}H)\cdot(\Gamma_0\cap A_FU_R);$$

and $\Gamma_0\cap H'$ is finite since H'_R is compact, hence, the image of Γ_0 under j_{D_F} is a finite group of roots of unity, say of order l_F.

Now let φ be a polynomial on F, let l be a "sufficiently large" positive integer, such that $l_F|l$, put $\Gamma_\infty=N(F)\cap\Gamma$, $\Gamma_0=B\cap\Gamma$, and for $w\in D_F$, define

$$E_{\varphi,l,F}(w)=\sum_{\gamma\in\Gamma/\Gamma_0}\varphi(\tau_F(w\cdot\gamma))j_{D_F}(w,\gamma)^l. \tag{72}$$

Since $B \cap G_R^0 \subset Z(F)$, we have $\tau_F(w \cdot \gamma \cdot \gamma_0) = \tau_F(w \cdot \gamma)$ for $\gamma_0 \in \Gamma_0$, and from $l_F | l$ it follows that $j_{D_F}(w, \gamma\gamma_0)^l = j_{D_F}(w, \gamma)^l$. Therefore, the series in (72) is well-defined.

Now assume for simplicity that all components of the multi-index $\boldsymbol{n}_F$ in (36) are equal. (This restriction may be removed easily.)

Let $w_{0F} \in D_F$ be the point fixed by the maximal compact subgroup K of G_R^0 and let $\zeta_F : G_R^0 \to D_F$ be defined by $\zeta_F(g) = \zeta(g \cdot c_F) = w_{0F} \cdot g$. Then to apply our previous criterion for convergence to $E_{\varphi, l, F}$, let $f : G_R^0 \to \boldsymbol{C}$ be defined by

$$f(g) = \varphi(\tau(\zeta_F(g))) j_{D_F}(w_{0F}, g)^l.$$

Clearly, $|f(k \cdot g)| = |f(g)|$, and for $b \in B \cap G_R^0$, we obtain easily

$$|f(g \cdot b)| = |f(g)| \cdot |\Delta(b, l)| = |f(g)| \cdot |\chi_F(b)|^{-lq_F/m_F}$$

(note that in this case, θ' consists of a single element α_F for which $\chi_F = e_{\alpha_F} \Lambda_{\alpha_F}$, $e_{\alpha_F} > 0$). Since

$$|j_{D_F}(w, g)| = |j_F(\tau_F(w), \pi(g))|^r \cdot |\Delta(b, 1)|, \quad g \in N(F),$$

for a certain rational number $r > 0$, we have

$$|f(g)| \cdot |\Delta(g, l)|^{-1} = |\varphi(\tau(\zeta(g)))| \cdot |j_F(\tau(w_{0F}), \pi(g))|^{lr}, \quad g \in N(F), \tag{73}$$

and the right side is bounded since 1) φ is a polynomial function on the bounded domain F, and 2) $|j_F(\tau(w_{0F}), h)|$ is bounded for $h \in \mathrm{Hol}(F)$. Moreover, the series

$$\sum_{\gamma \in \Gamma_\infty / \Gamma_0} |j_F(\tau(w_{0F}), \pi(g\gamma))|^{lr} \tag{74}$$

converges normally if $lr \geqq 2$. Or, if as before we put

$$C_1 = (P \cap H)/(B \cap H),$$

and if f' denotes the function obtained by removing the absolute value symbols from the right side of (73), then $f' \in L^1(C_1)$ if $lr \geqq 2$. Since f' arises by lifting a holomorphic function on F to C_1, we see that f' and f satisfy the hypotheses (i) and (ii) of Theorem 28. Hence

$$E_f(g) = \sum_{\gamma \in \Gamma / \Gamma_0} f(g \cdot \gamma)$$

converges normally if l is large enough, and since

$$E_{\varphi, l, F}(w_{0F} \cdot g) = j_{D_F}(w_{0F}, g)^{-l} E_f(g),$$

we have shown that the series (72) for $E_{\varphi, l, F}$ converges normally on

D_F.

Since $E_{\varphi, l, F}$ is defined by a series that combines features of both Poincaré series and Eisenstein series, that series is called a Poincaré-Eisenstein series. It is easily seen to be an automorphic form on D_F of weight l with respect to Γ.

§7. The Satake compactification

We sketch here results for which proofs may be found elsewhere [50b; 2c; 52; 3].

Let the relationships between G, $D\subset \boldsymbol{C}^n$, Γ, etc., be as in the preceding section. In particular, $G_R^0=\mathrm{Hol}(D)^0$. We denote by D^* the union of D and of all rational boundary components of D, and by $\bar{D}$, the closure of D in $\boldsymbol{C}^n$. If $\mathfrak{S}$ is, as before, a Siegel set in G_R^0, we denote by $\bar{\mathfrak{S}}$ its canonical image $\zeta(\mathfrak{S})$ as a subset of D and by $\mathfrak{S}^*$ the closure of $\bar{\mathfrak{S}}$ in $\bar{D}$ (or in $\boldsymbol{C}^n$). The group $G_Q^0=G_R^0\cap G_Q$ operates on the set D^* and $\Gamma\subset G_Q^0$ by assumption.

One may prove [3 : § 4] that there is one and only one topology $\mathcal{T}$ on D^* satisfying the following conditions:

1) $\mathcal{T}$ induces the natural topology as subsets of $\boldsymbol{C}^n$ on D and on any set of the form $\mathfrak{S}^*\cdot g$, $g\in G_Q^0$, $\mathfrak{S}$ a Siegel set.
2) Every $g\in G_Q^0$ operates continuously on D^* with respect to $\mathcal{T}$.
3) If $x, x'\in D^*$, $x\notin x'\cdot\Gamma$, then there exist neighborhoods $\mathcal{N}$ of x and $\mathcal{N}'$ of x' such that $\mathcal{N}\cdot\Gamma\cap\mathcal{N}'=\emptyset$, the empty set.
4) If $x\in D^*$, there exists a fundamental system of neighborhoods $\mathcal{N}$ of x such that $\mathcal{N}\cdot\gamma=\mathcal{N}$ if $\gamma\in\Gamma_x$ (the stability group of x in Γ) and such that $\mathcal{N}\cdot\gamma\cap\mathcal{N}=\emptyset$ if $\gamma\in\Gamma-\Gamma_x$.

Supplying D^* with this topology, one may show that the quotient space $X^*=D^*/\Gamma$ is a compact Hausdorff space on which $X=D/\Gamma$ is a dense open subset. One calls X^* the Satake compactification of X.

One supplies X^* with a ringed structure $\mathcal{U}\to\mathcal{R}_{\mathcal{U}}$ as follows: Let π be the canonical mapping of D^* onto X^*. If $\mathcal{U}$ is an open subset of X^* and if f is a complex-valued continuous function on $\mathcal{U}$, then $f\in\mathcal{R}_{\mathcal{U}}$ if and only if the restriction of $f\circ\pi$ to $D\cap\pi^{-1}(\mathcal{U})$ is a complex analytic function on the latter set.

The main problem in showing that $(X^*, \mathcal{R})$ is a normal analytic space is to be able to exhibit enough "analytic" functions on it. In fact, these may be constructed using Poincaré-Eisenstein series. To see how this is done, it is sufficient to consider points in the closure $\mathfrak{S}^*$ of the image of a sufficiently large Siegel set. Let $\mathfrak{F}$ be the flag of *rational* boundary components having non-empty intersection with $\mathfrak{S}^*$. If $F \in \mathfrak{F}$ and x is an interior point of $F \cap \mathfrak{S}^*$, we consider an automorphic form f on D, transform it to the associated automorphic form f_F on D_F, and develop f_F in a series (46). Let $\mathfrak{S}_F = \bar{\mathfrak{S}} \cdot c_F$; then $\mathfrak{S}_F \subset D_F$. Now f_F may be extended to a continuous function on $\mathfrak{S}_F \cup F$ by defining its values on F to be those of the function ψ_0^F, and by viewing F as the set of limit points of sequences of points $(z, u, t) \in D_F$ such that u and t remain bounded and $y = \operatorname{Im} z$ goes to infinity interiorly in V. Define $\psi_0^F = \Phi_F(f)$. If $f_F = E_{\varphi, l, F}$ is a Poincaré-Eisenstein series, then $\psi_0^F = \Phi_F(f)$ is a Poincaré series on F; moreover, one may show that one has

(75) $$\Phi_{F'}(f) = 0, \text{ if } F' \in \mathfrak{F},\ F' \subset \partial F.$$

One exploits the properties of Poincaré series proved in Chapter 5, section 2. One result is that if $x \in X^*$ is a point of the rational boundary component F, and if l is large enough, then we may find an automorphic form f of weight l such that $\Phi_F(f)(x) \neq 0$. If f_1 is another automorphic form of weight l, f_1/f is a well-defined element of $\mathcal{R}_{\mathcal{U}}$ for some suitably small neighborhood $\mathcal{U}$ of $\pi(x)$. One may then use the point-separating properties of Poincaré series together with (75) to show the existence of enough functions in $\mathcal{R}_{\mathcal{U}}$, if $\mathcal{U}$ is small enough, to separate the points of $\mathcal{U}$. It is not hard to show that the local ring $\mathcal{R}_x$ of X^* at x is integrally closed. Hence, applying a criterion of [2c ; 15d], it follows that $(X^*, \mathcal{R})$ is a normal analytic space.

To show that the normal analytic space X^* is an algebraic variety, one again utilizes the point-separating properties of Poincaré series (Chapter 5, section 2), which in turn imply point-separating properties of Poincaré-Eisenstein series, or rather of quotients of such, on X^*. Now using the diagonal device of Chapter 5, section 2, in conjunction with the Noetherian property of analytic subsets of a compact, complex analytic space, one may show, just as in the case when D/Γ was compact (cf. Chapter 5, section 2) that X^* is in fact isomorphic (as a complex analytic space) to a projective, algebraic variety. This is one of the main results of [3].

PART III

SOME SPECIAL TOPICS

CHAPTER 12
FOURIER COEFFICIENTS OF EISENSTEIN SERIES

§1. Generalized gamma integrals [56b; 52; 20: Thm. 24.6]

Let V be the space of $n\times n$ real symmetric matrices $Y=(y_{ij})$, $y_{ij}=y_{ji}\in\boldsymbol{R}$, and let $\mathfrak{P}$ be the cone of those which are positive-definite, i.e., those $Y\in V$ such that $Y=X^2$ for some $X\in V$ with $\det X\neq 0$. In this chapter, we denote by H_n the space of $n\times n$ complex symmetric matrices $Z=X+iY$, X, Y real, $Y\in\mathfrak{P}$. Let dv be the ordinary Euclidean measure on V, $dv=dy_{11}dy_{12}\cdots dy_{nn}$. We want to calculate the value of the integral

$$I=\int_{\mathfrak{P}}|Y|^{\rho-(n+1)/2}e^{-\operatorname{tr}Y}dv,\quad \rho\geqq(n+1)/2,$$

where $|Y|=\det Y$. For this purpose, we make a change of variables as follows: Each $Y\in\mathfrak{P}$ can be written uniquely in the form $Y={}^tTAT$, where $T=(t_{ij})$, $t_{ii}=1$, $t_{ij}=0$ for $i>j$, and A is an $n\times n$ diagonal matrix, with diagonal entries $a_1,\cdots,a_n$. Then

$$\operatorname{tr}Y=\sum_{i=1}^{n}a_i+\sum_{i=1}^{n}\Big(\sum_{j<i}a_jt_{ji}^2\Big),$$

$$\det Y=\det A=a_1\cdots a_n,$$

and the Jacobian determinant for the change of variables $Y\mapsto(A,T)$ is given by

$$\det\Big(\frac{\partial Y}{\partial(A,T)}\Big)=a_1^{n-1}\cdot a_2^{n-2}\cdots a_{n-1},$$

as is obvious from appropriate ordering of the variables (y_{ij}) such that the Jacobian matrix becomes triangular in form. Then we have

$$\begin{aligned}I&=\int_0^{+\infty}\cdots\int_0^{+\infty}\Big(\int_{-\infty}^{+\infty}\cdots\int_{-\infty}^{+\infty}\prod_{i=1}^{n}a_i^{\rho+(n-1)/2-i}e^{-\sum_i a_i-\sum_i(\sum_{j<i}a_jt_{ij}^2)}\prod_{j<i}dt_{ij}\Big)\prod_i da_i\\&=\prod_{i=1}^{n}\pi^{(n-i)/2}\cdot\prod_{i=1}^{n}\int_0^{\infty}a_i^{\rho-(i+1)/2}e^{-a_i}\cdot da_i\end{aligned}\qquad(1)$$

$$=\pi^{n(n-1)/4}\prod_{i=1}^{n}\Gamma(\rho-(i-1)/2),$$

where Γ denotes the gamma function.

Now if $X\in\mathfrak{P}$ is fixed, then the preceding result enables us to calculate

$$I_1=\int_{\mathfrak{P}}|Y|^{\rho-(n+1)/2}e^{-\operatorname{tr}(XY)}dv;$$

in fact, $X=A^2$, and if we write $Z=AYA$, an easy calculation gives

$$dv_Y=\det\left(\frac{\partial Y}{\partial Z}\right)dv_Z=(\det A)^{-(n+1)}dv_Z$$
$$=(\det X)^{-(n+1)/2}dv_Z,$$

and $\det Y=(\det X)^{-1}\det Z$, hence, $I_1=(\det X)^{-\rho}I$, i.e.,

$$\int_{\mathfrak{P}}|Y|^{\rho-(n+1)/2}e^{-\operatorname{tr}(XY)}dY=|X|^{-\rho}\pi^{n(n-1)/4}\prod_{i=1}^{n}\Gamma(\rho-(i-1)/2). \tag{2}$$

But now both sides of (2) are analytic functions of $X\in\mathfrak{P}$, hence are identical in the complex domain

$$\{Z\,|\,{}^tZ=Z,\ Z\text{ is }n\times n,\ \operatorname{Re}Z\in\mathfrak{P}\}.$$

Therefore, we have for any $T\in V$,

$$\begin{aligned}&\int_{\mathfrak{P}}|Y|^{\rho-(n+1)/2}e^{-\operatorname{tr}\{(X+2\pi iT)Y\}}dv_Y\\ &=\pi^{n(n-1)/4}\Gamma(\rho)\cdots\Gamma(\rho-(n-1)/2)\,|X+2\pi iT|^{-\rho},\quad X\in\mathfrak{P}.\end{aligned} \tag{3}$$

Results analogous to (2) and (3) may be obtained when V is replaced by the exceptional Jordan algebra $\mathfrak{J}$ of Chapter 6, section 6, and $\mathfrak{P}$ is replaced by the open convex cone $\mathfrak{R}^+$ of loc. cit. [2i : § 8]. The result for $\mathfrak{R}^+$ itself is

LEMMA 1. *Let $l>8$ be a positive integer. Then*

$$\int_{\mathfrak{R}^+}(\det X)^{l-9}e^{-\operatorname{tr}(X)}dX=\pi^{12}\cdot\Gamma(l-8)\Gamma(l-4)\Gamma(l)\,. \tag{4}$$

This has as a corollary

LEMMA 2. *Let $\mathfrak{T}$ be the tube domain of Chapter 6, section 6, and let $Z\in\mathfrak{T}$. Then we have, again for $l>8$,*

$$\int_{\mathfrak{R}^+}|Y|^{l-9}e((Y,Z))\cdot dY=\pi^{12}(2\pi i)^{-3l}\prod_{j=0}^{2}\Gamma(l-4j)\cdot|Z|^{-l}, \tag{5}$$

where $|Z| = \det Z$.

Now the boundary $\partial\mathfrak{R}^+ = \bar{\mathfrak{R}} - \mathfrak{R}^+$ of $\mathfrak{R}$ is a union of cones in linear subspaces of $\mathfrak{J}$ of lower dimensions, and for these one has analogs of (4) and (5) [2i : § 8].

The most general result known to us along these lines, of which the above are special cases, is [20 : Theorem 24.6].

§ 2. Application of the Poisson summation formula

We consider for $t > 0$ the function defined on V by

$$f(X) = \begin{cases} |X|^t e^{-\operatorname{tr}(XY)}, & X \in \mathfrak{P}, \\ 0 & \text{otherwise}, \end{cases}$$

where Y is a fixed element of $\mathfrak{P}$. Clearly f has continuous partial derivatives of orders $\leqq t-1$, since in any case the partial derivatives of all orders are evidently continuous except on $\partial\mathfrak{P}$, which is a subset of the hypersurface $\det X = 0$, and on $\partial\mathfrak{P}$, f vanishes to order t.

Let Λ be the lattice of integral, symmetric $n \times n$ matrices and Λ', the lattice of symmetric matrices $M = (m_{ij})$ such that $2m_{ij}$ and m_{ii} are integers for all i, j (i.e., the matrices of integral, quadratic forms in n variables). Then Λ and Λ' are duals of each other with respect to the inner product $(X, Y) = \operatorname{tr}(XY)$. Moreover, $\sum_{T \in \Lambda'} f(T)$ converges absolutely for any fixed $Y \in \mathfrak{P}$, and uniformly so when Y varies on a compact subset of $\mathfrak{P}$. Hence, by section 1 of Chapter 13, for sufficiently large t the series $\sum_{F \in \Lambda} \hat{f}(F)$ converges normally, where $\hat{f}$ is the Fourier transform of f, and by (3) we have

$$\hat{f}(F) = \pi^{n(n-1)/4} \Gamma(\rho) \cdots \Gamma(\rho - (n-1)/2) \cdot |Y + 2\pi i F|^{-\rho}, \tag{6}$$

if $t = \rho - (n+1)/2$. Since the volume of a fundamental parallelopiped of Λ' is $2^{-n(n-1)/2}$, the Poisson summation formula (q.v. infra, Chapter 13) gives for sufficiently large t, i.e., for sufficiently large ρ, with $2\pi i Z = -Y$ real, $Y \in \mathfrak{P}$, and $\Lambda^* = \Lambda' \cap \mathfrak{P}$,

$$\begin{aligned} &\sum_{T \in \Lambda^*} |T|^{\rho-(n+1)/2} e^{2\pi i \operatorname{tr}(TZ)} \\ &= (4\pi)^{n(n-1)/4} (2\pi i)^{-n\rho} \Gamma(\rho) \cdots \Gamma(\rho - (n-1)/2) \sum_{F \in \Lambda} |Z + F|^{-\rho} \end{aligned} \tag{7}$$

(since $4^{n(n-1)/4}=2^{n(n-1)/2}$). For fixed Z, each side is an analytic function of ρ, so (7) must hold for all real values of ρ equal to or greater than any ρ_0 for which both sides converge absolutely. The sum on the right side is a subseries of the Eisenstein series (28) of Chapter 11, section 2 (but for the factor $f(g)$), hence converges for $\rho>n+1$; on the other hand, it is known, and not hard to show, that the left side converges for $\rho>(n+1)/2$, hence (7) holds for $\rho>n+1$.

A similar formula may be obtained by similar methods for the lattice $\Lambda=\mathfrak{J}_0\subset\mathfrak{J}$ of Chapter 6, section 6. If we let $\Lambda^*=\Lambda\cap\mathfrak{R}^+$, it becomes

(8) $$\sum_{\lambda\in\Lambda}|Z+\lambda|^{-18l}=\Gamma_3(l)\cdot\sum_{T\in\Lambda^*}|T|^{18l-9}e((T,Z)),$$

where

(9) $$\Gamma_3(l)=2^{54l}\pi^{54l}\cdot\prod_{j=0}^{2}\Gamma(18l-4j).$$

Moreover, one has analogs of (8) for the lower-dimensional cones lying on $\partial\mathfrak{R}^+$, referred to in section 1 (cf. [2i : §§ 6, 9]).

§ 3. Fourier coefficients of Eisenstein series

In this section, we suppose the situation to be as in Chapter 11, section 6. Moreover, we assume that D has a rational boundary component of dimension zero, which we denote by F_0, let φ be the constant 1, and put $E_l=E_{1,l,F_0}$, in the notation of Chapter 11, section 6. Then the series in (72) of Chapter 11 becomes, with $j_0=j_{D_{F_0}}$,

(10) $$E_l(Z)=\sum_{\gamma\in\Gamma/\Gamma_0}j_0(Z,\gamma)^l.$$

Since $Z(F_0)=N(F_0)$, we have $\Gamma_0=\Gamma_\infty=\Gamma\cap P$. Now we assume that D_{F_0} is a tube domain; according to [39a : §§ 3, 4], this implies that in the real root system ${}_{R}\Sigma$ of G, no $\boldsymbol{R}$-root is the double of another $\boldsymbol{R}$-root, and hence that the unipotent radical U of P is Abelian. Accordingly (Chapter 11, section 5), any automorphic form on D_{F_0} with respect to Γ has a Fourier expansion

(11) $$f(Z)=\sum_{\lambda\in\Lambda^+}a(\lambda)e((\lambda,Z)),\quad \Lambda^+=\Lambda'\cap\bar{V}.$$

In particular, we have

(12) $$E_l(Z) = \sum_{\lambda \in \Lambda^+} a_l(\lambda) e((\lambda, Z)).$$

It turns out, in some cases at least, that the Fourier coefficients $a_l(\lambda)$ have interesting arithmetic properties. For example, in the case of the Siegel modular group $\Gamma = Sp(n, \mathbf{Z})$ (Chapter 6, section 5), they are rational numbers with denominators that are bounded for fixed l [42; 56f]. The same can be proved for the Fourier series attached to the exceptional arithmetic group Γ considered in Chapter 6, section 6 [2i; 35]. We now use the formulae of the preceding section to obtain an expression for the Fourier coefficients in the case of the Siegel modular group.

We should like to emphasize that the method outlined here seems to be rather general [58] and in particular may also be applied to the Eisenstein series for the exceptional arithmetic group referred to above. We hope some suggestion of the more general situation may become apparent to the reader from the developments below.

If $g = \begin{pmatrix} A & B \\ C & D \end{pmatrix} \in Sp(n, \mathbf{R})$, one may prove from the developments of Chapter 6, section 5, and of Chapter 7, section 7, that the Jacobian determinant of g, regarded as a transformation of H_n onto itself, is given by

(13) $$j_0(Z, g) = \det(ZB + D)^{-n-1}.$$

In fact, the non-singular matrix $ZB+D$ is the element k_2 of $K_C = GL(n, \mathbf{C})$ appearing in the decomposition (25) of Chapter 7, section 7; the proof that the determinant of k_2 in the adjoint representation of K_C on $\mathfrak{p}^+$ is $\det(ZB+D)^{n+1}$ may easily be carried out in the special case when $B=0$ and D is a scalar multiple of the $n \times n$ identity matrix; and the truth of (13) in general follows from this, since calculation of the exponent $n+1$ is sufficient to characterize the one-dimensional rational representation $\det(Ad_{\mathfrak{p}^+})$ of K_C.

Henceforth, we denote by m a positive integer divisible by $2(n+1)$ (hence, automatically, $> n+1$). However, it is to be observed that series of the form (14) below make sense for any large even m, and it is only because of the convenience of writing out certain subsequent formulas in terms of the functional determinant that we impose the condition that m be divisible by $2(n+1)$. Later, in Chapter 13, we shall have occasion to refer to the results of this chapter as though we had only assumed m to be even and $>n+1$.

In this case, with $G=Sp(n, \boldsymbol{C})$, we have $\Gamma=Sp(n, \boldsymbol{Z})$, P is the set of matrices in G of the form $\begin{pmatrix} A & 0 \\ C & D \end{pmatrix}$, and $\Gamma_0=\Gamma\cap P$ consists of such matrices with A, C, and D integral and A and D unimodular; then the series E_l becomes (with $l\equiv 0\ (2)$)

$$\sum_{\{B,\, D\}} \det(ZB+D)^{-m}, \quad m=l(n+1), \tag{14}$$

where the summation on B and D is as in (28) of Chapter 11.

Let

$$\iota_r=\begin{pmatrix} 0 & 0 & E_r & 0 \\ 0 & E_{n-r} & 0 & 0 \\ -E_r & 0 & 0 & 0 \\ 0 & 0 & 0 & E_{n-r} \end{pmatrix}, \quad 0\leqq r\leqq n, \tag{15}$$

and put $\iota_n=\iota$. Let L be the subgroup of $\begin{pmatrix} A & 0 \\ C & D \end{pmatrix}\in P$ for which $C=0$; if $\begin{pmatrix} A & 0 \\ 0 & D \end{pmatrix}\in L$, then $D={}^tA^{-1}$. Let U be the subgroup of P characterized by $A=D=E_n$, $C=S$, where ${}^tS=S$. Then P is the semi-direct product of U by L and P_Q is the semi-direct product of U_Q by L_Q. We note that ι normalizes L and L_Q, and that ι_r, $0\leqq r\leqq n$, normalizes the subgroup of L for which both A and D are diagonal. Thus, $\{\iota_r\}$ represent certain elements of the Weyl group W of G. Moreover, the subgroup W' of L consisting of the elements for which A is a permutation matrix may, by Chapter 6, section 2, be canonically identified with the Weyl group of L. From the known structure of the Weyl group of $G=Sp(n, \boldsymbol{C})$, that may be established easily along the lines of Chapter 6, one sees that

$$W=\bigcup_{0\leqq r\leqq n} W'\iota_r W' \quad \text{(disjoint union)}. \tag{16}$$

It follows then from the Bruhat decomposition of G_Q (Chapter 7, section 4) that

$$G_Q=\bigcup_{0\leqq r\leqq n} P_Q\iota_r P_Q; \tag{17}$$

we put $\omega_r=P_Q\iota_r P_Q$. In particular,

(18) $$\omega_n = P_Q \iota P_Q = U_Q \iota P_Q,$$

and

(19) $$\omega_r = \left\{ \begin{pmatrix} A & B \\ C & D \end{pmatrix} \in G_Q \mid rk(B) = r \right\}.$$

If $g \in \omega_r$, then there exist $l, l' \in L_Z = \Gamma \cap L$ such that, for $r > 0$,

(20) $$lgl' = \begin{pmatrix} A' & B' \\ C' & D' \end{pmatrix}, \quad B' = \begin{pmatrix} B'' & 0 \\ 0 & 0 \end{pmatrix}, \quad \det(B'') \neq 0,$$

where B'' is r by r. Moreover, one may prove in elementary fashion that

(21) $$G_Q = \Gamma \cdot P_Q = P_Q \cdot \Gamma.$$

We identify U_Q with the additive group of symmetric, $n \times n$ rational matrices and the lattice Λ with

$$U_Z = \left\{ \begin{pmatrix} E & 0 \\ S & E \end{pmatrix} \mid {}^tS = S, \ S \text{ integral} \right\};$$

let $\Gamma_r = \Gamma \cap \omega_r$, $\Gamma_n = \Gamma^*$, and note that $\Gamma_r \Gamma_0 = \Gamma_r$. Then define

(22) $$E_l^{(r)}(Z) = \sum_{\gamma \in \Gamma_r / \Gamma_0} j_0(Z, \gamma)^l, \quad E_l^{(n)} = E_l^*.$$

It follows from (18) that every element γ of Γ^* can be written in the form $u_\gamma \iota p_\gamma$, $u_\gamma \in U_Q$, $p_\gamma \in P_Q$. We now treat the Fourier expansion of E_l^* and simply note that the Fourier expansions of $E_l^{(r)}$, $0 \leqq r \leqq n$, may be treated in similar fashion [56d : § 7]. It follows from (21) that if $u \in U_Q$, then there exists $p \in P_Q$ such that $u\iota p \in \Gamma^*$, and if also $p' \in P_Q$ is such that $u\iota p' \in \Gamma^*$, then $p'p^{-1} \in P \cap \Gamma = \Gamma_0$. It is easily verified that if $u''\iota p'' = \iota$, then $p'' = u'' =$ identity; hence, if $u\iota p = u'\iota p'\gamma_0$ with $u, u' \in U_Q$, $p, p' \in P_Q$, $\gamma_0 \in \Gamma_0$, then $u = u'$ and $p^{-1}p' \in \Gamma_0$. Therefore, E_l^* may be rewritten as

(23) $$E_l^*(Z) = \sum_{u \in U_Q} j_0(Z, \gamma_u)^l,$$

where $\gamma_u \in \Gamma$ is such that for some $p = p_u \in P_Q$, we have $u\iota p = \gamma_u$; thus,

(24) $$\begin{aligned} E_l^*(Z) &= \sum_{u \in U_Q} j_0(Z, u\iota p)^l = \sum_{u \in U_Q} j_0(Z+u, \iota)^l j_0(*, p)^l \\ &= \sum_{u \in U_Q/\Lambda} \Big(\sum_{\lambda \in \Lambda} j_0(Z+u+\lambda, \iota)^l \Big) j_0(*, p)^l, \end{aligned}$$

since if $u^{-1}u' \in \Lambda$, we may use the same p for both u and u'. Now we are able to apply (7) to the inner sum. We have $j_0(Z,\iota)^l=(\det Z)^{-l(n+1)}$, hence

$$\sum_{\lambda\in\Lambda} j_0(Z+u+\lambda,\iota)^l=\sum_{\lambda\in\Lambda}|Z+u+\lambda|^{-l(n+1)} \tag{25}$$

$$=\pi^a\cdot r_l\cdot\sum_{T\in\Lambda^*}|T|^{(n+1)(l-1/2)}e((T,Z+u)),$$

where

$$\pi^a\cdot r_l=(4\pi)^{-n(n-1)/4}(2\pi)^{ln(n+1)}\prod_{j=0}^{n-1}\Gamma(l(n+1)-j/2)^{-1}, \tag{26}$$

so that for even n, $a=ln(n+1)-\frac{1}{4}n^2$, and for odd n, $a=ln(n+1)-\frac{1}{4}(n^2-1)$; therefore, in either case, a is an integer and r_l is a rational number depending only on l and n. Then we see that E_l^* has a Fourier expansion

$$E_l^*(Z)=\sum_{T\in\Lambda^*}a_l(T)e((T,Z)), \tag{27}$$

where $a_l(T)=\pi^a\cdot r_l\cdot|T|^{(n+1)(l-1/2)}\cdot\sigma_T$ and

$$\sigma_T=\sum_{u\in V_Q/\Lambda} j_0(*,p)^l e((T,u)). \tag{28}$$

§4. Euler product expansion of the Fourier coefficients

We now consider the relationship between u and p occurring in a single term of (28). Since $p\in P_Q$, $j_0(*,p)$ is a rational number $=\pm\prod_{\boldsymbol{p}:\ \text{rational prime}}\boldsymbol{p}^{\kappa_{\boldsymbol{p}}(u)}$, where $\kappa_{\boldsymbol{p}}(u)$ are rational integers, and are zero for all but a finite number of $\boldsymbol{p}$; we shall soon see that $\kappa_{\boldsymbol{p}}(u)\leqq 0$. In any case, $u\equiv\sum_{\boldsymbol{p}}u_{\boldsymbol{p}} \bmod \boldsymbol{Z}$, where $u_{\boldsymbol{p}}$ is integral at all primes but $\boldsymbol{p}$, and the power of $\boldsymbol{p}$ dividing $j(*,p)$, as we shall see also, depends only on the residue class of u modulo $\boldsymbol{Z}_{\boldsymbol{p}}$, i.e., only on $u_{\boldsymbol{p}}$. We start from the relation

$$g=\begin{pmatrix}E&0\\S&E\end{pmatrix}\begin{pmatrix}0&E\\-E&0\end{pmatrix}\begin{pmatrix}A&0\\C&D\end{pmatrix}=\begin{pmatrix}C&D\\-A+SC&SD\end{pmatrix}. \tag{29}$$

Let $\Lambda_{\boldsymbol{p}}=\Lambda\otimes_{\boldsymbol{Z}}\boldsymbol{Z}_{\boldsymbol{p}}=V_{\boldsymbol{Z}_{\boldsymbol{p}}}$. Given $u_{\boldsymbol{p}}$ at a finite number of $\boldsymbol{p}$, there is one and only one $u\in V_Q/\Lambda$ such that $u\equiv u_{\boldsymbol{p}} \bmod \Lambda_{\boldsymbol{p}}$ for the given primes $\boldsymbol{p}$ and such that $u\in\Lambda_{\boldsymbol{p}}$ for all other $\boldsymbol{p}$.

Now it is clear that if $l\in L_{\boldsymbol{Z}_{\boldsymbol{p}}}$, $l=\begin{pmatrix}M^{-1}&0\\0&{}^tM\end{pmatrix}$, $M\in GL(n,\boldsymbol{Z}_{\boldsymbol{p}})$, then

(30) $$\kappa_p(l\cdot u_p)=\kappa_p(u_p[M])=\kappa_p(u_p), \quad l\cdot =Ad\, l^{-1}.$$

In fact, the relation between u and p implies that for a suitable $n\times n$ matrix B over Q_p and a suitable $A\in GL(n, \boldsymbol{Q}_p)$, we have (29) with $D={}^tA^{-1}$, $S=u$, and $g\in Sp(n, \boldsymbol{Z}_p)$; and for this A and u, the power of $\boldsymbol{p}$ in $\det(A)^{n+1}$ is $\boldsymbol{p}^{\kappa_p(u)}$. On the other hand, if $M\in GL(n, \boldsymbol{Z}_p)$, then $lgl^{-1}=g'\in Sp(n, \boldsymbol{Z}_p)$ and we have again (29) with g replaced by g', u by $u'=u[M]$, and A by $A'=A[M]$, while $\det A'=\eta\cdot\det A$, where η is a $\boldsymbol{p}$-adic unit, so that $\kappa_p(u)=\kappa_p(u')$. Moreover, if $u_p{}^0\in\Lambda_p$, we clearly have $\kappa_p(u_p{}^0)=0$ and $\kappa_p(u_p+u_p{}^0)=\kappa_p(u_p)$ for any $u_p\in V_{Q_p}$, as one sees in similar fashion. Hence, κ_p is constant on orbits of L_{Z_p} in V_{Q_p} and on (additive) cosets of Λ_p. It is clear that if $\boldsymbol{p}^\nu$ is the smallest *non-negative* power of $\boldsymbol{p}$ such that $\boldsymbol{p}^\nu u_p\in\Lambda_p$, then it is also the smallest non-negative power of $\boldsymbol{p}$ such that $\boldsymbol{p}^\nu l\cdot u_p\in\Lambda_p$. In these circumstances, we write $\nu=\nu(u_p)$. From the coarsest sort of estimates based on the definitions and simple matrix calculations, one sees that there are strictly positive fixed constants δ and δ' such that

(31) $$\delta\nu(u_p)\leqq-\kappa_p(u_p)\leqq\delta'\nu(u_p)$$

for all $u_p\in V_{Q_p}$. (In fact, one may show [56a: §7] that $\kappa_p(u_p)$ is the product of the reduced *denominators* of the elementary divisors of u_p.) It follows from (30) and (31) that if we define

(32) $$V_{p,\kappa}=\{u\in V_{Q_p}\mid \kappa_p(u_p)=\kappa\}/\Lambda_p,$$

then $V_{p,\kappa}$ is a finite set on which L_{Z_p} operates. Let $\xi_{\kappa 1},\cdots,\xi_{\kappa m}$ be the orbits of L_{Z_p} in $V_{p,\kappa}$.

Let ε_p be the character on $\boldsymbol{Q}_p/\boldsymbol{Z}_p$ defined by: If $a\in\boldsymbol{Q}_p$ and $a\equiv a_{-\nu}\boldsymbol{p}^{-\nu}+\cdots+a_{-1}\boldsymbol{p}^{-1} \bmod \boldsymbol{Z}_p$, where $a_{-i}\in\boldsymbol{Z}$, then $\varepsilon_p(a)=e(a_{-\nu}\boldsymbol{p}^{-\nu}+\cdots+a_{-1}\boldsymbol{p}^{-1})$. Then we have $e((T,u))=\prod_p\varepsilon_p((T,u_p))$ (where only a finite number of factors are different from one). Hence

(33) $$\sigma_T=\prod_p\sigma_{T,p},$$

where

(34) $$\sigma_{T,p}=\sum_{u_p\in V_{Q_p}/\Lambda_p}\boldsymbol{p}^{\kappa_p(u_p)l}\varepsilon_p((T,u_p)).$$

It follows from a straightforward generalization of Hensel's lemma [2h] that if $|\kappa|$ is large enough, then for each $j=1,\cdots,m$, we have

$$\sum_{x \in \xi_{\kappa j}} \varepsilon_p((T, x)) = 0. \tag{35}$$

The proof of this uses the known fact that if $t \in \boldsymbol{Z}_p - \{0\}$, then for a sufficiently large positive integer N we have

$$\sum_{a \in \boldsymbol{Z}_p \bmod \boldsymbol{p}^N} \varepsilon_p(\boldsymbol{p}^{-N} at) = 0.$$

Thus, for each $\boldsymbol{p}$, the sum in (34) is actually a finite sum (absolute convergence and thus the legitimacy of grouping terms follows from elementary estimates based on (31)), and its value lies in the field F_N of $\boldsymbol{p}^N$-th roots of unity for some sufficiently large N. If η is a $\boldsymbol{p}$-adic unit, clearly, by (29), $\kappa_p(\eta u) = \kappa_p(u)$ for all $u \in V_{Q_p}$. On the other hand, η determines naturally a Galois automorphism of F_N by $\varepsilon_p(\boldsymbol{p}^{-N})^\eta = \varepsilon_p(\eta \boldsymbol{p}^{-N})$. Hence, the sum (34) which gives the value of $\sigma_{T,p}$ is invariant under all such Galois automorphisms and, consequently, $\sigma_{T,p}$ is a rational number for *every* $\boldsymbol{p}$.

Thus, to get part of the information we desire about σ_T, it will be sufficient to find a formula giving the value of each factor $\sigma_{T,p}$ for all but a finite number of $\boldsymbol{p}$ (the finite set of excluded $\boldsymbol{p}$ depending on T). Lacking space for a treatment of the general situation (for which we refer the reader to [58]), we compute a few simple examples by way of illustration.

EXAMPLE 1. $n=1$, $\boldsymbol{p}>2$, $T=1$.

If $s \in \boldsymbol{Q}_p$, then the conditions for

$$\begin{pmatrix} 1 & 0 \\ s & 1 \end{pmatrix} \begin{pmatrix} 0 & 1 \\ -1 & 0 \end{pmatrix} \begin{pmatrix} a & 0 \\ c & d \end{pmatrix} = \begin{pmatrix} c & d \\ -a+sc & sd \end{pmatrix}, \quad \text{where } p = \begin{pmatrix} a & 0 \\ c & d \end{pmatrix},$$

to belong to $Sp(1, \boldsymbol{Z}_p) = SL(2, \boldsymbol{Z}_p)$ are

$$sd,\ c,\ d,\ sc-a \in \boldsymbol{Z}_p, \quad ad=1. \tag{36}$$

Write $a = \varepsilon_a \boldsymbol{p}^{\nu_a}$, $c = \varepsilon_c \boldsymbol{p}^{\nu_c}$, $s = \varepsilon_s \boldsymbol{p}^{\nu_s}$ (then $d = \varepsilon_a^{-1} \boldsymbol{p}^{-\nu_a}$). Then $\nu_d, \nu_c \geqq 0$, $\nu_d + \nu_s \geqq 0$, so $\nu_s \geqq -\nu_d$. Moreover, $sc - a \in \boldsymbol{Z}_p$.

Case I. Suppose $s \in \boldsymbol{Z}_p$. Then a, c, and $d \in \boldsymbol{Z}_p$, $a, d \in \boldsymbol{Z}_p^*$ (the group of $\boldsymbol{p}$-adic units), and $\kappa_p(s)=0$, while $\det(Ad_{\mathfrak{u}} p) = d^2$ is a $\boldsymbol{p}$-adic unit.

Case II. Suppose $s \notin \boldsymbol{Z}_p$, so $\nu_s < 0$. Then $\nu_a = -\nu_d \leqq \nu_s$, so $\nu_a < 0$; but then $sc \equiv a \bmod \boldsymbol{Z}_p$ implies that $\nu_s + \nu_c = \nu_a \leqq \nu_s$, i.e., $\nu_c \leqq 0$. Since $\nu_c \geqq 0$, we conclude that $\nu_c = 0$, i.e., c is a unit and $\nu_a = \nu_s = -\nu_d$, so $\det(Ad_{\mathfrak{u}}p) = \varepsilon \cdot \boldsymbol{p}^{-2\nu_s}$, where ε is a $\boldsymbol{p}$-adic unit. Thus $\kappa_p(s) = 2\nu_s$ is always even. Now, $V_{Z_p} = \Lambda_p$ and $\Lambda_p/\Lambda_p = \{0\}$, so that $\sum_{s \in V_{p,0}} \varepsilon_p(s) = 1$. Moreover,

$$V_{p^{-1}Z_p}/V_{Z_p} - \{0\} = V_{p,-2}$$

consists of $\boldsymbol{p}-1$ elements representing the elements of $F_p^* = F_p - \{0\}$, and $\sum_{s \in V_{p,-2}} \varepsilon_p(s) = -1$. If $-\nu > 1$, then

$$V_{p,2\nu} = \{a_\nu \boldsymbol{p}^\nu + \cdots + a_{-1}\boldsymbol{p}^{-1} \mid a_j \bmod \boldsymbol{p},\ a_\nu \not\equiv 0\ (\boldsymbol{p})\},$$

from which it follows that $\sum_{s \in V_{p,2\nu}} \varepsilon_p(s) = 0$. Hence,

$$\sigma_{1,p} = (1 - \boldsymbol{p}^{-2l}),$$

and

$$\sigma_1 = \prod_p \sigma_{1,p}$$

is therefore a rational multiple of $\zeta(2l)^{-1}$, hence is a rational multiple of π^{-2l}; it follows that (for $l > 1$) the Fourier coefficients of E_l are rational numbers.

EXAMPLE 2. $n = 2$, $\boldsymbol{p} > 2$, $T = E = \begin{pmatrix} 1 & 0 \\ 0 & 1 \end{pmatrix}$.

In this case, as before, $\Lambda_p/\Lambda_p = \{0\}$, so that $\sum_{u \in V_{p,0}} \varepsilon_p(\mathrm{tr}(u)) = 1$. Now the elements of $\Lambda_p/\boldsymbol{p}\Lambda_p$, different from zero, are divided into two classes: Let $\bar{u} = \begin{pmatrix} \bar{a} & \bar{b} \\ \bar{b} & \bar{c} \end{pmatrix}$, $\bar{a}, \bar{b}, \bar{c} \in F_p$ (the field of $\boldsymbol{p}$ elements); then $\bar{u}$, if not zero, is of rank 1 or of rank 2. Since $\boldsymbol{p} > 2$, any symmetric, 2×2, $\boldsymbol{p}$-adic matrix is congruent by a $\boldsymbol{p}$-adic unimodular transformation to a diagonal 2×2 $\boldsymbol{p}$-adic matrix. Then with u in diagonal form $\begin{pmatrix} s_1 & 0 \\ 0 & s_2 \end{pmatrix}$, a matrix $p \in P_{Q_p}$ satisfying $u\iota p \in G_{Z_p}$ may be found in the form

$$\begin{pmatrix} a_1 & 0 & 0 & 0 \\ 0 & a_2 & 0 & 0 \\ c_1 & 0 & d_1 & 0 \\ 0 & c_2 & 0 & d_2 \end{pmatrix},$$

where $\begin{pmatrix} a_j & 0 \\ c_j & d_j \end{pmatrix}$ is related to $u_j = \begin{pmatrix} 1 & 0 \\ s_j & 1 \end{pmatrix}$ in the manner determined in the calculation of Example 1. It follows that the power of $\boldsymbol{p}$ dividing $j(*, p)$ is $\boldsymbol{p}^{3(o_1+o_2)}$, where $o_j = \min(0, \operatorname{ord}_p s_j)$, thus $\kappa_p(u) = 3(o_1+o_2)$. We may always assume $o_1 \leqq o_2$. If $o_1 < -1$ and if ξ is the finite orbit under L_{Z_p} of u viewed as an element of V_{Q_p}/Λ_p, then simple calculations show that $\sum_{u \in \xi} \varepsilon_p(\operatorname{tr}(u)) = 0$. So we may assume $-1 \leqq o_1 \leqq o_2$; thus $\boldsymbol{p}u = \begin{pmatrix} a & b \\ b & c \end{pmatrix} \in \Lambda_p$. We denote the residue classes modulo $\boldsymbol{p}$ of $\boldsymbol{p}u$, a, b, and c by $\bar{u}$, $\bar{a}$, $\bar{b}$, and $\bar{c}$ respectively. Then $\kappa_p(u) = -3rk(\bar{u})$.

Case I. $rk(\bar{u}) = 0$. It is easy to see, as already remarked, that the exponential sum in this case has the value one.

Case II. $rk(\bar{u}) = 1$. Then $\overline{ac} - \bar{b}^2 = 0$, while $\bar{a}$, $\bar{b}$, and $\bar{c}$ are not all zero.

Subcase A. $\bar{a} \neq 0$. Then $\bar{c} = \bar{a}^{-1}\bar{b}^2$, and $\operatorname{tr}(\bar{u}) = \bar{a} + \bar{a}^{-1}\bar{b}^2$, and the sum for such $\bar{u}$ becomes

$$\sum_{u \in V_{p,-3}} \varepsilon_p(\operatorname{tr}(u)) = \sum_{a,\, b \bmod p,\; a \not\equiv 0(p)} \varepsilon_p(\boldsymbol{p}^{-1}(a + a^{-1}b^2)),$$

and the last sum may be evaluated by well-known properties of Gaussian sums; the result is that that sum has the value

$$\left(\frac{-1}{\boldsymbol{p}}\right)\boldsymbol{p},$$

$\left(\frac{\quad}{\quad}\right)$ being the Legendre symbol.

Subcase B. $\bar{a} = 0$. Then $\bar{b} = 0$, so $\operatorname{tr}(\bar{u}) = \bar{c}$, $\bar{c} \neq 0$, and the sum for this case is equal to

$$\sum_{\bar{c} \in F_p^*} \varepsilon_p(\boldsymbol{p}^{-1}c) = -1.$$

Case III. $rk(\bar{u}) = 2$. Then $\overline{ac} - \bar{b}^2 \neq 0$. We could proceed just as in the preceding cases; however, everything is much simplified if we note that

$$\sum_{\bar{u}} \varepsilon_p(a + c) = 0, \tag{37}$$

where the summation on $\bar{u}$ is over all symmetric 2×2 matrices over F_p and the relations between a, b, c, and u are as previously stated. It follows that the sub-sum of (37), where $\bar{u}$ is restricted to have rank 2, has the value

$$-\left(\frac{-1}{p}\right)p.$$

We conclude, finally, that for $p>2$, one has

$$\begin{aligned}\sigma_{E,p} &= 1+\left(\left(\frac{-1}{p}\right)p-1\right)p^{-3l}-\left(\frac{-1}{p}\right)p^{1-6l} \\ &= \left(1+\left(\frac{-1}{p}\right)p^{1-3l}\right)(1-p^{-3l}).\end{aligned} \tag{38}$$

Of course,

$$\prod_p (1-p^{-3l})=\zeta(3l)^{-1},$$

where the product is over all primes p. On the other hand,

$$\begin{aligned}\prod_p{}' \left(1+\left(\frac{-1}{p}\right)p^{1-3l}\right) &= \prod_p{}'\left[(1-p^{2-6l})\left(1-\left(\frac{-1}{p}\right)p^{1-3l}\right)^{-1}\right] \\ &= (1-2^{2-6l})^{-1}\zeta(6l-2)^{-1}L(\chi, 3l-1),\end{aligned}$$

where $\prod_p{}'$ indicates the product over all *odd* primes and $L(\chi, s)$ is the L-function of the Gaussian number field $\boldsymbol{Q}(i)$ associated to the non-trivial character of $\mathrm{Gal}(\boldsymbol{Q}(i)/\boldsymbol{Q})$, when we identify the latter with the multiplicative group of principal odd ideals (n), $n>0$, modulo those for which $n\equiv 1$ (4), via the Artin homomorphism. As for the value of $L(\chi, s)$ for odd integral values of $s>1$, one may compare the Mittag-Leffler partial fraction expansion of *sech* x with the power series expansion of the same function at 0 and obtain the result that $L(\chi, 3l-1)$ is a rational multiple, involving an Euler number, of π^{3l-1} (recall that $l\equiv 0 \pmod 2$).

More generally, if $t=\det(2T)\neq 0$, let k_t be the imaginary quadratic number field $\boldsymbol{Q}(\sqrt{-t})$ and let χ_t be the non-trivial character of $\mathrm{Gal}(k_t/\boldsymbol{Q})$. Denote by $L(\chi_t, s)$ the L-function of k_t associated to χ_t. Then one may prove that for odd $m>1$,

$$L(\chi_t, m)=\pi^m\cdot t^{1/2}\cdot \text{(rational number)}, \quad [56c: p. 288]. \tag{39}$$

Using (39), one may again see for $n=2$ that all the Fourier coefficients of E_l are rational numbers, the cases when $\det T=0$ being already included in Example 1, since the Fourier coefficients for such T are easily seen to be equal to Fourier coefficients of Eisenstein series in the one-variable case.

In fact, one may prove [56d: §7] that for any n the Fourier coefficients of E_l are rational numbers. In [56d] this is proven using Siegel's main theorem on definite quadratic forms, in particular the result of [56c : p. 288] just referred to, and some facts about quadratic exponential sums. The method outlined above appears, at first glance, to be more special and clumsy than Siegel's methods. However, it may in fact be applied in a broad range of cases [58] and in particular may be applied [2i] to the case of Eisenstein series for the exceptional arithmetic group Γ acting on the tube domain $\mathfrak{T}$ of Chapter 6, section 6. One employs the formula (8) of section 2, among others, to show that, in the exceptional case, if $\det T \neq 0$, the Fourier coefficient $a_l(T)$ is a product of "Euler factors",

$$(40) \qquad a_l(T) = \prod_{\boldsymbol{p}} \sigma_{T,\boldsymbol{p},l},$$

and for all but a finite number of $\boldsymbol{p}$ one has

$$(41) \qquad \sigma_{T,\boldsymbol{p},l} = (1-\boldsymbol{p}^{-18l})(1-\boldsymbol{p}^{4-18l})(1-\boldsymbol{p}^{8-18l}).$$

For all $\boldsymbol{p}$, $\sigma_{T,\boldsymbol{p},l}$ is a rational number. Using (9), (40), and (41) one sees again that $a_l(T)$ is a rational number. The cases when $\det T = 0$ are handled similarly.

In the case of the Siegel modular group, it has been shown [56f; 42] that for fixed l, the rational numbers $a_l(T)$ have bounded denominators, a common denominator being given as the product of a power of 2 and of the numerators of certain Bernoulli numbers. A similar result has been obtained in the case of the Fourier coefficients for the Eisenstein series of the exceptional arithmetic group [35]. However, it should be noted again that in proving this result also, the methods employed in [35] are different from those of Siegel.

Going back to the last section of Chapter 11, we consider the special case when D is of tube type and satisfies the special assumptions of section 3 of this chapter (which include among others the assumption that D have a zero-dimensional rational boundary component——a condition which really refers to the group G and its $\boldsymbol{Q}$-structure). When we are able to prove that the Eisenstein series have rational Fourier coefficients and when we can show that the Eisenstein series generate the full field of automorphic functions for the group Γ, then it follows from the results of [2g] that the algebraic variety X^* (of Chapter 11, section 7) may be assumed to be defined over the rational number field $\boldsymbol{Q}$. It would be interesting to see what other consequences may follow from the special properties of the Fourier coefficients of Eisenstein series, such as their being rational, having bounded denominators, etc.

In [2g], mentioned above, a sufficient criterion was developed in order for suitable combinations of Eisenstein series to generate the field of all automorphic functions with respect to a maximal arithmetic group Γ acting on a tube domain. Recently, Mr. L.-C. Tsao has directed the author's attention to an oversight in that paper, which we now take the opportunity of correcting. Namely, let $\mathfrak{T}=\prod_j \mathfrak{T}_j$ be a tube domain, where $\mathfrak{T}_j$ are its simple factors, and let $\mathrm{Hol}(\mathfrak{T}_j)^0=G_{j\mathbf{R}}^0=H_j$, where $G=\prod_j G_j$ may be realized as an algebraic group defined over $\boldsymbol{Q}$ with simple factors G_j. Let $\hat{H}_j$ be the group $\mathrm{Hol}(\mathfrak{T}_j)$ and put $\hat{H}=\mathrm{Hol}(\mathfrak{T})$. The index $[\hat{H}_j : H_j]$ is either one or two, but the full group $\hat{H}$ is a group extension by the direct product $\prod_j \hat{H}_j$ of the permutation group Ω of isomorphic factors $\mathfrak{T}_j$; the group Ω was ignored in [2g]. This may be rectified as we now indicate. Let $F_0=\prod_j F_{0j}$ be the representation of the rational, zero-dimensional boundary component F_0 as a product of zero-dimensional boundary components F_{0j} of the simple factors $\mathfrak{T}_j$. Then, for *each* maximal set of mutually holomorphically isomorphic simple factors, we choose and fix a holomorphic isomorphism of one of them, say $\mathfrak{T}_1$, onto each of the others such that F_{01} is identified with the corresponding F_{0j} in the factor $\mathfrak{T}_j$, if $\mathfrak{T}_1$ is isomorphic to $\mathfrak{T}_j$. In this way, we obtain an explicit realization of the group Ω (as a cross-section in the group extension) in the form of a finite group permuting the simple factors. If $\omega\in\Omega$, then $F_0\cdot\omega=F_0$ and therefore $\Omega\subset N_h$, the normalizer of $N(F_0)$ in $\mathrm{Hol}(\mathfrak{T})$, and since any automorphism of a symmetric tube domain leaving the zero-dimensional boundary component at infinity fixed has to be a linear affine transformation with constant functional determinant, it follows that we may assume ω has constant (non-zero) functional determinant. On the other hand, N_h is generated by $N(F_0)$, by Ω, and by elements τ_j ($=\tau$ in the notation of [2g]) when $[\hat{H}_j : H_j]=2$. Clearly $j_0(*,\omega)=\text{const.}$ for every $\omega\in\Omega$, and one easily sees, using the same kind of arguments as in [2g], that for $g\in G_h=\hat{H}$, we have $j_0(*,g)\equiv\text{const.}$ if and only if $g\in N_h$. Therefore, $g\mapsto j_0(*,g)$ is a homomorphism of N_h into $\boldsymbol{C}^*$ that maps $\Gamma\cap N(F_0)$ into a finite group of roots of unity [3 : § 3.14]; therefore, the image of $\Gamma \cap N_h$ is also a finite group (of course, of roots of unity), because $[N_h : N(F_0)]$ is finite. Consequently, the rest of the arguments of [2g], including the appropriately corrected first paragraph of section 5 of [2g] with $2d_a$ replaced by possibly some other suitable multiple of d_a, now continue to hold without change. (N.B. The possible temptation to choose $\omega \in \Omega$ in some way compatible with the $\boldsymbol{Q}$-structure on G should be avoided as an unnecessary complication; the main point is to verify from the $\boldsymbol{R}$-structure that N_h consists precisely of the elements of G_h having constant functional determinant.)

§ 5. Eisenstein series on the adele group

Let D, G, F_0, j_0, etc., be as at the beginning of section 3, *except* that now we assume G to be simply connected; the latter assumption entails the minor technical inconvenience that now $G_R = G_R^0$ is no longer identifiable with $\mathrm{Hol}(D)^0$, but is only isogenous to the latter (cf. Chapter 7, section 5). However, there is a natural correspondence between discrete subgroups of G_R and those of $\mathrm{Hol}(D)^0$ in that if π is the natural mapping of G_R onto $\mathrm{Hol}(D)^0$, then the image under π of a discrete group as well as the inverse image of a discrete group are discrete. Moreover, the center of G_R is the center of G (cf. [3 : proof of 11.5]), and so is contained in every parabolic subgroup of G.

We first note what the main idea in [2g] is for the construction of linear combinations of Eisenstein series. It is that one attaches an Eisenstein series of the type (10) to each vertex of a set of representatives of the Γ-orbits of zero-dimensional rational boundary components, and then takes a suitable linear combination of these. We now indicate how to get the desired Eisenstein series, or linear combinations of Eisenstein series, in another, seemingly more natural, way.

We proceed, namely, to construct Eisenstein series on the adele group G_A of G, for the definition and properties of which we refer the reader to [60d]. For our purposes here, we introduce some notation. If v is any place, finite or infinite, of $\boldsymbol{Q}$, denote by $\boldsymbol{Q}_v$ the completion of $\boldsymbol{Q}$ at v; in particular, $\boldsymbol{Q}_\infty = \boldsymbol{R}$. For each v, let $G_v = G_{\boldsymbol{Q}_v}$. We assume that G is a subgroup of $GL(V)$ and choose an $\boldsymbol{R}$-lattice Λ of V_R, $\Lambda \subset V_Q$. We use in part the notation of Chapter 7, section 6, and in particular may assume that $\Gamma=\{g\in G \mid g\cdot\Lambda=\Lambda\}$ is a special arithmetic group and that $K_p=\{g\in G_p \mid g\cdot\Lambda_p=\Lambda_p\}$ satisfies $G_p=K_p\cdot P_{Q_p}$ for every finite $\boldsymbol{p}$. For the case $v=\infty$, there exists a maximal compact subgroup $K=K_\infty$ of G_R such that $G_R=K\cdot P_R^0$. Write $P_{Q_p}=P_p$, $P_R^0=P_\infty$.

Now, the adele group G_A is the restricted direct product

$$\prod_v{}' G_v$$

with respect to the system of compact subgroups $\{K_v\}$; i.e., each $g\in G_A$ is of the form $g=(g_v)$, where $g_p\in K_p$ for all but a finite number of $\boldsymbol{p}$. For each v, let $\chi_v=\det(Ad_{\mathfrak{U}_v})$, where $\mathfrak{U}_v$ is the Lie algebra of U_v, U being the unipotent radical of P. Moreover, let ν_v be a character of absolute value one on K_v which we assume to be trivial for all but a finite number of v and to take the value one on $K_v\cap P_v$. Let s be a complex number. We now define a complex-valued function $F=F(s,\chi,\nu)$ on G_A. Namely, if $g=(g_v)\in G_A$, we may write each g_v in the form $k_v\cdot p_v$, $k_v\in K_v$, $p_v\in P_v$, and we define

$$F_v(g_v)=\nu_v(k_v)\,|\chi_v(p_v)|_v{}^s,$$

where $|\ |_v$ denotes the normalized absolute value at v appearing in the so-called product formula [60d]. We define

$$F(g)=\prod_v F_v(g_v).$$

Since $F_v(g_v)=1$ for all but a finite number of v, F is well-defined.

We now view G_Q as being diagonally imbedded in $G_A : G_Q\ni g_0\to(g_v)$, where $g_v=g_0$ for each v. Then G_Q is a discrete subgroup of G_A [60d]. If $p\in P_A$, we have

$$F(gp)=F(g)\prod_v |\chi_v(p_v)|_v{}^s,$$

and in particular we have from the product formula

$$F(gp)=F(g),\quad g\in G_A,\ p\in P_Q.$$

Thus, if the real part of s is large enough, Godement's criterion (cf. Chapter 11, section 2; virtually the same proof as for Theorem 25 works for Eisenstein series on the adele group [22c]) shows that the series

$$E(g)=\sum_{\gamma\in G_Q/P_Q} F(g\gamma) \tag{42}$$

converges normally on G_A. If now we take l to be a sufficiently large, positive, even integer and let

$$\nu_\infty(k)=j_0(w_{0F_0}, k)^l \quad \text{(notation of Chapter 11 and of section 3)},$$

$$\nu_{\boldsymbol{p}}(k_{\boldsymbol{p}})=1 \quad \text{(all finite } \boldsymbol{p}\text{)},$$

and if we view G_R as naturally imbedded in G_A by the mapping $g\to(g, 1, 1, \cdots)$, then the linear combinations of Eisenstein series considered in [2g] may be recovered with certain choices for the coefficients c_a. In particular, if Γ is unicuspidal, so that $G=\Gamma\cdot P_Q$, we obtain the Eisenstein series considered in earlier sections of this chapter (after, of course, lifting them to the group G_R). The details are left as an exercise.

Series of the type (42) were considered by Godement (unpublished manuscript (cf. [22*d*])), with the characters ν_v all trivial, and also in [22c] (with fairly general characters of K_v). We wonder what results might be proved about a field of rationality for the Fourier coefficients of such Eisenstein series (cf. [58; 2*j*]).

CHAPTER 13
THETA FUNCTIONS AND AUTOMORPHIC FORMS

§1. The Poisson summation formula

We first derive a simple classical version of the Poisson summation formula. Let f be a function of differentiability class C^ν on $\boldsymbol{R}^n$. We consider the series

(1) $$g(x)=\sum_{m_1,\cdots,m_n\in \boldsymbol{Z}} f(x_1+m_1,\cdots,x_n+m_n),$$

and assume that it and all of the series obtained by replacing f by any of its mixed partial derivatives of orders $\leqq\nu$ converge normally on $\boldsymbol{R}^n$, so that also $g\in C^\nu$. Clearly $g(x+m)=g(x)$ for all $x\in\boldsymbol{R}^n$, $m\in\boldsymbol{Z}^n$; then if ν is large enough, it is known (Chapter 9, section 6) that g has a Fourier expansion

(2) $$g(x)=\sum_{l_1,\cdots,l_n\in \boldsymbol{Z}} a_{l_1\cdots l_n}e(l_1x_1+\cdots+l_nx_n)$$

converging normally on $\boldsymbol{R}^n$, where

$$\begin{aligned}
a_{l_1\cdots l_n}&=\int_0^1\cdots\int_0^1 g(x)e(-(l_1x_1+\cdots+l_nx_n))\cdot dx_1\cdots dx_n\\
&=\sum_{m_1,\cdots,m_n\in\boldsymbol{Z}}\int_0^1\cdots\int_0^1 f(x_1+m_1,\cdots,x_n+m_n)e(-(l_1x_1+\cdots+l_nx_n))\cdot dx_1\cdots dx_n\\
&=\int_{-\infty}^{+\infty}\cdots\int_{-\infty}^{+\infty} f(x_1,\cdots,x_n)e(-(l_1x_1+\cdots+l_nx_n))\cdot dx_1\cdots dx_n\\
&=\hat{f}(l_1,\cdots,l_n),
\end{aligned}$$

where $\hat{f}$ is the Fourier transform of f. Then writing an expression for $g(0)$ in two different ways, we get

(3) $$\sum_{m_1,\cdots,m_n\in\boldsymbol{Z}} f(m_1,\cdots,m_n)=\sum_{l_1,\cdots,l_n\in\boldsymbol{Z}}\hat{f}(l_1,\cdots,l_n).$$

Now replace the lattice $\boldsymbol{Z}^n$ of integral points by another commensurable lattice Λ in $\boldsymbol{R}^n$, let Λ' denote the dual lattice with respect to the inner product $(x, y)=\sum_j x_j y_j$, and let $\varDelta$ be the volume of a fundamental period parallelogram of $\varLambda$. If f satisfies the same hypotheses as before, then (3) is replaced by

$$\varDelta\cdot\sum_{\lambda\in\varLambda} f(\lambda)=\sum_{\lambda'\in\varLambda'}\hat{f}(\lambda'). \tag{4}$$

If S is a symmetric $n\times n$ matrix, and if X is an $n\times m$ matrix, let $S[X]$ be the $m\times m$ matrix ${}^tXSX$. If S is positive-definite and real, then there exists a non-singular, $n\times n$, real, symmetric matrix A such that $S=A^2$. Let $\operatorname{tr}(M)$ be the trace of a square matrix M. We need the following integration formulae $(c>0)$:

$$\int_{-\infty}^{+\infty} e^{-c\xi^2}d\xi=\int_{-\infty+ia}^{+\infty+ia} e^{-c\xi^2}d\xi=\int_{-\infty}^{+\infty} e^{-c(\xi+ia)^2}d\xi=\sqrt{\frac{\pi}{c}} \tag{5}$$

and

$$\int_{-\infty}^{+\infty} e^{-\pi x^2}dx=1. \tag{6}$$

Formula (6) is well-known, and (5) follows from it by applying Cauchy's theorem and elementary estimates to a rectangular contour with two sides parallel to and two sides perpendicular to the real axis, the x-coordinates of the latter two being allowed to tend to infinity.

Let S and Y be positive-definite, real matrices, S being $m\times m$ and Y being $n\times n$, and let $S=A^2$, $Y=B^2$, where A and B are real, symmetric, and non-singular. Let V be a fixed, real, $m\times n$ matrix and X a variable, real, $m\times n$ matrix; then

$$\begin{aligned}&\int_{R^{mn}} e^{-\pi\operatorname{tr}(S[X]\cdot Y)-2\pi i\operatorname{tr}({}^tVX)}\cdot dX\\ &=\nu^{-1}\cdot\int_{-\infty}^{+\infty}\cdots\int_{-\infty}^{+\infty} e^{-\pi\operatorname{tr}({}^t\xi\xi)-2\pi i\operatorname{tr}({}^tVA^{-1}\xi B^{-1})}\cdot d\xi,\end{aligned} \tag{7}$$

where $\xi=AXB$ is also an $m\times n$ matric variable and ν is the determinant of the transformation $X\mapsto\xi$, that is,

$$\nu=(\det(A))^n(\det(B))^m=(\det(S))^{n/2}(\det(Y))^{m/2}, \tag{8}$$

and the right side of equation (7) is equal to

$$\nu^{-1}\cdot\int_{-\infty}^{+\infty}\cdots\int_{-\infty}^{+\infty} e^{-\pi\operatorname{tr}\{{}^t(\xi+i\eta)(\xi+i\eta)+B^{-1}\cdot{}^tVA^{-2}VB^{-1}\}}\cdot d\xi,$$

where $\eta = A^{-1}VB^{-1}$. If we denote by I the common value of the two sides of (7), then by (5) and (6) we have

$$I = \nu^{-1} \cdot e^{-\pi \operatorname{tr}(Y^{-1} \cdot S^{-1}[V])}. \tag{9}$$

Therefore, if Λ denotes the lattice of integral $m \times n$ matrices, we have

$$\begin{aligned} &\sum_{M \in \Lambda} e^{-\pi \operatorname{tr}(S[M] \cdot Y)} \\ &\quad = (\det(S))^{-n/2}(\det(Y))^{-m/2} \cdot \sum_{L \in \Lambda} e^{-\pi \operatorname{tr}(S^{-1}[L] \cdot Y^{-1})}, \end{aligned} \tag{10}$$

since Λ is self-dual. Let $Z = X + iY \in H_n$, so that Y is positive-definite. If $X = 0$, we have from (10)

$$\begin{aligned} &\sum_{M \in \Lambda} e\left(\frac{1}{2} \operatorname{tr}(S[M] \cdot Z)\right) \\ &\quad = i^{mn/2}(\det(S))^{-n/2}(\det(Z))^{-m/2} \cdot \sum_{L \in \Lambda} e\left(\frac{1}{2} \operatorname{tr}(S^{-1}[L] \cdot (-Z^{-1}))\right). \end{aligned} \tag{11}$$

However, both sides of (11) converge normally in H_n to analytic functions there, and since by (10) they are equal there for purely imaginary Z, they are identical in all of H_n. In other words, (11) holds for all $Z \in H_n$.

§2. Quadratic forms and Siegel's main formula (definite case)

Let S be $m \times m$ and T be $n \times n$ symmetric, integral, positive-definite matrices. Let $A(S, T) \geqq 0$ be the (finite) number of $m \times n$ integral matrices X such that $S[X] = T$. If q is a positive integer, let $A_q(S, T)$ be the number of incongruent integral solutions X mod q of the congruence $S[X] \equiv T \pmod q$.

We now proceed to define $A_\infty(S, T)$. Let $\mathfrak{N}$ be a relatively compact neighborhood of T in the cone $\mathfrak{P}$ of positive-definite $n \times n$ matrices, which in turn is an open subset of $\boldsymbol{R}^{n(n+1)/2}$. We define a mapping φ of $\boldsymbol{R}^{mn}$ into $\boldsymbol{R}^{n(n+1)/2}$ by setting $\varphi(X) = S[X]$, where X is an $m \times n$ real matrix viewed as a point in $\boldsymbol{R}^{mn}$. It is easy to see that φ is a proper mapping because S is definite; therefore, $\varphi^{-1}(\mathfrak{N})$ has finite volume. Let $A_\infty(S, \mathfrak{N}) = vol(\varphi^{-1}(\mathfrak{N})) \cdot (vol(\mathfrak{N}))^{-1}$. It can be shown by elementary integral calculus (cf. [56b : pp. 352–354]) that if $\mathfrak{N}$ runs through a sequence of neighborhoods shrinking down to T, then the limit

$$\lim_{\mathfrak{N}\to T} A_\infty(S, \mathfrak{N})$$

exists. This limit is denoted by $A_\infty(S, T)$. If there exists no X such that $S[X]=T$, which occurs for example when $m<n$, of course we define $A_\infty(S, T)=0$.

We use a special notation to cover the case $T=S$: $E(S)=A(S, S)$ (the order of the group of integral units of S), $E_\infty(S)=A_\infty(S, S)$, and $A_q(S, S)=E_q(S)$.

One may prove by counting arguments [56b: pp. 338–346] that if $\boldsymbol{p}$ is a prime integer, then

$$\boldsymbol{p}^{-a(mn-n(n+1)/2)} \cdot A_{\boldsymbol{p}^a}(S, T) \tag{12}$$

has a limit as $a\to+\infty$, and in fact is constant for $a\geqq a_0(\boldsymbol{p})$, where $a_0(\boldsymbol{p})=1$ for all but a finite number of $\boldsymbol{p}$. We denote this limit by $d_{\boldsymbol{p}}(S, T)$.

Now use the notation $q\to\infty$ to mean that q runs through a sequence $\{q_\nu\}$ of positive integers such that 1) $q_\nu \mid q_{\nu+1}$ for all ν, and 2) given any positive integer l, there exists $\nu>0$ such that $l \mid q_\nu$. Siegel has shown [56a] that

$$\lim_{q\to\infty} q^{n(n+1)/2-mn} \cdot A_q(S, T) \tag{13}$$

exists if $m>n+1$ (and is independent of the choice of sequence $\{q_\nu\}$). In fact, if q and r are relatively prime, one may see, in like manner as the Chinese remainder theorem, that

$$A_{qr}(S, T)=A_q(S, T)A_r(S, T)$$

and one then derives the result that the limit in (13) is equal to

$$\prod_{\boldsymbol{p}} d_{\boldsymbol{p}}(S, T), \tag{14}$$

where $\boldsymbol{p}$ runs over the set $\mathscr{P}$ of all finite primes.

Let S' be another non-singular, symmetric, $m\times m$ integral matrix. We write $S\sim S'$ and say that S and S' are in the same genus if $A_\infty(S, S')A_\infty(S', S)\neq 0$ and if $A_q(S, S')A_q(S', S)\neq 0$ for all positive integers q; in particular, S' is also positive-definite. We write $S\approx S'$ and say that S and S' are in the same class if there exist $m\times m$ integral matrices X and X' such that $S[X]=S'$ and $S'[X']=S$; of course, this implies that X and X' are unimodular and that XX' is a unit of S. The number of classes in the genus of S is finite; we denote this number by h and let $S_1, \cdots, S_h$ be representatives of the different classes in the genus of S. It is easy to see on geometric grounds that $A_\infty(S_i, T)=A_\infty(S, T)$ for all $i=1, \cdots, h$. We let $\mu(S)=\sum_{i=1}^{h} E(S_i)^{-1}$. Then Siegel's main theorem in this case [56b: Satz 1, p. 354] says that the limit in (13) is equal to

$$(A_\infty(S, T)\mu(S))^{-1} \sum_{i=1}^{h} E(S_i)^{-1}A(S_i, T),$$

which of course is equivalent to saying that

$$\mu(S)^{-1} \sum_{i=1}^{h} E(S_i)^{-1}A(S_i, T)=A_\infty(S, T) \prod_{p\in\mathscr{P}} d_p(S, T). \tag{15}$$

We omit the proof of this.

If, now, T is only semi-definite, $T\geqq 0$, $\det T=0$, then T has a rational null-vector, hence has an integral null-vector, and, finally, a primitive integral null-vector $x\neq 0$. Then x may be complemented to a unimodular integral matrix. Proceeding in this way, one obtains an integral, unimodular $n\times n$ matrix U such that

$$T[U]=\begin{pmatrix} T_1 & 0 \\ 0 & 0 \end{pmatrix}=T', \tag{16}$$

where T_1 is $r\times r$ and non-singular. Suppose that X is an $m\times n$ integral matrix such that $S[X]=T'$; since S is positive-definite, one sees that the last $(n-r)$ columns of X are zero and that the number of such X is just $A(S, T_1)$. In similar fashion, one sees that the "density factor" $A_\infty(S, T)$ is well-defined and equal to $A_\infty(S, T_1)$ if the limiting process is properly interpreted. Thus the formula (15) also holds when T is only semi-definite if we replace T everywhere that it occurs in (15) by a positive-definite $r\times r$ matrix T_1 given by an equation of the form (16).

We denote the function given by the series on the left-hand side of (11) by $\theta_S(Z)$. Then

$$\theta_S(Z)=\sum_{T\geqq 0} A(S, T)e\left(\frac{1}{2}\operatorname{tr}(TZ)\right), \tag{17}$$

since $A(S, T)$ is equal to the number of $M\in\Lambda$ such that $S[M]=T$, whether T is positive-definite or only positive semi-definite. Combining (15) and (17), we obtain [56f]

$$\mu(S)^{-1} \sum_{i=1}^{h} E(S_i)^{-1}\theta_{S_i}(Z) = \sum_{T \geqq 0} d(S, T) e\left(\frac{1}{2}\operatorname{tr}(TZ)\right), \tag{18}$$

where

$$d(S, T) = A_\infty(S, T) \prod_{p \in \mathcal{P}} d_p(S, T) \tag{19}$$

depends only on the genera of S and T.

We now want to explain connections of the terms in (18) with automorphic forms. First of all, equation (11) becomes

$$\theta_S(Z) = i^{mn/2}(\det(S))^{-n/2}(\det(Z))^{-m/2}\theta_{S^{-1}}(-Z^{-1}). \tag{20}$$

It is clear that if $S \approx S'$ then $\theta_S = \theta_{S'}$; this follows from the series definition of θ_S. In particular, if S is symmetric, integral, and unimodular, then $S^{-1} \approx S$ because $S = SS^{-1}S$; hence, in this case, $\theta_S = \theta_{S^{-1}}$. Suppose, in addition to S being integral and unimodular, that $m \equiv 0 \pmod 8$. Then (20) becomes

$$\theta_S(Z) = \det(Z)^{-m/2}\theta_S(-Z^{-1}). \tag{21}$$

Thus, if $S = E = E_{(m)}$, the $m \times m$ identity matrix, if $m = 8q$, and if we write $\theta_E = \theta_m$, then (21) becomes

$$\theta_m(Z) = \det(Z)^{-4q}\theta_m(-Z^{-1}). \tag{22}$$

If U is an integral, $n \times n$ unimodular matrix, then

$$\theta_S(Z[U]) = \theta_S(Z), \tag{23}$$

because $\operatorname{tr}(S[M] \cdot Z[U]) = \operatorname{tr}(S[M^tU] \cdot Z)$. On the other hand, if A is a symmetric, integral, $n \times n$ matrix, then

$$\begin{aligned} \theta_m(Z+A) &= \sum_{M \in \Lambda} e\left(\frac{1}{2}(Z+A)[M]\right) \\ &= \sum_{M \in \Lambda} \varepsilon_M \cdot e\left(\frac{1}{2}Z[M]\right), \end{aligned} \tag{24}$$

where $\varepsilon_M = e\left(\frac{1}{2}A[M]\right) = \pm 1$. If $A[x] \equiv 0 \pmod 2$ for every integral m-vector x, or, what amounts to the same thing, if the diagonal entries of A are all even, then $\varepsilon_M = 1$ for all $M \in \Lambda$ and $\theta_m(Z+A) = \theta_m(Z)$; call such a matrix A, simply, *even*. Let Γ_2 be the subgroup of $\Gamma=Sp(n, \boldsymbol{Z})$ generated by all the matrices $\begin{pmatrix} E & A \\ 0 & E \end{pmatrix}$, where A is even, by all the matrices $\begin{pmatrix} U & 0 \\ 0 & {}^tU^{-1} \end{pmatrix}$, where U is integral and unimodular, and by $\iota=\begin{pmatrix} 0 & E \\ -E & 0 \end{pmatrix}$. Then Γ_2 is a subgroup of finite index in Γ. In fact, Γ_2 contains the principal congruence subgroup $\Gamma(2)$ consisting of all elements γ of Γ which are congruent to the identity modulo 2; for Γ_2 obviously contains all $\gamma\in\Gamma(2)$ of the forms $\begin{pmatrix} A & B \\ 0 & D \end{pmatrix}$ and $\begin{pmatrix} A & 0 \\ C & D \end{pmatrix}$, and the elements of $\Gamma(2)$ of these forms are known [30] to generate $\Gamma(2)$.

For purposes of *this chapter*, a modular form of weight l with respect to an arithmetic subgroup Γ' of $Sp(n, \boldsymbol{R})$ commensurable with Γ will mean a holomorphic function f on H_n satisfying

$$f(Z\cdot\gamma)=f(Z)\cdot\det(ZB+D)^{2l}, \quad Z\in H_n, \tag{25}$$

for all $\gamma=\begin{pmatrix} A & B \\ C & D \end{pmatrix}\in\Gamma'$. Thus, if f is an automorphic form of weight $2g$ with respect to Γ', then f is a modular form of weight $g(n+1)$. Now for $m\equiv 0 \pmod 8$, it is clear from (22), (23), and (24) that θ_m is a modular form of weight $\frac{1}{4}m$ with respect to Γ_2. If $\gamma=\begin{pmatrix} A & B \\ C & D \end{pmatrix}\in Sp(n, \boldsymbol{R})$, define, for any holomorphic function f on H_n,

$$(T_m(\gamma)f)(Z)=\det(ZB+D)^{-m/2}f(Z\cdot\gamma).$$

Then f is a modular form of weight $\frac{1}{4}m$ with respect to Γ' if and only if $T_m(\gamma)f=f$ for all $\gamma\in\Gamma'$. If Γ'' is a *normal* subgroup of Γ', let $\gamma_1, \cdots, \gamma_k$ be coset representatives of Γ'' in Γ' and define

$$T_{\Gamma'/\Gamma''}f=\sum_{i=1}^{k} T_m(\gamma_i)f.$$

If f is a modular form of weight $\frac{1}{4}m$ with respect to Γ'', it follows easily that $T_{\Gamma'/\Gamma''}f$ is a modular form of weight $\frac{1}{4}m$ with respect to Γ'. Applying this in the case when $\Gamma''=\Gamma(2)$, $\Gamma'=\Gamma=Sp(n, \boldsymbol{Z})$, and $f=\theta_m$, $m\equiv 0 \pmod 8$, we obtain a method of constructing modular forms for Γ out of linear combinations of transforms of

θ-series. The modular forms obtained in this way will not be $\equiv 0$; this may be seen by examining their behavior for purely imaginary values of the argument Z.

There are other choices of S which give a more "economical" construction of modular forms and at the same time bring out the arithmetical significance of (18) in another way. Namely, we now let S be unimodular and integral as before, we assume $m \equiv 0 \pmod 8$, and assume, moreover, that S itself is now even (which, by the way, in itself implies $m \equiv 0 \pmod 8$). That such an S exists whenever $m = 8q$ is guaranteed by the known form of the Cartan matrix for the exceptional Lie algebra E_8 (cf. Chapter 1). Under these assumptions we have in addition to (21) and (23) that

$$\theta_S(Z+A) = \theta_S(Z) \tag{26}$$

for every symmetric, integral $n \times n$ matrix A. It follows that θ_S is itself a modular form of weight $2q$, since the transformations represented in (21), (23), and (26) generate the discrete group of holomorphic isometries of H_n coming from $\Gamma = Sp(n, \boldsymbol{Z})$. (In fact, (21) and (26) alone suffice for this.) If S' is in the same genus as S, then $\det(S') = \det(S) = 1$ and the diagonal entries of S' must also be even; therefore, $\theta_{S'}$ is also a modular form of weight $2q$ for Γ. Consequently, the left side of (18) is a modular form of weight $2q$ for Γ.

We return to the Eisenstein series given by the series in (14) of Chapter 12, where now the only restriction on m is that it be an even integer $> n+1$; denote that series by s_m. Then as for the right side of (18), it may be proven, following in part the ideas of Chapter 12, sections 3 and 4, or those of [56b, d : § 7], that the Fourier coefficient $d(S, T)$ is just the Fourier coefficient $a_{4q}\left(\frac{1}{2}T\right)$ in the expansion

$$s_{4q}(Z) = \sum_{\lambda \in \Lambda'} a_{4q}(\lambda) e((\lambda, Z)), \tag{27}$$

where Λ' is the same lattice as in Chapter 12, section 2, noting that since S is even, every matrix $T = S[M]$, where M is $n \times n$ and integral, is the double of an element of Λ'. The proof that $d(S, T) = a_{4q}\left(\frac{1}{2}T\right)$ is also given explicitly in [56f]. Thus (18) takes the form

$$\mu(S)^{-1} \sum_{i=1}^{h} E(S_i)^{-1} \theta_{S_i}(Z) = s_{4q}(Z). \tag{28}$$

Since the Fourier coefficients of θ_S are always integers, it follows that the denominators of the Fourier coefficients of s_{4q} are bounded. This, together with an *expression* for a bound, is one of the main results of [56f]. As mentioned earlier (Chapter 12, section 3), a better result for the special case $n=2$ is obtained in [42] using more direct methods, and a result analogous to that of [42] is obtained in [35] for Eisenstein series associated to an exceptional arithmetic group. It would be most interesting to obtain an extensive generalization of the intriguing results of [56f; 42; 35].

BIBLIOGRAPHY

[1] Allan, N. daS., Maximality of some arithmetic groups, Thesis, University of Chicago, 1965.

[2] Baily, W. L., Jr., a) The decomposition theorem for V-manifolds, Amer. J. Math. 78 (1956), 862–888.
b) On the imbedding of V-manifolds in projective space, Amer. J. Math. 79 (1957), 403–430.
c) Satake's compactification of V_n, Amer. J. Math. 80 (1958), 348–364.
d) On the theory of θ-functions, the moduli of Abelian varieties, and the moduli of curves, Ann. of Math. 75 (1962), 342–381.
e) Fourier-Jacobi Series, in Proc. Symp. Pure Math. IX, Amer. Math. Soc., Providence, 1966, 296–300.
f) Classical Theory of θ-functions, in Proc. Symp. Pure Math. IX, Amer. Math. Soc., Providence, 1966, 306–311.
g) Eisenstein Series on Tube Domains, Problems in Analysis, in A Symposium in Honor of Salomon Bochner, Ed. R. C. Gunning, Princeton University Press, Princeton, 1970, 139–156.
h) On Hensel's lemma and exponential sums, in Global Analysis——Papers in honor of K. Kodaira, Ed. D. C. Spencer and S. Iyanaga, Tokyo University Press and Princeton University Press, 1969, 85–100.
i) An exceptional arithmetic group and its Eisenstein series, Ann. of Math. 91 (1970), 512–549.
j) On the Fourier coefficients of certain Eisenstein series on the adele group, in Number Theory, in honor of Y. Akizuki, Kinokuniya, Tokyo, 1973.

[3] Baily, W. L., Jr., and Borel, A., Compactification of arithmetic quotients of bounded symmetric domains, Ann. of Math. 84 (1966), 442–528.

[4] Bieberbach, L., Lehrbuch der Funktionentheorie, Chelsea Publishing Co., New York, 1945.

[5] Bochner, S., and Martin, W. T., Several Complex Variables, Princeton University Press, Princeton, 1948.

[6] Borel, A., a) Groupes linéaires algébriques, Ann. of Math. 64 (1956), 20–80.
b) Density and maximality of arithmetic groups, J. Reine Angew. Math. 224 (1966), 78–89.
c) Linear algebraic groups, Notes by H. Bass, Benjamin, New York, 1969.
d) Introduction aux groupes arithmétiques, Actualités Sci. Ind. 1341, Hermann, Paris, 1969.
e) Introduction to Automorphic Forms, in Proc. Symp. Pure Math. IX, Amer. Math. Soc., Providence, 1966, 199–210.
f) Some finiteness properties of adele groups over number fields, Publ. Math. Inst. HES No. 16, pp. 101–126, Presses Univ. de France, Paris, 1963.

[7] Borel, A., Chowla, S., Herz, C. S., Iwasawa, K., and Serre, J.-P., Seminar on Complex Multiplication, Lecture Notes in Mathematics 21, Springer-Verlag, 1966.

[8] Borel, A., and Harish-Chandra, Arithmetic subgroups of algebraic groups, Ann. of Math. 75 (1962), 485–535.

[9] Borel, A., and Tits, J., Groupes réductifs, Publ. Math. Inst. HES No. 27, pp. 55–151, Presses Univ. de France, Paris, 1965.

[10] Borevich, Z. I., and Shafarevich, I. R., Number Theory, Academic Press, New York, 1966.

[11] Bourbaki, N., a) Éléments de Mathématique, Livre VI, Intégration, Hermann, Paris.
b) Éléments de Mathématique, Groupes et algèbres de Lie, Chaps. IV, V, VI, Hermann, Paris, 1968.

[12] Bruhat, F., and Tits, J., a) Un théorème de point fixe, mimeographed, Paris, 1966.
b) Groupes algébriques simples sur un corps local, in Proceedings of a Conference on Local Fields, Ed. T. A. Springer, Springer-Verlag, 1967.
c) Four articles in C. R. Acad. Sci. Paris, Série A, v. 263 (1966), 598–601, 766–768, 822–825, 867–869.

[13] Burnside, W., Theory of Groups of Finite Order, Dover Publications, 1955.

[14] Cartan, É., Sur les domaines bornés homogènes de l'espace de n variables, Abh. Math. Sem. Univ. Hamburg 11 (1935), 116–162.

[15] Cartan, H., a) Sur les groupes de transformations analytiques, Actualités Sci. Ind., Exposés Math. IX, Hermann, Paris, 1935.
b) Quotient d'un espace analytique par un groupe d'automorphismes, in Symposium in Honor of S. Lefschetz, Princeton University Press, Princeton, 1957.
c) Fonctions automorphes et séries de Poincaré, J. Analyse Math. 6 (1958), 169–175.
d) Prolongement des espaces analytiques normaux, Math. Ann. 136 (1958), 97–110.

[16] Chevalley, C., a) Theory of Lie Groups, Princeton University Press, Princeton, 1946.
b) Théorie des groupes de Lie, II, III, Hermann, Paris, 1951, 1955.
c) Séminaire sur la classification des groupes de Lie algébriques, 2 vol., Paris, 1958 (mimeographed).

[17] Coxeter, H. S. M., Integral Cayley Numbers, Duke Math. J., 13 (1946), 561–578.

[18] Dieudonné, J., La dualité dans les espaces vectoriels topologiques, Ann. Sci. École Norm. Sup. 59 (1942), 107–139.

[19] Freudenthal, H., a) Oktaven, Ausnahmegruppen, Oktavengeometrie, Math. Inst. der Rijksuniv. te Utrecht, 1951.
b) Beziehungen der E_7 und E_8 zur Oktavenebene I, Proc. Konkl. ned. Akad. Wet., Series A, 57 (1954), 218–230.

[20] Fuks, B. A., Special Chapters of the Theory of Analytic Functions of Several Complex Variables (Russian), Fizmatgiz, Moscow, 1963.

[21] Gelfand, I. M., Graev, M. I., and Pyateckii-Shapiro, I. I., Theory of Representations and Automorphic Functions (Generalized Functions v. 6) (Russian), Publ. "Nauka", Ed. A. A. Kirillov, Moscow, 1966.

[22] Godement, R., a) Sur la théorie des représentations unitaires, Ann. of Math. 53 (1951), 68–124.

b) A theory of spherical functions I, Trans. Amer. Math. Soc. 73 (1952), 496-556.

c) Introduction à la théorie de Langlands, Sém. Bourbaki 19 (1966/67), Exposé 321.

d) Formes automorphes et produits Euleriennes, d'après R. P. Langlands, Sém. Bourbaki 21 (1968/69), Exposé 349.

[23] Gunning, R. C., Lectures on Modular Forms, notes by A. Brumer, Ann. of Math. Studies No. 48, Princeton University Press, Princeton, 1962.

[24] Gunning, R. C., and Rossi, H., Analytic Functions of Several Complex Variables, Prentice Hall, Englewood Cliffs, N. J., 1965.

[25] Gutnik, L., On the extension of integral subgroups of some groups (Russian), Vestnik Leningrad Univ., Ser. Math., Mech., Astr. 19 (1957), 51-79.

[26] Harish-Chandra, a) Representations of a semi-simple Lie group on a Banach space I, Trans. Amer. Math. Soc. 75 (1953), 185-243.

b) Lie algebras and the Tannaka duality theorem, Ann. of Math. 51 (1950), 299-330.

c) Representations of semi-simple Lie groups VI, Amer. J. Math. 78 (1956), 564-628.

d) Discrete series for semi-simple Lie groups II, Acta Math. 116 (1966), 1-111.

e) Automorphic Forms on Semi-simple Lie Groups, notes by J. G. M. Mars, Lecture Notes in Mathematics 62, Springer-Verlag, 1968.

[27] Helgason, S., Differential Geometry and Symmetric Spaces, Academic Press, New York, 1962.

[28] Hua, L.-K., a) On the theory of Fuchsian functions of several variables, Ann. of Math. 47 (1946), 117-191.

b) On the theory of functions of several complex variables, I, II, III, Trans. Amer. Math. Soc. (2) 32 (1963), 163-263.

[29] Huppert, B., Endliche Gruppen I, Springer-Verlag, 1967.

[30] Igusa, J., On the graded ring of theta constants, Amer. J. Math. 86 (1964), 219-246.

[31] Iwahori, N., and Matsumoto, H., On some Bruhat decomposition and the structure of the Hecke rings of p-adic Chevalley groups, Publ. Math. Inst. HES No. 25, pp. 5-48, Presses Univ. de France, Paris, 1965.

[32] Jacobson, N., a) Structure theory of simple rings without finiteness assumptions, Trans. Amer. Math. Soc. 57 (1945), 228-245.

b) Lie algebras, Interscience Publishers, New York, 1962.

c) Some Groups of Transformations defined by Jordan Algebras I, J. Reine Angew. Math. 201 (1959), 178-195; II, ibid. 204 (1960), 74-98; III, ibid. 207 (1961), 61-85.

[33] Jensen, K. L., Om talteoretiske Egenskaber ved de Bernoulliske Tal, Nyt Tidsskr. for Mat. B 26 (1915), 73-83.

[34] Kaplansky, I., Groups with representations of bounded degree, Canad. J. Math. 1 (1949), 105-112.

[35] Karel, M. L., On Certain Eisenstein Series and their Fourier Coefficients, Thesis, University of Chicago, 1972.

[36] Klingen, H., Über den arithmetischen Charakter der Fourierkoeffizienten von Modulformen, Math. Ann. 147 (1962), 176-188.

[37] Kneser, M., a) Starke Approximation in algebraischen Gruppen I, J. Reine Angew. Math. 218 (1965), 190-203.
b) Strong Approximation, in Proc. Sym. Pure Math. IX, Amer. Math. Soc., Providence, 1966, 187-198.

[38] Koecher, M., Zur Theorie der Modulformen n-ten Grades I, Math. Z. 59 (1954), 399-416.

[39] Korányi, A., and Wolf, J., a) Realization of Hermitian Symmetric Spaces as Generalized Half-planes, Ann. of Math. 81 (1965), 265-288.
b) Generalized Cayley transformations of bounded symmetric domains, Amer. J. Math. 87 (1965), 899-939.

[40] Langlands, R. P., Euler Products, Whittemore lectures in mathematics, Yale University, 1967.

[41] Loomis, L. H., An Introduction to Abstract Harmonic Analysis, van Nostrand, New York, 1953.

[42] Maass, H., Die Fourierkoeffizienten der Eisensteinreihen zweiten Grades, Mat.-fys. Med. Kong. Danske Vid. Selskab, 34, No. 7 (1964).

[43] Montgomery, D., and Zippin, L., Topological Transformation Groups, Interscience, New York, 1955.

[44] Narasimhan, R., a) Introduction to the Theory of Analytic Spaces, Lecture Notes in Mathematics 25, Springer-Verlag, 1966.
b) Analysis on Real and Complex Manifolds, North Holland, Amsterdam, 1968.
c) Several Complex Variables, University of Chicago Press, Chicago, 1971.

[45] Neumann, J. v., Zur algebra der Funktionaloperatoren und Theorie der normalen Operatoren, Math. Ann. 102 (1930), 370-427.

[46] Pyateckii-Shapiro, I. I., Geometry of classical domains and theory of automorphic functions (Russian), Fizmatgiz, Moscow, 1961.

[47] Remmert, R. and Stein, K., Über die wesentliche Singularitäten analytischer Mengen, Math. Ann. 126 (1953), 263-306.

[48] Riesz, F., and Sz.-Nagy, B., Functional Analysis, Frederick Ungar, New York, 1955.

[49] Rossi, H., Analytic Spaces II, lectures at Princeton University (notes), 1960.

[50] Satake, I., a) On a generalization of the notion of manifolds, Proc. Nat. Acad. Sci. USA 42 (1956), 359-363.
b) On the Compactification of the Siegel Space, J. Indian Math. Soc. 20 (1956), 259-281.

[51] Séminaire H. Cartan, E.N.S., 1953-54, hectographed notes.

[52] Séminaire H. Cartan, E.N.S., 1957-58, mimeographed notes.

[53] Séminaire C. Chevalley (Classification des groupes de Lie algébriques), E.N.S., 1956-58.

[54] Séminaire Sophus Lie, E.N.S. (Théorie des algèbres de Lie, Topologie des groupes de Lie) mimeographed notes, 1954-55.

[55] Shimura, G., Introduction to the Arithmetic Theory of Automorphic Functions, Publications of the Math. Soc. of Japan 11, Iwanami Shoten, Publishers and Princeton University Press, 1971.

[56] Siegel, C. L., a) Lectures on the Analytical Theory of Quadratic Forms, notes by Morgan Ward, Institute for Advanced Study, Princeton, 1949.
b) Über die analytische Theorie der quadratischen Formen, Ann. of Math.

36 (1935), 527-606.
c) Do. III, ibid. 38 (1937), 212-291.
d) Einführung in die Theorie der Modulfunktionen n-ten Grades, Math. Ann. 116 (1939), 617-657.
e) Analytic Functions of Several Complex Variables, notes by P. T. Bateman, Institute for Advanced Study, Princeton, 1948-49.
f) Über die Fourierschen Koeffizienten der Eisensteinschen Reihen, Mat.-fys. Med. Kong. Danske Vid. Selskab, 34, No. 6, 1964.

[57] Tits, J., a) Théorème de Bruhat et sous-groupes paraboliques, C. R. Acad. Sci. Paris 249 (1959), 1438-1440.
b) Algebraic and abstract simple groups, Ann. of Math. 80 (1964), 313-329.

[58] Tsao, L.-C., Thesis, Univ. of Chicago, 1972.

[59] Waerden, B. L. v.d., Modern Algebra, Frederick Ungar, New York, 1950.

[60] Weil, A., a) Intégration dans les groupes topologiques et ses applications, Actualités Sci. Ind. 1145, Hermann, Paris, 1965.
b) On algebraic groups of transformations, Amer. J. Math. 77 (1955), 355-391.
c) On algebraic groups and homogeneous spaces, Amer. J. Math. 77 (1955), 493-512.
d) Adeles and algebraic groups, notes by M. Demazure and T. Ono, Institute for Advanced Study, Princeton, 1961.
e) Basic Number Theory, Springer-Verlag, New York, 1967.

[61] Whitney, H., Elementary structure of real algebraic varieties, Ann. of Math. 66 (1957), 545-556.

[62] Wolf, J., Spaces of Constant Curvature, McGraw-Hill, 1967.

[63] Wyler, A., On the Conformal Groups in the Theory of Relativity and their Unitary Representations, Archive for Rational Mechanics and Analysis 31 (1968), 35-50.

[64] Yoshida, K., Functional Analysis, Springer-Verlag, Berlin, 1965.

[65] Zariski, O., and Samuel, P., Commutative Algebra, van Nostrand, Princeton, 1958.

Index

(the numbers refer to pages)